Microelectronic Test Structures for CMOS Technology

Manjul Bhushan · Mark B. Ketchen

Microelectronic Test Structures for CMOS Technology

Springer

Manjul Bhushan
IBM Systems & Technology Group
Hopewell Junction, NY 12533, USA
bhushan@us.ibm.com

Mark B. Ketchen
IBM T.J. Watson Research Center
Yorktown Heights, NY 10598, USA
mketchen@us.ibm.com

ISBN 978-1-4419-9376-2 e-ISBN 978-1-4419-9377-9
DOI 10.1007/978-1-4419-9377-9
Springer New York Dordrecht Heidelberg London

Library of Congress Control Number: 2011934043

Printed on acid-free paper

Springer is part of Springer Science+Business Media (www.springer.com)

Preface

Test structures are widely used in all manner of scientific and technological endeavors. Their role in the development and monitoring of technology and its applications in the specific area of microelectronics has become ever more important as the technology becomes more complex and expensive, while market demands exert downward pressure on product prices. Scaling of complementary metal oxide semiconductor (CMOS) technology to ever smaller dimensions, large investments in silicon manufacturing lines, and the race to follow Moore's law to its ultimate limits have driven special attention to test structures. In addition to sessions at many conferences with themes of microelectronic design, development, and manufacturing, the IEEE sponsors a conference fully devoted to microelectronic test structures, the International Conference on Microelectronic Test Structures (ICMTS), with participants from industry, universities, and research laboratories from across the Americas, Europe, and Asia. In recent years, several small companies marketing test chips along with data analysis services have emerged to assist larger companies engaged in CMOS technology development and manufacturing.

With backgrounds in low temperature physics, much of our earlier experience in research and development was in the area of superconducting devices and circuits for magnetic and high-frequency digital and analog applications. Prior to our work on test structures for CMOS technologies, we collectively worked in various capacities on a wide range of other large and small projects, from optoelectronics and photovoltaics to silicon bipolar and CMOS devices and fabrication, with some continued involvement in basic and applied research in superconductivity. Work on smaller projects often involved several or all aspects of technology development from modeling and fabrication to circuit design and test. In larger projects such as mainstream CMOS silicon technology development, on the other hand, the distribution of technical work and responsibilities is much more modular, although some management and advisory roles remain broad, but with less depth.

In late 1990s we started to work on the evaluation of IBM's CMOS technology performance based on compact models used in circuit simulations. This work was initially driven by the need to elucidate the trade-offs between SOI and bulk CMOS for IBM applications. Based on our past experience in physics and technology, an approach incorporating mode-to-hardware correlation was a natural consequence.

We started by designing high-speed test structures, sometimes resurrecting ideas that were used in superconducting technology many years earlier. Bridging high-speed circuit and product performance to MOSFET characteristics and process recipes was critical. Acquiring precious silicon real estate on product wafers, meeting test time budgets, and providing rapid feedback to technology and product teams added an interesting business spin to the underlying technical effort. Many of our designs are now in routine use in IBM's CMOS technology development and manufacturing and embedded in all of IBM's high-performance products. A number of these designs are also shared with IBM's partner industries in CMOS technology development.

Information on different aspects of design and test of test structures is spread out in bits and pieces across many textbooks and scientific publications. We have written this book to provide a single source covering a wide range of the basic concepts of test structures for digital CMOS technology. Our text features an integrated view of design, test, and data analysis and includes numerous examples throughout, to serve as a handbook to those working on test structures. While the focus is on digital CMOS technology applications, many of the concepts described here are applicable to other electronic devices and material systems as well.

Our knowledge on this topic was enhanced through work with our colleagues in IBM Research, in IBM's Server and Technology Group, and especially from our collaborations within IBM's Microelectronics division. Carl Anderson gave us strong encouragement and support to couple with microprocessor design and test teams. The manuscript was carefully reviewed by Tak Ning and we are grateful for his insights into the technical content. We would also like to thank all the individuals who helped in reviewing the manuscript. Chin Kim gave us input from a test structure design engineer's view and Richard Rouse provided a perspective from semiconductor industry in general. Selected sections of the manuscript were reviewed by Brian Ji, Mary Lanzerotti, and Jayant Sogani.

Hopewell Junction, NY — Manjul Bhushan
Yorktown Heights, NY — Mark B. Ketchen

Contents

List of Figures

List of Tables

Chapter 1
Introduction

Contents

Since ancient times it has been the nature of man to build physical structures, ranging from the simple to the very sophisticated and complex. It has also been the nature of man to devise experiments to explain how and why things behave the way as they do in the physical world. As an end goal or as intermediate steps along the way, such activities often involve the design, fabrication, and characterization of test structures of one kind or another from which scientific principles and behavioral patterns are derived, generalized, and applied on a wider scale.

An early example of a scientific experiment to prove a hypothesis is attributed to Archimedes, circa 250 BC. In the popular story of the "golden crown," Archimedes was able to determine the purity of gold in the emperor's crown by comparing its density, as determined in part by the volume of water displaced, to that of lumps of pure gold and silver [1]. Although there is no known written record by Archimedes of this event, the story serves well to illustrate the concept of non-destructively evaluating the properties of an object by learning derived from test samples.

A natural concern in conducting scientific experiments has been the validity and accuracy of measurements. As far back as 1020 AD, Persian scientist Al-Biruni wrote a commentary on systematic and random errors in measurement. "He argued that if instruments produce random errors because of their imperfections or idiosyncratic qualities, then multiple observations must be taken, analyzed qualitatively, and on this basis, arrive at a common-sense single value for the constant sought, whether an arithmetic mean or a reliable estimate" [2]. Many sophisticated statistical techniques to account for measurement errors and variability in the data collected from test samples or structures have been developed since the time of Al-Biruni. The use of these statistical methods has become ever more important as large volumes of data are collected and analyzed.

M. Bhushan, M.B. Ketchen, *Microelectronic Test Structures for CMOS Technology*,
DOI 10.1007/978-1-4419-9377-9_1,

The integrity of the design and characterization of test structures has been instrumental in separating the winners from the losers in pursuit of some of the great technological feats of all time. One such example is the invaluable contribution of the Wright brothers to the field of aviation. In 1901 while Orville and Wilbur Wright were conducting glider trials in Kitty Hawk, North Carolina, they came to realize that the lift tables used by them and many of their immediate predecessors to design and size wings were significantly in error. Subsequently they very carefully constructed a wind tunnel and over a period of several months beginning in late 1901 made thousands of precise measurements on a wide variety of aerodynamic test structures. The data from these experiments formed the basis of the design of the wings and propellers of the first airplane which the Wright brothers constructed and successfully demonstrated on December 17, 1903. Orville's comment on the errors made in earlier designs is noteworthy [3]: "From one way of looking at it, you might even have called it encouraging, that the data others had used could not be relied upon. It suggested that maybe the reason others had failed to fly was not because the thing couldn't be done."

Test structures have played an essential and indispensible role in the unprecedented growth and success of the semiconductor industry from the invention of the bipolar transistor in 1947 through today's widespread use of silicon chips. Complementary metal oxide semiconductor (CMOS) technology based on Frank Wanlass's invention in 1963 has now become the dominant silicon technology for digital applications. Low power consumption, noise immunity, and scalability to smaller dimensions leading to very large-scale integration (VLSI) circuits are some of its attractive features. This technology is now utilized in products ranging from low-cost consumer electronics to high-performance computer systems.

1.1 Role of Test Structures in CMOS Technology

CMOS technology follows the now-standard principles of planar silicon microfabrication [4–5]. Metal oxide semiconductor field effect transistors (MOSFETs) with n-type or p-type channel conduction, generally referred to as n-FET (NMOS) and p-FET (PMOS), are formed in the silicon substrate. A vertical stack of metal and dielectric isolation layers is patterned on top of silicon for electrical interconnections. A schematic cross section of a part of a CMOS circuit with an n-FET and a p-FET and a five-metal-layer stack is shown in Fig. 1.1. The number of metal layers may vary from three to ten or more. Other active and passive circuit elements such as p–n diodes, resistors, capacitors, and inductors are also simultaneously fabricated.

Following Moore's original projection on semiconductor technology scaling trends in 1965 [6] and MOSFET scaling rules set forth by Dennard and co-workers in 1974 [7], there has been an exponential growth in the device count per integrated circuit. This trend of approximately doubling the transistor count every 2 years, corresponding to the reduction in minimum feature size by $0.7\times$ is shown in Fig. 1.2 for CMOS technologies [8]. Every 2–3 years, products in a new CMOS technology

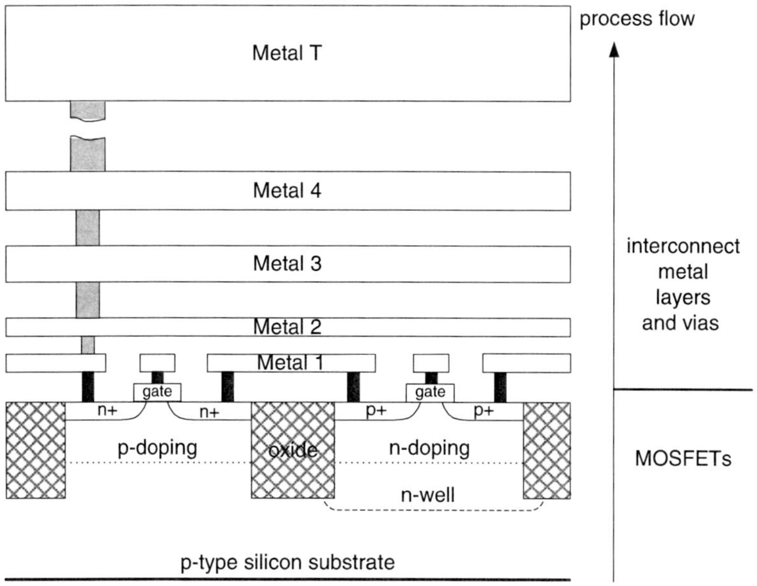

Fig. 1.1 Schematic cross section of a part of CMOS circuit with five metal layers, indicating the process flow beginning with MOSFET delineation in silicon followed by metal interconnect layers and inter-level dielectric isolation

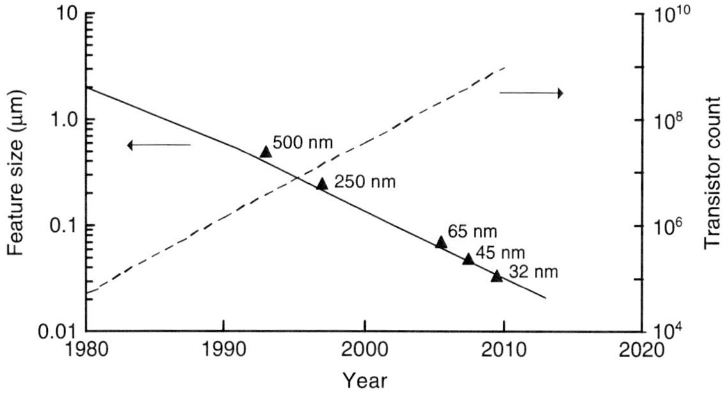

Fig. 1.2 Trends in minimum feature size and transistor count for integrated circuit chips since 1980. CMOS technologies corresponding to 500, 250, 65, 45, and 32 nm nodes are indicated [8]

generation are introduced in the marketplace. Each technology generation or node is referred to by the minimum feature size in nm. These nodes are indicated in Fig. 1.2, from 500 nm in 1993 to 32 nm in 2010. Development is currently under way on the 22 and 14 nm nodes.

During the development of a new CMOS technology node, test structures are used in both the development of the process technology and the optimization of device behavior. As the technology transitions into manufacturing, test structures play an important diagnostic role in climbing the yield learning curve and in providing valuable data for process monitoring and feedback for centering the fabrication line. Careful characterization of device and circuit test structures provides input to models used in designing CMOS products. Test structures both alongside and embedded within the product facilitate tracking the technology, aid in debugging product functionality, and enable early learning on reliability.

In manufacturing for mass production, quality control is generally enforced by monitoring a small set of representative samples at different steps in the production process. This ensures that the final product delivered to the customer is within prescribed specifications. If the test samples indicate the process to be outside the specified limits at any step, corrective actions are taken to re-center the process in a timely fashion. Hence, any deviation from the normal process window affects only a small fraction of the output and helps keep the manufacturing cost down. Statistical Process Control (SPC) [9], Taguchi [10], and Six Sigma [11] are some of the commonly used statistical control methods for improving manufacturing efficiency. Quality control is facilitated by design of experiments (DOE) and with the aid of variability analysis and control charts.

Development and manufacturing of CMOS products follow these same general principles of process monitoring and quality control with some unique requirements. The number of process steps being large (>100), manufacturing time of a CMOS product is typically several weeks. Investment in a state-of-the-art CMOS fabrication facility (fab) is now running into several billion dollars. At the same time, the market price of silicon chips and consumer products that utilize these chips has been steadily declining. For an acceptable return on investment (ROI) from a silicon fab, a steady flow of silicon wafers must be maintained through the manufacturing line with nearly full capacity utilization to generate a continuous supply of CMOS products. In this scenario, it is increasingly important that the process be monitored at each step as the silicon wafers flow through the pipeline and that the fraction of rejected parts be kept to a minimum.

Characterization in CMOS fabrication is carried out at several different levels of process integration. At a lower level, processes such as ion implantation, photoresist patterning, etching, and chemical mechanical polishing are locally characterized on individual tools. Techniques for such characterization involve, for example, thin-film resistivity monitoring, line-width measurements, ellipsometry, and optical and electron microscopy [12]. At a higher integration level, electrical test structures are used for determining the characteristics and yield of CMOS circuit elements and functional circuit blocks for logic and memory. These can be tested after completion of many process steps as illustrated in Fig. 1.3. The MOSFET device and simple circuit blocks can be tested only after at least the first metal layer, M1, has been delineated. Hence, the first major test stop for electrical tests occurs immediately post M1. The processing time to M1 is typically half or more of the total fabrication time. Following M1, there may be several intermediate test stops prior to testing

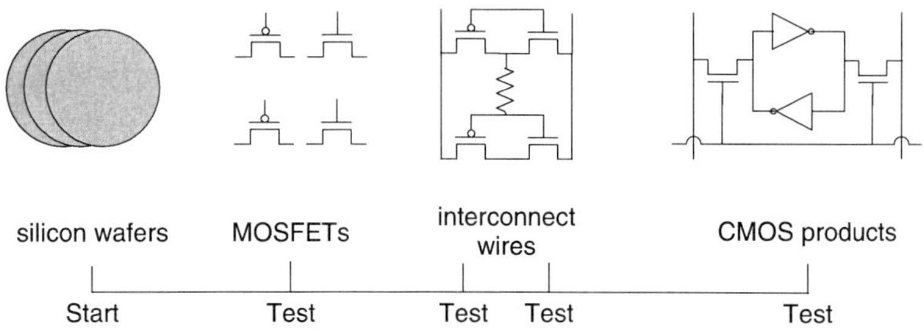

Fig. 1.3 Locations of electrical test stops in a CMOS fabrication line

the final product. The test structures described in this book fall into this category of electrical tests and cover parametric, functional, and yield characterization.

Electrical test structures provide input for building compact models for MOSFETs and other circuit elements. These semi-empirical models are based on known theoretical properties of circuit elements coupled with fitting parameters derived from hardware characterization. Such models are used in a simulation program with integrated circuit emphasis (SPICE) simulator for circuit analysis. Silicon foundries offer several different types of MOSFETs, from low power to high performance for digital and mixed signal applications. In addition, MOSFET properties may be sensitive to physical dimensions and layout. A large number of test structures are therefore required to cover this multidimensional space for building models that can accurately represent the properties and functional behavior of all circuit elements. New models are generated as the technology matures to accommodate significant changes in the process and device behavior.

Increasing complexity of VLSI circuits has made the use of small test structures for diagnostics and debug of product functionality, reliability, and yield ever more important. Signal propagation delays and power are measured on logic gates and complex circuit blocks at different voltage and temperature application corners. Impact of process variations, such as of physical dimensions and material properties, on circuit elements is monitored to determine a manufacturing process window with acceptable yield. Test structures for product reliability are subjected to accelerated stress, and reliability models are developed to predict product end-of-life. The measured data from test structures are correlated with model predictions derived from circuit simulations.

The role of test structures in increasing product yield and in building more accurate models to get the right product design in place is frequently hidden and the impact of test structures on the revenue stream, although great, is indirect. The large number of test structures needed to measure the parameters of interest takes up precious silicon space which could otherwise be used to produce revenue-generating product chips. Measurement time decreases wafer throughput in the manufacturing line. Adding more test equipment to increase throughput adds to the capital equipment cost. Finally, a team of engineers is devoted to data mining and analysis and correlation of electrical measurements to physical measurements and metrology. The information gathered in this way needs to be fed back to the technology and design teams in a timely fashion for appropriate action.

While test structures are critical in maintaining a robust and profitable manufacturing line, there is a cost associated with their use. The time taken for testing and the unavoidable waiting time for test equipment availability add to the total process time and impact the supply chain. The cost of test structure design, cost associated with silicon area, cost of test, and cost of data analysis and feedback must be compensated by the savings derived from reduced technology development time or product rejection rate. Devising and executing an effective and efficient test structure plan is a complex exercise that plays an important role in determining who profits the most in the silicon technology game.

1.2 Placement of Test Structures

CMOS fabrication facilities require an ultra–clean environment and use partial or fully automated wafer-handling systems and tools. These fabs are each dedicated to a fixed silicon wafer diameter ranging from 100 mm (4 in.) to 300 mm (12 in.), and many product chips are processed on each wafer. A silicon wafer map with rectangular areas denoting lithographic exposure fields is shown in Fig. 1.4a. The pattern for each process layer is delineated on a glass plate or mask known as a reticle. The exposure field (or reticle field) is equal to some fixed reduction of the reticle pattern. Generally all field sizes are identical on a wafer and at or near the maximum for the exposure tool in use. In an automated fabrication facility, the whole wafer is processed and then diced with a diamond or a laser saw to separate each chip. The width of the saw cut, which is of the order of 30–100 μm, is called the kerf. The area designated for the kerf or the scribe line, indicated in Fig. 1.4b for a

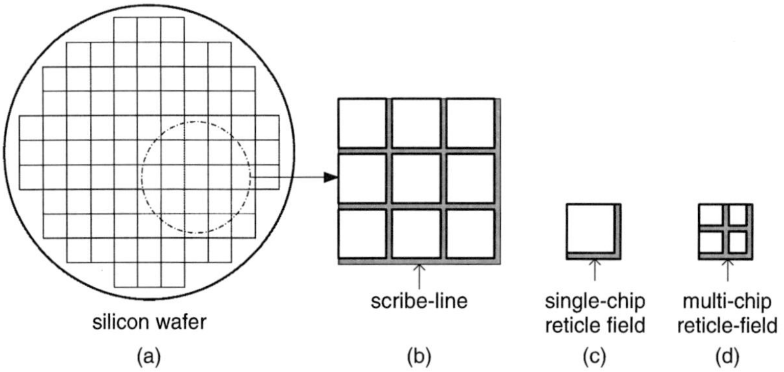

Fig. 1.4 Planer views **a** of reticle exposure fields on a silicon wafer. **b** Kerf or scribe lines for sawing the wafer into chips. **c** Single-chip reticle field. **d** Multi-chip reticle field with increased scribe-line area

single-chip exposure field, is outside the product chip area, and it is removed during sawing. The scribe line is utilized for placement of test structures for process monitoring in the manufacturing line without any additional demand for silicon space. The scribe-line area is very limited if there is only one product chip per reticle field as shown in Fig. 1.4c but significantly larger for multi-chip reticle fields as shown in Fig. 1.4d.

Silicon technology research, development, and manufacturing organizations take different approaches to the placement of test structures. A graphical illustration of representative approaches is shown in Fig. 1.5. In early technology development, the full reticle field design contains only test structures. At this stage, the "test vehicle" or "characterization vehicle" is an assembly of test structures, each with its own independent inputs/outputs (I/Os) to interface with the test equipment. These test structures usually require partial processing and are designated as short-loop test vehicles. The designs are relatively simple and the development of different process modules is carried out in parallel. MOSFET devices are ideally characterized with a single layer of metal and characterization of metal layers may be carried out without any active MOSFET elements. Only limited space on the reticle field may be ascribed to more complex, product-like designs. As there may be multiple test vehicles fully devoted to a sub-set of process steps, there is ample space for placement of test structures.

In the second phase of the technology cycle, full qualification for manufacturing takes place. In this phase, product-like designs utilizing all process steps and structures for yield learning with adequate critical area are included on the reticle field. The space allocated for simpler designs used in the first phase is reduced.

In the third phase where the technology reaches full-scale manufacturing, the reticle field area is primarily occupied by the product chips. At this point, most test structures are accommodated in the scribe line, although some are embedded in the product chip itself for diagnostic purposes. For electrical measurements, the embedded structures may either have independent I/Os and be tested only at the wafer level, or share the power supply and I/Os with the product and be tested at wafer, package, or system level. In either case, they are placed in the unused silicon space on the product which is typically very limited. Short-loop test vehicles may be run periodically for full process characterization.

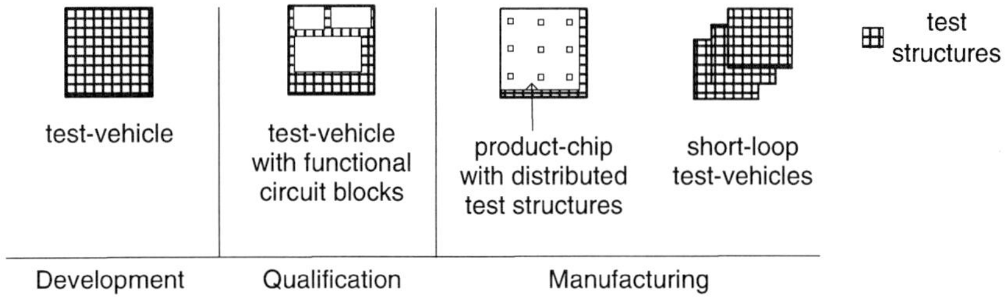

Fig. 1.5 Placement of electrical test structures on a reticle field, from technology development through manufacturing of CMOS products

From process development through manufacturing, a sub-set of common test structures is placed on all the wafers for tracking key aspects of the technology. These test structures are maintained and faithfully migrated to other technology nodes so that technology comparison in different phases, among different fabrication facilities, and across technology nodes can be carried out in an unambiguous fashion.

1.3 Classification of Electrical Test Structures

Electrical test structures are classified in a number of ways based on their function, content, placement, and test requirements [13]. Frequently used categories (followed by some examples of each in parenthesis) are listed below. There is considerable overlap among the different categories. As we work our way through the electrical test structure landscape in this book, the significance of these different categories and their interrelationships will become more apparent:

A. Function (silicon fabrication process, yield, model build, product debug, and reliability)
B. Circuit elements or circuit blocks under test (resistors, capacitors, MOSFETs, ring oscillators, logic circuits, and memory arrays)
C. Placement on silicon (scribe line, test vehicle, on-product)
D. Test metal layer (first, fifth, and last)
E. Test type (DC parametric, low frequency (<10 MHz), high frequency, and digital)
F. Test equipment location (manufacturing in-line and laboratory)

1.4 Scope of the Book

In this book we cover the basic ideas and concepts of electrical test structure design, characterization, and analysis for digital CMOS applications. The implementation of these ideas and concepts is demonstrated with numerous examples. The importance of differential design and measurement techniques, compact designs using a single layer of metal, and test time reduction techniques are emphasized throughout the book. References at the end of each chapter include text books on related topics, selected articles from the scientific literature, and application notes from equipment manufacturers.

Chapter 2 begins with a brief description of the circuit elements in CMOS technology and scaling rules for circuit parameters. An overview of electrical tests, from DC to high frequency, is followed by a discussion of the different types of test equipment and techniques for interfacing to silicon. The nuts and bolts of test structure macros including I/O pads, signal propagation and delay rules, power distribution, and some commonly used logic circuit blocks are reviewed. A macro template concept for achieving high efficiency in design and test is introduced and tips on migrating designs from one technology generation to next are given. We recommend that readers not familiar with test structure design read through Chapter 2 before delving into the other chapters.

Test structures for resistors, capacitors, and MOSFET characterization are covered in Chapters 3, 4, and 5 respectively. Ring oscillator-based test structures are described in Chapter 6. In each of these four chapters, five test structure macro design examples are presented. The first example contains basic design principles implemented in a single metal layer. The design complexity is gradually increased in the subsequent examples.

High-speed test structures for measurements of circuit delays with sub-ps precision are described in Chapter 7. In Chapter 8, test structures specific to partially depleted silicon-on-insulator (PD-SOI) technology are described. These test structures are designed to measure the influence of floating-body effects on circuit delays and the impact of thermal effects on MOSFET characteristics. As in Chapters 3, 4, 5, and 6, Chapters 7 and 8 each contain five macro design examples.

Test equipment for in-line characterization and laboratory bench tests is covered in Chapter 9. In Chapter 10, standard statistical data analysis methods are introduced followed by data visualization techniques specific to CMOS technology. Numerous examples are included to illustrate and expand upon key concepts. Our experience has convinced us that developing a sound plan for test and data analysis must be an integral part of the test structure design process.

The subject of microelectronic test structures for CMOS technology is very broad and multi-faceted. While it is impossible in a book of finite length to address all of the important issues, we describe a methodology and an approach to this subject that transcends the content of the book. This approach is more generally applicable to other electronic technologies as well, such as Si bipolar, GaAs, and photovoltaics. We hope this book will be of interest to those both with and without previous microelectronic test structure design experience and that it will help all to create first-time correct designs with a high degree of efficiency.

References

1. Chowdhury R 'Eureka!' the stories of Archimedes and the golden crown. http://www.longlongtimeago.com/llta_greatdiscoveries_archimedes_eureka.html. Accessed 15 Mar 2011
2. Biruni AR. In: Wikipedia. http://en.wikipedia.org/wiki/Abu_Rayhan_Biruni#Scientific_method. Accessed 15 Mar 2011
3. Wright O (1953) How we invented the airplane: an illustrated history. Dover, New York, NY, p 55
4. Jaeger RC (2001) Introduction to microelectronic fabrication, vol 5, Modular series on solid state devices, 2nd edn. Prentice Hall, Upper Saddle River, NJ
5. Campbell SA (2001) The science and engineering of microelectronic fabrication, 2nd edn. Oxford University Press, Oxford
6. Moore GE (1965) Cramming more components onto integrated circuits. Electronics 38:114–117

7. Dennard RH, Gaensslen FH, Yu H-N, Rideout VL, Bassous E, LeBlanc AR (1974) Design of ion-implanted MOSFET's with very small physical dimensions. IEEE J Solid-State Circuits SC-9:256–268
8. Bohr M (2009) The new era of scaling in an SoC world. International solid-state circuit conference (ISSCC) proceedings, 2009:23–28
9. Shewhart WA (1986) Statistical method from the viewpoint of quality control. Dover, New York, NY
10. Taguchi G, Chowdhury S, Wu Y (2004) Taguchi's quality engineering handbook. Wiley, Hoboken, NJ
11. Tennant G (2001) Six sigma: SPC and TQM in manufacturing and services. Gower, Aldershot, England
12. Schroder DK (2006) Semiconductor material and device characterization, 3rd edn. Wiley, Hoboken, NJ
13. Orshansky M, Nassif S, Boning D (2008) Test structures for variability. In: Chandrakasan A (ed) Design for manufacturability and statistical design: a constructive approach. Springer

Chapter 2
Test Structure Basics

Contents

M. Bhushan, M.B. Ketchen, *Microelectronic Test Structures for CMOS Technology*,
DOI 10.1007/978-1-4419-9377-9_2,

Electrical measurements are made either by directly contacting the test structure on a silicon wafer to the test equipment as shown in Fig. 2.1 or after dicing and packaging the test structure. The design of test structures is therefore closely coupled to the method of interfacing to the test equipment and the capabilities and limitations of the test equipment itself. Design and test efficiency is improved by standardizing the footprints of each class of test structures and maintaining a high degree of commonality among them. Many of the test structures for MOSFET and CMOS circuit characterization may be implemented at the first level of metal. During technology development and manufacturing, feedback from such test structures early in the process cycle can help reduce the fabrication cost of CMOS products.

In this chapter, the essential elements of electrical test structures in CMOS technology are introduced. Special emphasis is placed on integrating the design effort with silicon area usage constraints and test requirements. The concepts presented here for achieving efficiency in design, data acquisition, and analysis are followed throughout the book.

The following terminology is used loosely throughout the book. A device under test (DUT) is the circuit element or the circuit block to be characterized. A DUT may be a MOSFET, a resistor, a ring oscillator, a CMOS circuit block, or a number of these circuit elements and blocks connected together to form a single unit such as an array of MOSFETs. A *test structure* comprises a DUT and peripheral circuitry required for carrying out the measurements. A *macro* contains one or more test structures, assembled together as a single unit to be placed on silicon. The test structures within a macro may share peripheral circuits and I/O pads to improve silicon area utilization and test efficiency.

In Section 2.1, a brief description of CMOS circuit elements and basic logic gates is given followed by CMOS technology scaling rules. The types of test equipment used in the manufacturing line and in the laboratory are introduced in Section 2.2. Silicon-to-test equipment interfaces, which include metal probes and packages, are discussed in Section 2.3. Nuts and bolts of macro designs covered in Section 2.4 include I/O pads, signal propagation in logic gates and interconnects,

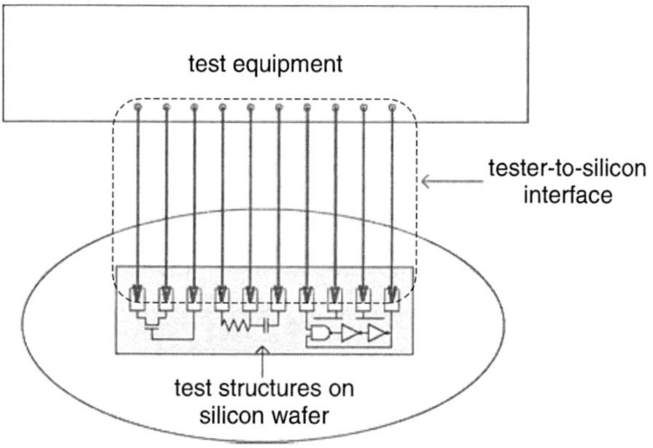

Fig. 2.1 Interfacing test equipment to test structures on silicon

on-chip buffers, I/O drivers, power distribution, and commonly used logic circuit blocks. The concept of macro templates, design methodology, and tips on migrating macros from one technology node to another are described in Section 2.5.

We have assumed that the reader is familiar with at least some aspects of CMOS technology, circuit design, and test. For an in-depth treatment of these topics, we recommend standard textbooks on CMOS fundamentals [1–3], semiconductor devices including MOSFETs [4, 5], microelectronic fabrication [6, 7], and semiconductor characterization [8].

2.1 CMOS Circuit Elements and Scaling

The two key constituents of digital CMOS circuits are the complementary MOSFETs, n-FET (NMOS), and p-FET (PMOS), configured to consume power primarily during switching. A p/n diode is another active element used in I/O protection, voltage reference, and isolation circuits. Precision resistors are used in CMOS analog circuit applications such as amplifiers, analog-to-digital (A/D) converters, and phase-locked loops for internal clock generation. Metal oxide semiconductor (MOS) capacitors serve as decoupling capacitors for power supply stabilization. Parasitic resistances and capacitances associated with MOSFETs and metal interconnects have a measurable impact on circuit behavior. These parasitic elements are characterized and modeled along with the other active and passive circuit elements.

2.1.1 MOSFETs and Diodes

Physical schematic cross sections of an n-FET and a p-FET in a conventional bulk silicon technology are shown in Fig. 2.2a. The source, S, and drain, D, electrodes or terminals of the n-FET are formed with heavily doped n-type diffusion, or n+, regions in a p-type body or p-well. Similarly, S and D electrodes of the p-FET are formed with p+ diffusion in an n-type body or n-well. The voltage on the gate electrode, G, which is separated by a thin oxide layer (not shown) from the channel region between S and D terminals controls the source-to-drain current of the MOSFET. Contact to the body, B, of the MOSFET is made through a p+ diffusion layer for a p-type body and through an n+ diffusion layer for an n-type body. Both n-well and p-well may be electrically isolated from the substrate in a twin-well (twin-tub) process.

Physical layout schematics of an n-FET and a p-FET are shown in Fig. 2.2b in a standardized format used throughout the book. Only the key design layers are shown to preserve clarity in the drawings. The drawings are not to scale. Outline of the n-well is shown for the p-FET, which distinguishes it from the layout of the n-FET. Layer DF denotes the n+ and p+ diffusion layers, PS is the gate layer, and H0 squares represent vias connecting DF and PS to the first metal layer M1 through a dielectric isolation layer. Each MOSFET has two PS electrodes connected

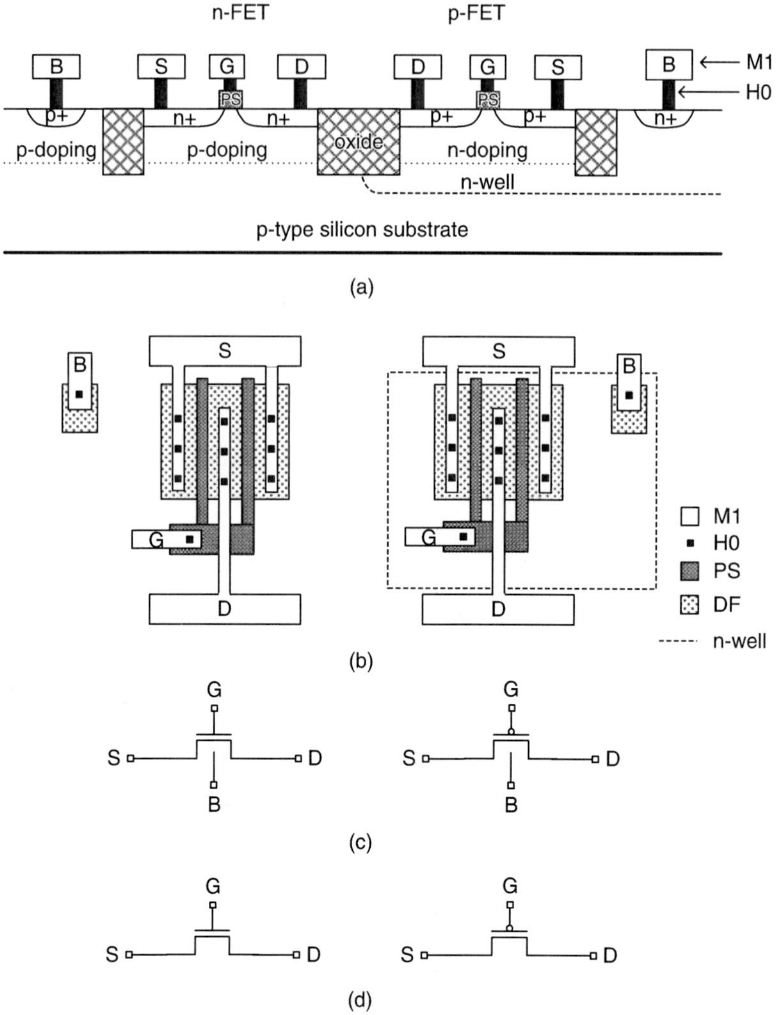

Fig. 2.2 An n-FET and a p-FET: **a** schematic cross sections, **b** physical layouts, **c** circuit symbols with S, D, G, and B terminals, and **d** circuit symbols with S, D, and G terminals

in parallel (two-finger design), the outer DF regions form the S terminal and the region between the fingers is the D terminal. The drain-to-source current I_{ds} is modulated by the voltages on the G and D terminals with respect to the S terminal, V_{gs} and V_{ds}, respectively. The I_{ds} increases as the width W of the MOSFET is increased or its channel length L_p is reduced.

Circuit symbols for the n-FET and the p-FET with terminals S, D, G, and B are shown in Fig. 2.2c. The B terminal is connected to an independent power supply to control the body-to-source voltage V_{bs} if a twin-well process is used. In most circuits, terminal B is connected to S ($V_{bs} = 0$) and three terminal representations

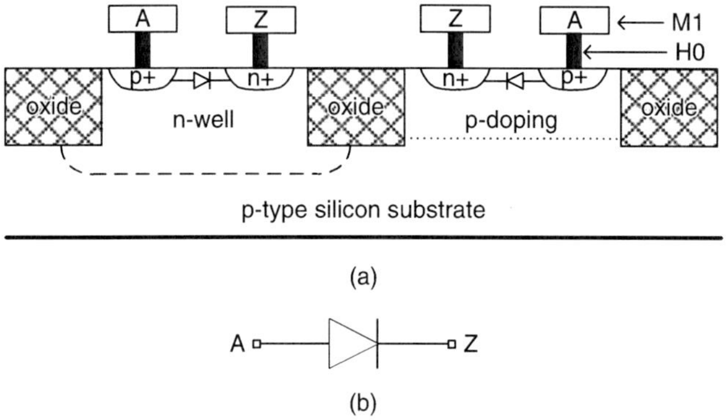

Fig. 2.3 **a** Schematic cross sections of p+/n and n+/p diodes. **b** Circuit symbol of a diode

of the MOSFETs shown in Fig. 2.2d are used instead. MOSFET properties are described in more detail in Chapter 5.

The cross sections and the circuit symbol of p–n diodes are shown in Fig. 2.3. One application of p–n diodes is in electrostatic discharge (ESD) protection circuits. An ESD circuit provides protection to the thin gate oxide layer of MOSFETs from damage by a high-voltage build-up on the I/O pads (Section 2.4.5). Diodes are also used in temperature sensing and temperature-insensitive bandgap voltage reference circuits.

2.1.2 Precision Resistors and Capacitors

The schematic cross section of a resistor formed in the silicon n-well is shown in Fig. 2.4. Contact resistance is reduced with n+ diffusion implants. Resistors may also be defined in the PS layer. These precision resistors are primarily used in analog circuit designs.

The MOS capacitor provides a higher capacitance per unit area than do metal capacitors. These capacitors, placed across the power supply, stabilize the voltage spikes generated during switching of CMOS circuits. The schematic cross section of a decoupling capacitor (DECAP), is shown in Fig. 2.5a. Contact to the n-well is made through an n+ diffusion layer, and the PS layer forms the second electrode of

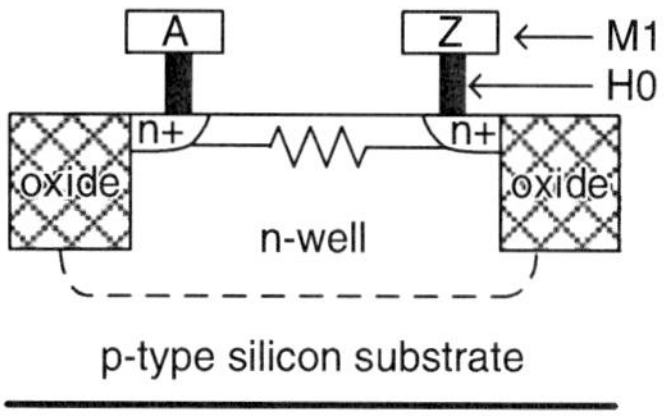

Fig. 2.4 Schematic cross section of a resistor element in silicon, formed in an n-well with n+ diffused regions for low resistance contacts

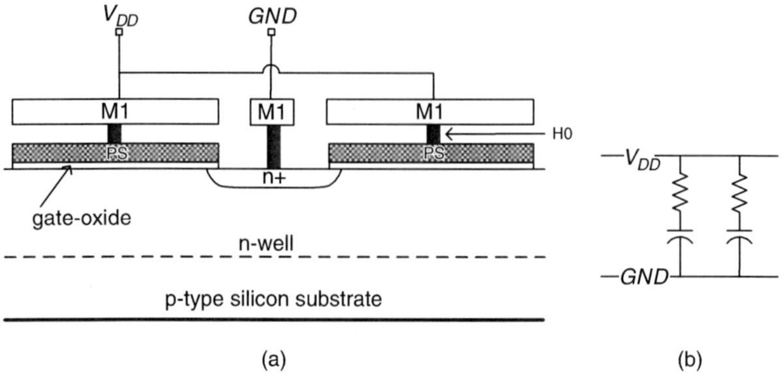

Fig. 2.5 **a** Schematic cross section of a gate oxide capacitor (DECAP) with two adjacent sections. **b** Equivalent circuit of a DECAP

the capacitor, separated by the gate oxide layer from the n-well. An equivalent circuit of the DECAP is shown in Fig. 2.5b. There is a parasitic resistance contribution from the PS and diffusion layers and via contacts in series with the capacitor. When using DECAPs for voltage stabilization, an upper limit on the application frequency is set by the *RC* time constant of the DECAP.

Silicon film resistances and p/n junction capacitances add parasitic elements to CMOS circuits. Parasitic bipolar transistors are formed by n–p–n and p–n–p layers in the n-FET and p-FET, respectively. By design, these transistors are prevented from turning on and rendering a CMOS circuit inoperative (latch up).

2.1.3 Interconnects

The active and passive devices defined in silicon are electrically connected together with metal layers deposited on top of silicon. In CMOS technologies, a planarized vertical stack of three to as many as 12 or more Cu or Al metal layers is offered for interconnects. The metal layers are vertically separated by an inter-level dielectric (ILD) material. Holes or vias are etched in the ILD layers, which are then filled with metal to serve as contacts between vertically adjacent metal layers. The lower metal layers are thinner and the allowed minimum linewidth is small to accommodate dense local wiring. The wire thicknesses and linewidths are increased in higher layers to meet the current-carrying requirements of clock and power distributions across the chip and to reduce parasitic resistances in long signal wires. Generally, product chip wiring is facilitated with wires travelling in orthogonal directions in vertically adjacent layers.

The schematic cross section of a planar five-layer metal stack with two thin ($1\times$) layers and two thick ($2\times$) layers with twice the minimum dimensions of thin layers is shown in Fig. 2.6a. The fifth metal layer, MT, which is also the top layer, is thicker and may have a protective coating of Al for ease of probing or packaging.

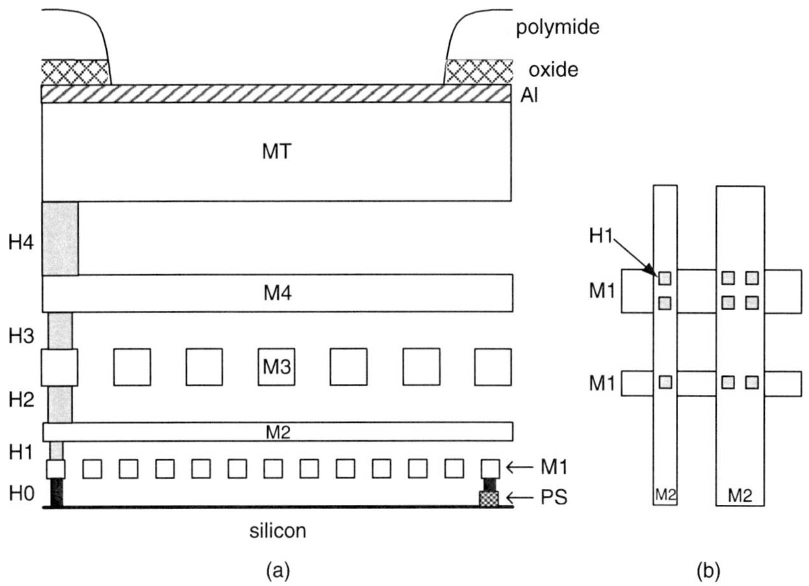

Fig. 2.6 **a** Metal wiring stack with two thin layers M1 and M2, two 2× thick layers M3 and M4, and a top layer MT. **b** Planar view of M1 and M2 layers with and without redundant vias. Via layer H0 connects DF and PS to M1; H1, H2, H3, and H4 inter-level vias above M1 are shown in a stacked configuration

The symbols for the metal layers are M*X*, where *X* is the layer number starting with 1 at the bottom, and the via layer above M*X* is H*X*. Other layers between M4 and MT may be present in CMOS product chips.

Via layers H0–H4 are square or rectangular in shape. The material for the lowest layer via, H0, contacting DF or PS layers to M1 wires is tungsten or tungsten alloy and has a higher resistance than Cu and Al. Via dimensions are smaller than the metal line dimensions and therefore vias are more susceptible to open-circuit failures. Process yield is improved with wider metal lines and redundant vias wherever possible. A planar view of metal interconnects between layers M1 and M2 with and without redundant vias is shown in Fig. 2.6b.

Resistances and capacitances associated with interconnect wires are measured and modeled for circuit simulations as they add to signal propagation delays and introduce coupling between signal lines. Spurious metal between wires may cause electrical shorts. Breaks in wires and insufficient metal in vias create electrical opens. These "shorts" and "opens" are detrimental to product yield. Test structures are designed to detect the presence of such defects.

In a CMOS manufacturing flow, the first electrical test stop for full active devices occurs immediately after M1 delineation. In Chapters 3, 4, 5, 6, 7, and 8, a number of macro design examples of moderate complexity, implemented at the M1 metal level, are provided. Such designs facilitate electrical characterization early in the process cycle.

2.1.4 Physical Layout and Ground Rules

Electrical characteristics of circuit elements in CMOS products are dependent on the properties and physical dimensions of various material layers. In the 65 nm CMOS technology node and beyond, charge carrier mobility enhancement in strained silicon layers and across wafer and local temperature gradients during processing of the silicon wafer impose physical layout dependency on MOSFET characteristics. Linewidth and layer thickness variations arising from photolithographic processes, reactive ion etching (RIE), and chemical mechanical polishing (CMP) are influenced by local pattern density of films, and location of the test structure on wafer. Parasitic resistances and capacitance of silicon diffusion layers, polysilicon or metal gate electrodes, and metal interconnects are therefore dependent on the process recipes as well as their physical layouts.

The ground rules (GRs) for physical design in each technology are provided to the designers by the technology development team or the silicon foundry. The GRs include minimum and maximum dimensions of shapes on all layers, inter-layer and intra-layer spacing and overlap between layers, allowed shapes in a layer, and their pattern densities. There are additional rules arising from design for manufacturability (DFM) which further restrict design styles in favor of higher product yield. The extent of incorporation of DFM GRs may vary from product to product in the same technology node. The physical layout styles for CMOS gates and circuit blocks may vary with the layout of the on-chip power distribution grid. Hence in designing test structures, it is important to follow the layout styles which best represent the product chips of interest.

We have adopted a standard physical layout style for illustrating the macro design concepts presented in this book. This style, with key design layers on a grid, is typical of the 65 nm CMOS technology node and beyond. In Appendix A, a standardized set of physical dimensions of layers and electrical parameters of circuits used in examples throughout the book are listed. These properties have been selected to simplify calculations and are in the range of the 65 and 45 nm CMOS technology nodes but do not represent any specific CMOS foundry. Description of a standard macro template with a linear array of 25 I/O pads (1×25 padset) is also included.

2.1.5 CMOS Logic Gates

In combinational logic, the most commonly used circuit blocks for carrying out simple logic operations are inverter, NAND, and NOR. Other circuits such as XOR, XNOR, AOI, and OAI are also available as single blocks in a circuit design library. These circuit blocks are referred to as static logic gates. Except during signal transition through a logic gate, the input and output voltage levels of these logic gates are held either high or "1" or low or "0," where "1" and "0" represent V_{DD} and GND potentials. This basic CMOS family of logic gates comprises n-FETs and p-FETs and one or two metal layers for interconnects. Single n-FETs and p-FETs, or both MOSFET types connected in parallel, act as switch elements in signal lines. The static logic gates and switches form the basic building blocks of more complex functional circuits such as decoders, multiplexers, adders, and multipliers. In this section, we describe the circuit schematics and logic operations of logic gates (inverter, NAND, NOR, and XNOR) and MOSFET switches.

A static logic gate may have one or more inputs and a single output. A signal transition occurs when the output voltage level changes in response to changes in one or more input levels. If the transition results in an output changing from "0" to "1," it is called a pull-up (PU) transition. A pull-down (PD) transition occurs when the output changes from "1" to "0." The signal propagation delay across a logic gate is commonly defined as time delay from the mid-point of the transition-inducing waveform to the mid-point of the output waveform.

The physical layout and the circuit schematic of a CMOS inverter comprising an n-FET and a p-FET are shown in Fig. 2.7a, b, respectively. The gates of the p-FET and the n-FET are connected at the PS level. The metal wires at the top and the bottom are connected to the V_{DD} and GND terminals of the power supply. An inverter circuit symbol with power supply connections is shown in Fig. 2.7c. This "mixed-mode" representation of a logic gate or a circuit is a useful way of combining its symbol with the physical arrangement of the power grid in the layout. The relationship between voltage levels at input node A and output node Z with A either at "0" or "1" is shown in the logic truth table in Fig. 2.7d.

The NAND, NOR, and XNOR logic gates have two or more inputs and a single output. Any one or multiple inputs may switch voltage states at a time. It is convenient to add the number of inputs to the name of the gate. For example, a two-input NAND gate is referred to as NAND2. Circuit schematics, symbols with

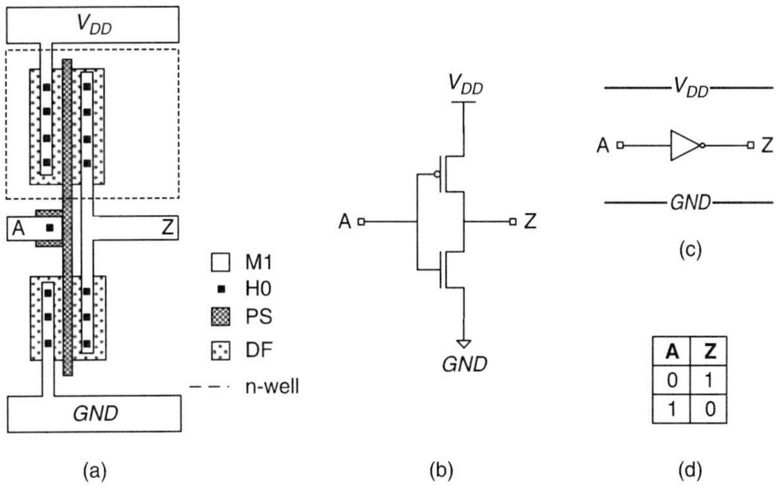

Fig. 2.7 An inverter: **a** physical layout, **b** circuit schematic, **c** a mixed circuit and physical representation with symbol and power busses, and **d** logic truth table

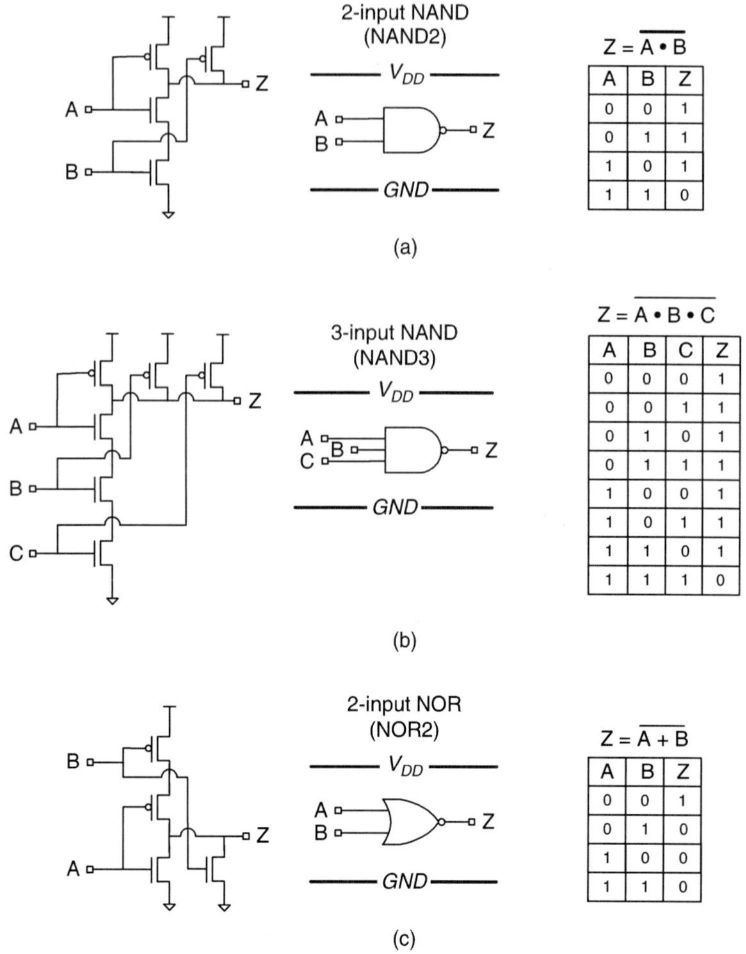

A	B	Z
0	0	1
0	1	1
1	0	1
1	1	0

A	B	C	Z
0	0	0	1
0	0	1	1
0	1	0	1
0	1	1	1
1	0	0	1
1	0	1	1
1	1	0	1
1	1	1	0

A	B	Z
0	0	1
0	1	0
1	0	0
1	1	0

Fig. 2.8 Circuit schematic, symbol with power busses, and logic truth table of **a** a NAND2, **b** a NAND3, and **c** a NOR2 logic gate

power busses, and truth tables for a NAND2, a NAND3, and a NOR2 are shown in Fig. 2.8a–c. Circuit schematic, symbol with power busses, and logical truth table for an XNOR2 are shown in Fig. 2.9.

In Fig. 2.10, the three commonly used MOSFET switch configurations for passing or blocking a signal are shown. The switch is opened or closed by turning the MOSFET channel current off or on with a control signal applied to its gate. An n-FET switch, referred to in this configuration as an n-passgate or n-passtransistor, is more effective in passing a "0." A p-FET switch (p-passgate or p-passtransistor) is more effective in passing a "1." A transmission gate shown in Fig. 2.10c with both MOSFET types connected in parallel is used for effectively passing either a "1" or a "0." If a single control signal is used for both the n-FET and the p-FET in

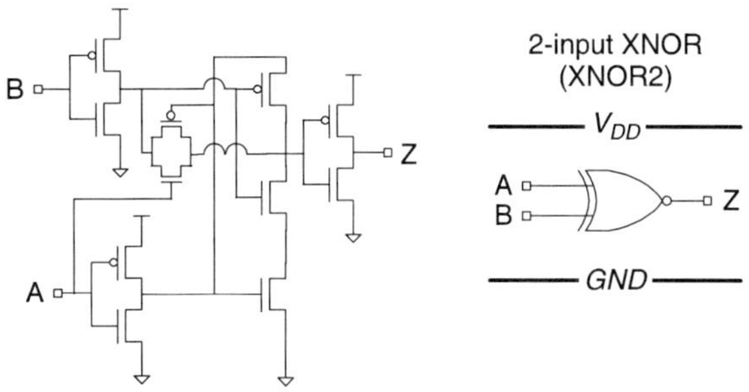

$Z = \overline{A \oplus B}$

A	B	Z
0	0	1
0	1	0
1	0	0
1	1	1

Fig. 2.9 Circuit schematic, symbol, and logic truth table of an XNOR2 logic gate

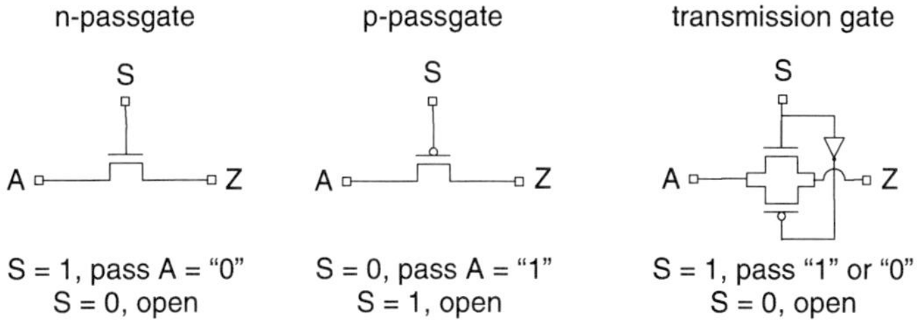

Fig. 2.10 MOSFET switches to pass or block a signal from A to Z with a control input S: **a** n-passgate, **b** p-passgate, and **c** transmission gate

the switch shown in Fig. 2.10c, an inverter is included to provide a complementary control signal to the gate of the p-FET.

For validation of models used in circuit simulations, the AC behavior of these logic gates is characterized to correlate the measured signal propagation delays, power, and other circuit properties with model predictions. Test structures for high-speed characterization of logic gates are covered in Chapters 6, 7, and 8.

2.1.6 CMOS Scaling Rules

With advances in lithography and semiconductor fabrication tools, the feature sizes in CMOS technologies have been following the trend predicted by Gordon Moore in 1965 [9]. Since then minimum feature size has been shrinking by a factor of ~0.7× every 2 years, with the potential of doubling the maximum transistor count on product chips. The technology node name reflects the approximate physical gate length (in nm) of the MOSFET, which is the minimum dimension of the PS layer in the direction of current flow in the channel. CMOS scaling is also described in the context of the minimum contacted gate pitch, which is the sum of PS-to-PS spacing and gate length with H0 contact via in-between the two PS shapes. The trends in minimum contacted pitch dimension and minimum SRAM cell area with technology nodes are shown in Fig. 2.11. Typically, other minimum dimensions in

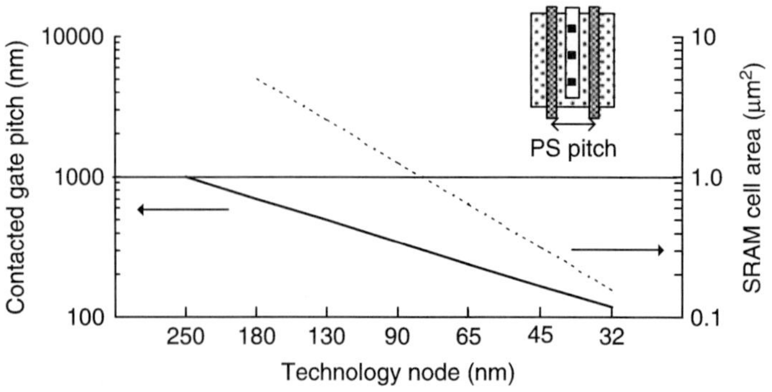

Fig. 2.11 Contacted gate pitch and SRAM cell area scaling trends of 0.7× and 0.5× per technology node, respectively [10]. *Inset* depicts PS-contacted pitch

silicon and in lower metal layers and vias are scaled by about the same factor as the contacted pitch.

In addition to feature sizes and circuit density, the power supply voltage is also scaled. In a constant electric field scaling scheme (constant V_{DD}/L_p and V_{DD}/t_{ox}), V_{DD} is scaled by the same factor as feature size. In practice, to maintain voltage compatibility between different generations, V_{DD} is scaled more gradually. In a generalized scaling scheme, the vertical and horizontal electric fields increase at the same rate, preserving the shape of the electric field pattern as the feature size is reduced. In Table 2.1, the multiplication factors for scaling key MOSFET, wire, and circuit parameters are listed [4], where S (<1) and $1/S_k$ ($S_k < 1$) are the multiplication factors for feature size and electric field, respectively. Prior to the introduction of high-dielectric constant (HK) materials for the gate oxide in the 45

Table 2.1 Generalized rules for scaling MOSFET, wire, and circuit parameters to smaller dimensions [4]

MOSFET or circuit parameter	Multiplication factor ($S < 1$, $S_k < 1$)
Contacted gate pitch	S
MOSFET dimensions (L_p, W, t_{ox})	S
MOSFET capacitance	S
V_{DD}	S/S_k
I_{on}	$\sim S/S_k$
Circuit delay (C–V/I)	S
Power density	$(1/S_k)^2$
Circuit density	$1/S^2$
Metal wire thickness (1× wire)	$\sim S$
Metal wire width (1× wire)	$\sim S$
Metal wire resistance per unit length (R_w)	$1/S^2$
Metal wire capacitance per unit length (C_w)	~ 1
Voltage drop per unit wire length ($I_{on}R_w$)	$1/(SS_k)$
Delay per unit wire length (R_wC_w)	$1/S^2$
Delay for wire length scaled by S	$1/S$

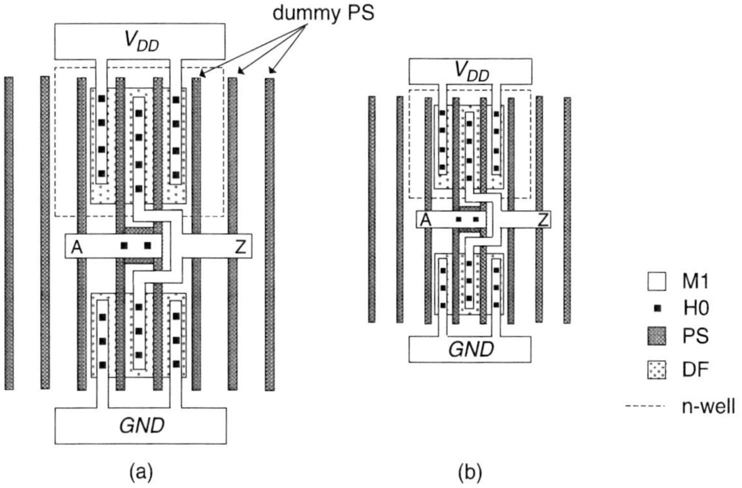

Fig. 2.12 **a** Inverter layout with two PS fingers. **b** Inverter layout scaled by 0.7×. Neighboring PS dummy shapes at the same pitch as in the inverter are also shown

and 32 nm technology nodes, to limit the gate oxide tunneling current, the gate oxide thickness t_{ox} was scaled more gradually than feature size.

Physical layout of an inverter (with two PS fingers) and its 0.7× scaled version, with a factor of two reduction in area, are shown in Fig. 2.12. Faithful printing of layer design dimensions smaller than the wavelength of the optical source in the photolithography exposure tool has led to many challenges. Prior to optical mask generation, the design dataset is modified using an algorithm for optical proximity correction. Assist features are placed on the mask to reproduce the design dimensions on the wafer. RIE and CMP processes are sensitive to local area coverage and it is important to keep the pattern densities of different layers nearly constant over a desired length scale. This is accomplished by either manually adding dummy shapes on each layer that are electrically isolated from the circuits or using a tool to automatically fill any available space with dummy shapes. The dummy shapes for the PS fingers are included in Fig. 2.12.

2.2 Electrical Measurements and Test Equipment

In early technology development and in process monitoring in the manufacturing line, prior to the availability of all active elements or metal interconnect layers, electrical characterization is primarily focused on DC or low-frequency tests. High-frequency and digital tests are performed on more complex test structures after partial or full process integration of the active circuit elements with interconnect layers. Automated test equipment (ATE) for DC parametric, digital, and RF tests are commercially available. A more detailed description of test equipment for CMOS devices and circuits is covered in Chapter 9.

Laboratory test equipment for DC tests comprise one or more independent power supplies; voltage, current, resistance, and capacitance meters; pulse generators; frequency counters; and other high-frequency test equipment. A semiconductor parameter analyzer (SPA) is used for full I–V and C–V characterization of MOSFETs and diodes. These SPAs are programmable and provide a user-friendly interface with built-in data analysis features.

In-line DC parametric ATE provide a more efficient means of data collection than laboratory test equipment. These testers typically comprise up to eight source measure units (SMUs) which can be configured as voltage and current sources or meters. An electronic switch matrix is used to connect any SMU to one or multiple locations allowing flexibility in I/O configuration. The two test modes for measuring I–V characteristics and resistances are voltage force current measure (VFIM) and current force voltage measure (IFVM). Parametric ATE with a low level of parallel test capability is also available. This parallelism provides significant reduction in test time.

An impedance meter is used for measuring stand-alone capacitors. Capacitance test structures, with CMOS circuit integration schemes, can be measured with a standard parametric tester. Frequency measurements from the kHz to the GHz range are made with a frequency counter, an oscilloscope, or a spectrum analyzer. Impedance meters and pulse generators for AC measurements may be incorporated in parametric ATE.

ATE for digital tests (at frequencies as high as a few GHz) are equipped with internal clocks and have the capability of handling programmable bit-stream logic inputs and outputs. These testers are fast and capable of making many measurements in parallel. Limited DC tests can also be carried out but the measurement accuracy in current ranges below 1 μA and in voltage ranges below 10 mV is poor compared to parametric ATE. Measurements at RF and microwave frequencies require special fixtures, cables, and careful impedance matching to couple the test equipment to the test structures.

Test structure efficiency may be enhanced by judicious use of the available features in the test equipment. Prior to designing a test structure, it is a good practice to become familiar with the characteristics and limitations of the designated test equipment. Some of the important factors to consider are number of independent source and sense elements, range and accuracy of measurements, frequency limits, measurement time, and parallelism. Whenever software packages or test algorithms are provided by the test equipment manufacturer, it is helpful to design test structures within the constraints of the existing code and avoid the overhead of custom software development.

2.3 Silicon Interface to Test Equipment

Electrical connections are made between the macroscopic test equipment and cables to the microscopic CMOS circuit elements of typical dimensions ~1 μm. These connections are facilitated with sharp metal needles or wires contacting large conducting I/O areas on silicon with dimensions of ~40 μm. The two primary physical configurations of these I/O contact areas in a macro are planar metal pads and solder bumps. Electrical tests at the wafer level are conducted with metal probes contacting the pads or bumps. Contacts to an individual chip in a package are made by wire bonding to planar metal pads or flip-chip bonding of solder bumps by reflow. I/O layout and contacting techniques are selected on the basis of the process step at which tests are conducted, test structure geometry, I/O count, current and power requirements, frequency of I/O signals, and number of test repetitions.

Example distributions of I/Os for a macro with planar pads and solder bumps are shown in Fig. 2.13a, b. Planar I/O pads have one or more layers of metal and are generally placed on the periphery of a macro or some other geometrical arrangement suitable for landing probe needles or wire bonds. Pb–Sn alloy solder bumps, also called controlled collapse chip connections (C4s), can be distributed across the surface of a macro or a chip.

For wafer-level tests, contact to I/O pads can be made with metal cantilever probes on planar metal pads or with vertical probes on solder bumps as shown in Fig. 2.13c, d. Cantilever probes are constructed from a hard material such as tungsten, rhenium–tungsten, beryllium–copper, and a palladium alloy. The probes are pressed down and a low resistance contact is made with a mechanical scrubbing action to break through the oxide layer on the metal surface.

Macros with a limited number (~10) of I/Os can be tested in a laboratory bench setup with individual probes mounted on a wafer probing station. The probes are

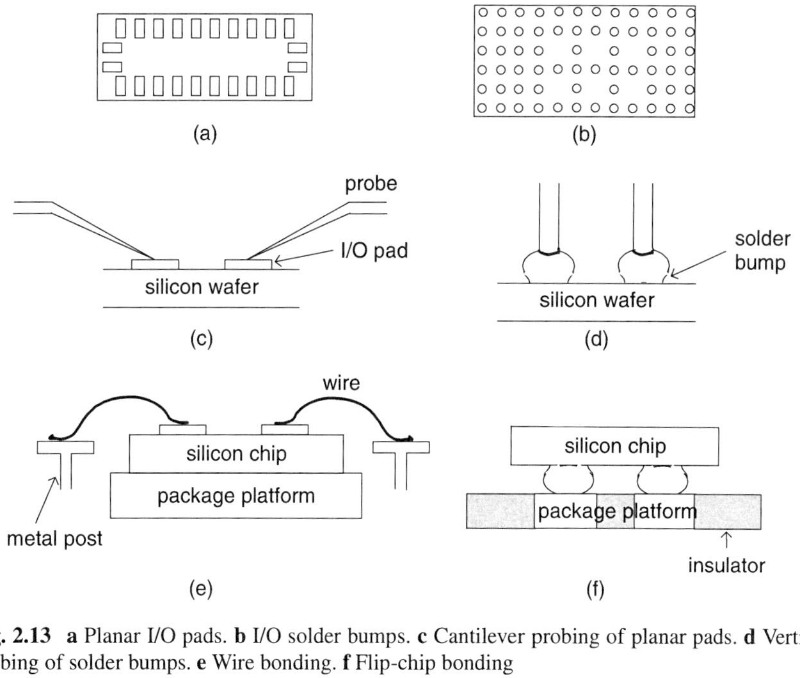

Fig. 2.13 **a** Planar I/O pads. **b** I/O solder bumps. **c** Cantilever probing of planar pads. **d** Vertical probing of solder bumps. **e** Wire bonding. **f** Flip-chip bonding

manually positioned and aligned one at a time to the I/O pads with the help of micromanipulator arms to control *X*, *Y*, and *Z* motions, and an optical microscope. Alternatively, a probe card with contact needles (up to ~100 or more) can be manipulated as a single unit and brought into contact with an underlying array of I/O pads on silicon. Motorized and programmable probing stations with micro-positioning capability for precision alignment are commercially available for test structures with a large number of I/O pads and smaller pad geometries. In silicon manufacturing, automated wafer probing stations and probe cards with a customized arrangement of probe tips to match a standardized footprint of I/O pads are used for a large set of macros to be tested. A more detailed discussion on the configuration of probe cards and the impact of their use on macro design and test are included in Section 2.3.1.

Individual macros for extended or repeated testing, for stress and reliability characterization, and for other non-standard tests are packaged in a chip carrier after dicing the silicon wafer. The diced chip is mounted on the package platform with an epoxy adhesive or by the formation of a metal eutectic bond. One end of a fine wire, 15–50 μm in diameter, is bonded to the I/O pad and the other to the metal post on the package as shown in Fig. 2.13e. There are two types of wire bonding processes. Ball bonding is used with copper and gold wires with good thermal and electrical conductivity. A ball of metal is formed at the tip of the wire by heat and then attached to a heated bonding surface by applying pressure and ultrasonic energy. Ultrasonic bonding alone is used with Al and Al alloy wires. In all cases, a protective coating of Al on the pads is preferred. Minimum pad area for wire bonding is ~1,600 μm^2. Both manual and automatic wire bonding machines are commercially available. With an automated setup, many thousands of reliable bonds can be made within an hour.

Macros with solder bumps (C4 s) are packaged with a flip-chip technique as shown in Fig. 2.13f. The package substrate has metalized areas matching the C4 pattern on the macro. The chip is placed upside down on the substrate and attached to it by applying heat and pressure to reflow the solder and create a bond. Dummy C4 s may be added on the macro to achieve uniform pressure during bonding. This technique is useful for large macros or chips with hundreds or thousands of I/Os for signals and for power distribution across the macro. With inherent low contact resistance, capacitance, and inductance, flip-chip techniques are well suited for measurements at high frequencies.

2.3.1 Probe Cards

A high density of I/O connections is achieved with the use of appropriately configured probe cards [11]. A probe card comprises metal probes mounted in a ring on a printed circuit board or ceramic substrate in a predefined geometrical arrangement. Simplified views of probe needles and a common probe card design are shown in Fig. 2.14. This mechanically robust card with very delicate probe needles can be manually aligned to the I/O pads on silicon in a laboratory environment. In an

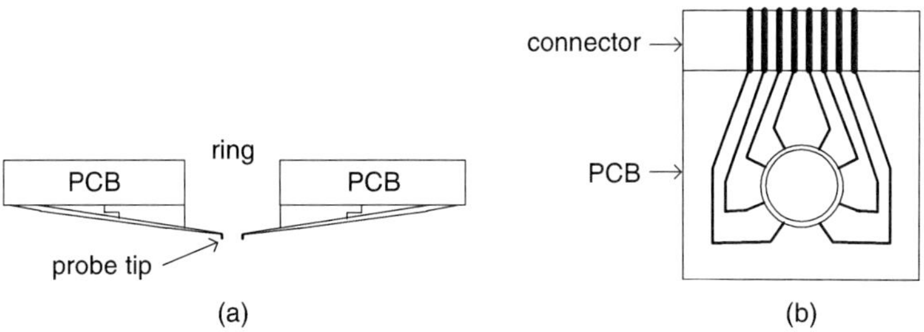

Fig. 2.14 **a** Side view of probes mounted around a ring on a printed circuit board (PCB). **b** Top view of a probe card schematic with eight probes mounted on a PCB with a plug-in connector

industrial setup, the probe card is mounted on a computer-controlled wafer probing station for automatic positioning and alignment to a specified macro location on the wafer.

The schematic of a test station in a manufacturing line is shown in Fig. 2.15. A wafer handler automatically loads and unloads wafers from a cassette which can hold 20–25 wafers. In a 300 mm silicon fabrication line, the cassette is replaced with a foup for keeping wafers in a clean and controlled environment. The wafer is placed on a temperature-controlled vacuum chuck. The probe card is mounted in a test head positioned above the wafer and a computer-controlled micromanipulator arm aligns the probe card to a specified X, Y location on the wafer with an alignment accuracy of $\pm 1\ \mu$m. The height of the probe card above the wafer is adjusted to facilitate loading and unloading of wafers and for applying desired pressure on the probes to establish low resistance ($\sim 1\ \Omega$) ohmic contacts.

Only a single probe card with <10–1,000 probe wires may be mounted on the probe station at a time. At the completion of the tests on one site, the wafer is stepped to the next site. With the step-and-repeat capability of the wafer-handling system, the probe card stays in alignment as the wafer is sequentially positioned to different macro locations on the wafer. Electrical wires connect the probe card to the test equipment. The entire wafer loading–unloading, probe card alignment to the test structure, test code for controlling the inputs and outputs of the test equipment, and data transfer to a database for characterization and analysis are automated using

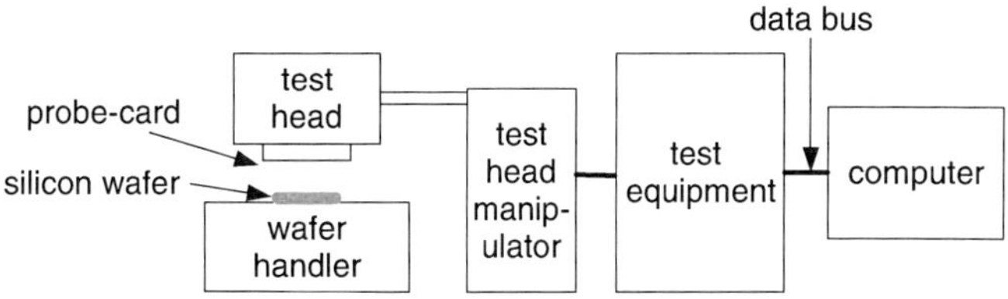

Fig. 2.15 Components of a test station, including wafer handler, test head with probe card assembly, a manipulator arm for positioning the test head, test equipment, and computer interface for data acquisition

computer control. The test station is housed in a controlled environment, as probe alignment precision and contact pressure are sensitive to fluctuations in ambient temperature and mechanical vibrations.

Custom probe cards to match the footprint of I/O pads of a macro can be purchased commercially. Probes are sufficiently electrically isolated from each other to reduce DC leakage currents and cross-coupling of high-frequency signals to acceptable levels. Probe cards for current measurements in the pA to fA range are constructed on ceramic substrates to improve DC electrical isolation. Probe card specifications include the maximum operating voltage and current-carrying capability of the probes. Currents exceeding the maximum limit may cause thermal heating and ultimately result in probe meltdown. High voltages may result in excessive leakage currents in the low current measurement ranges.

Probe cards for measurements at wafer temperatures exceeding 100°C over an extended period of time may include heat reflectors to keep the card temperature within an acceptable range. There are also constraints on the card materials used for measurements at temperatures below –25°C and if the probe card is placed in a refrigerated system. These special probe card designs ensure mechanical integrity of the low resistance electrical contact to the I/O pads under different conditions.

For high-speed measurements at frequencies above 1 GHz to as high as 50 GHz, signal I/O lines are shielded by using GND–signal–GND (G-S-G) pad arrangements shown in Fig. 2.16. Capacitors are incorporated in DC I/O lines to smooth out any noise spikes. These capacitors are mounted close to the probe tips and also on the printed circuit board.

Some mechanical rubbing motion is required for the probe tip to break through the thin native oxide layer to contact the metal I/O pad surface. Hence, probe tips tend to accumulate debris. The probes are cleaned with an abrasive material such as alumina or tungsten carbide after a predetermined number of touchdowns. The cleaning procedure, repeated after an assigned number of touchdowns, takes tens of seconds to perform and adds to the total test time. Repeated touchdowns and abrasive cleaning damage the probe tips, and the probe cards have to be replaced periodically. Qualifying different probe card designs and changing probe card on an ATE are time consuming. In a manufacturing environment, the time to test a macro must be minimized to increase wafer throughput and hence profit margins. It is therefore important to standardize the I/O pad geometry on as many macros as possible and avoid frequent exchanges of probe cards in a manufacturing line. New materials and probing techniques are evolving to improve probe lifetime.

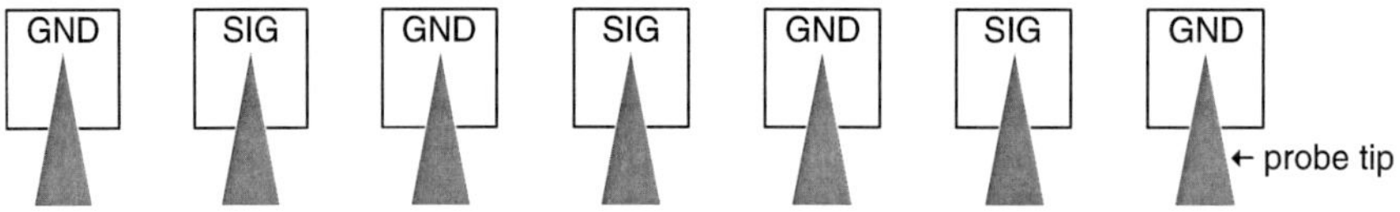

Fig. 2.16 I/O pad arrangement for high-frequency signals: GND-Signal-GND (G-S-G)

2.3.2 Advanced Probing Techniques

Thin flexible interposers (TFIs) have now largely replaced vertical probes for contacting very high-density C4s. Micro electro-mechanical system (MEMS) probe cards are available for full wafer testing. With a pad pitch of 50 μm, and low contact pressure allowing millions of touchdowns between probe card replacement, MEMS technology offers an attractive alternative to traditional needle probes.

In failure analysis, to contact circuit elements in the absence of I/O pads, extremely small probes (pico-probes) are used. These delicate pico-probes can make contact with a metal area less than 1 μm^2. The current-handling capability of such probes is of the order of 1 mA, and alignment is carried out via atomic force microscopy. The test capability is further limited as the maximum number of pico-probes in a test setup is $\sim$5.

In picosecond imaging circuit analysis (PICA), circuit delays are measured by detecting light emission during switching. Although this is a powerful technique for measuring circuit delays and relative signal arrival times in complex circuits, special sample preparation techniques are required, limiting its use to failure analysis applications. A number of other electron and optical emission techniques have been developed for failure analysis of CMOS circuits [8].

Development is underway to replace mechanical probing techniques with non-contact wireless methods. Each I/O pad is connected to an antenna and trans-receiver circuit and high-frequency wireless communication is established between the test structure and the test equipment. One application of the non-contact technique for measuring the frequency of a ring oscillator is mentioned in Section 6.5.

2.3.3 Macro Area and Test Time Efficiency

There are several factors to be considered in optimizing overall macro efficiency. The area occupied by a macro on silicon must be minimized so that either more macros can be placed on a test vehicle or more product chips can be produced per wafer. This must be done without compromising the information obtained from test structures for quality control of the manufacturing process and product debug. Compact macros with minimum available probe pitch and I/O pad area, shared I/O pads, and multiplexing schemes to increase the contents and tests per macro are some of the ways to improve macro efficiency.

The total test time is the sum of measurement time, indexing time which includes stepping and alignment, and probe cleaning time. Test efficiency, defined as (time per test $\times$ tests per macro $\times$ number of macros)/(total time), increases with the number of tests per probe touchdown. Test efficiency for an electrical test time of 0.01 s per test, indexing time of 0.5 s, and a cleaning time of 50 s per 100 touchdowns is shown as a function of number of tests per macro in Fig. 2.17. It is apparent that increasing the number of tests per touchdown improves test efficiency. This is facilitated by using multiplexing schemes to increase the number of

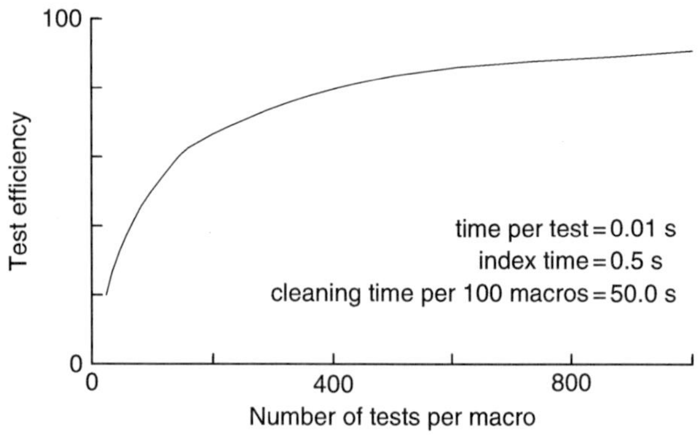

Fig. 2.17 Test efficiency as a function of number of tests per macro

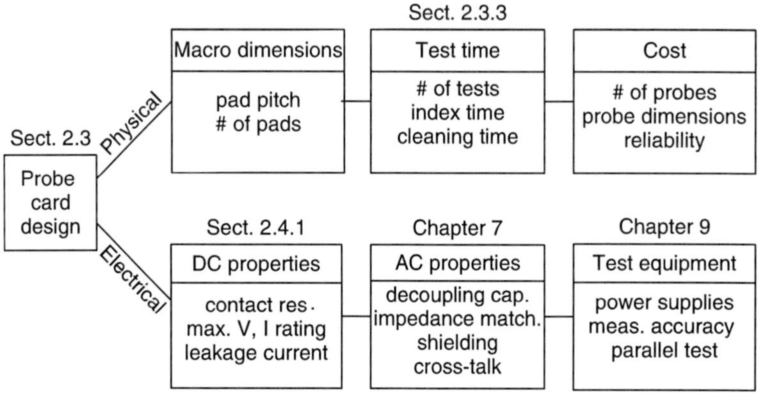

Fig. 2.18 Probe card design considerations

DUTs per macro and by increasing both the number of I/O pads within the macro and corresponding probes per probe card.

Test time can be significantly improved further by testing a number of DUTs in parallel. Design considerations for parallel testing for various types of DUTs are covered in the examples in Chapters 3, 4, 5, 6, 7, and 8. Probe card design and I/O pad layout contribute directly to macro area and test efficiency as mentioned above. In addition, probe cards impact measurement accuracy and limit current, voltage, and frequency ranges over which measurements are made. These considerations are discussed in a number of places in this book as indicated in Fig. 2.18.

2.4 Nuts and Bolts of Test Structure Macro Designs

The form factor (width/height) of macros placed in the scribe line is typically between 5 and 25, although large area macros may have a form factor approaching unity. In the macros designed for wafer-level probing, planar I/O pads serve as landing areas for cantilever probes mounted on probe cards. The I/O pads are placed in a linear array on one or both sides along the longer dimension of the macro. Pads frequently occupy a significant fraction of the total macro area and hence their utilization must be carefully optimized in creating area-efficient designs.

The constraints on macro geometry for scribe-line placement pose a greater challenge in signal and power distribution than does a large square or rectangular CMOS product chip with uniformly distributed I/Os. Computer-aided-design (CAD) tools developed for CMOS chip floorplanning and for power and timing optimization are typically not easily adapted to test structure designs. Often these tools may not be available to technology developers and test vehicle designers as the risk of a few malfunctioning test structures does not justify the cost of licensing and customizing these tools and providing training to a small design team. It is therefore prudent for a test structure macro designer to follow general circuit design principles and use rules of thumb described here and illustrated in examples throughout this book.

In this section, common features in macro designs which include I/O pads, electrical wiring, I/O drivers, ESD protection circuits, and power distribution and voltage stabilization are described. Placement of DUTs and peripheral circuits is dependent on the DUT types, measurement requirements, and I/O pad sharing schemes. Discrete circuit element DUTs, such as resistors and MOSFETs, and small-area circuit DUTs may be placed in the space between pads as shown in Fig. 2.19a or even under the pads. Large-area DUTs and peripheral circuits are placed along one or both sides of the I/O pad array as shown in Fig. 2.19b, c. The

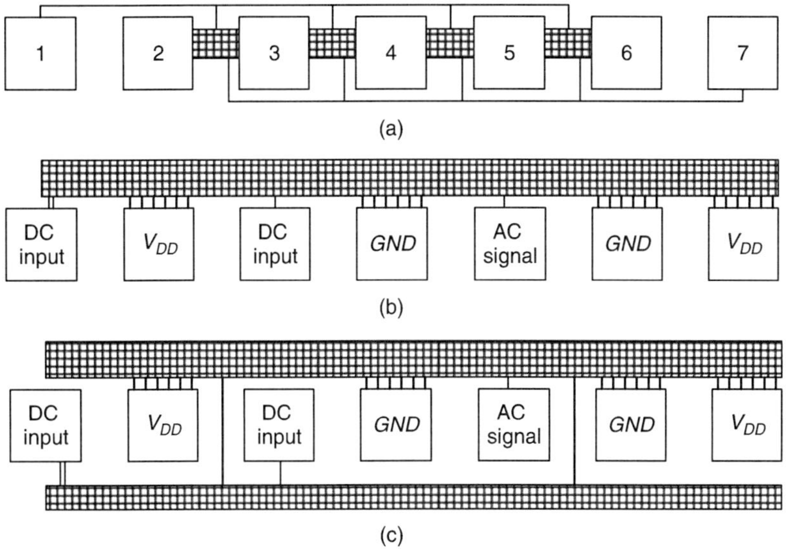

Fig. 2.19 Examples of placement of test structures (*hatched areas*) with respect to a linear array of I/O pads: **a** small-area DUTs placed in the space between pads. Large-area test structures placed **b** along one side and **c** along both sides of a linear pad array

majority of test structure macro designs examples in this book use a standard 1×25 padset described in Appendix A. The pads are numbered from left to right and in many cases the pad label indicates the I/O function.

2.4.1 I/O Pads

I/O pads are square or rectangular shapes of conducting material deposited on silicon to define landing areas for probe needles or wire bonds. In a regular linear array of pads, pad pitch is defined as the sum of pad width and pad spacing. The I/O pad resistance is minimized with the use of low resistance materials such as Cu and Al. During CMOS processing, wafer probing is done on Cu pads. At the final step of wafer processing, the pads are covered with a layer of Al which forms a thin oxide passivation layer on its surface when exposed to air. This oxide film is less susceptible to corrosion and mechanical damage than is Cu. It is easily punched through by the application of a mechanical scrubbing action when contact is made with probe needles or wire bonds. Al pads are also more suitable for measurements at high temperatures than are Cu pads. Prior to the M1 process step, however, DF or PS layers may be used for an early evaluation of silicon and MOSFET gate layers.

When a multi-layer metal stack is available, I/O pads may be patterned in any one or all the metal layers. If the I/O pads have several metal layers, interconnect vias between adjacent layers may be included in the pads. In Fig. 2.20a, the pads include stacked metal layers and interconnect vias. Contact to the DUTs can be made at any metal layer. Stacking metals in this way has the advantage of reducing the pad resistance. This arrangement is also useful when a DUT is first tested at metal M1 and subsequently re-tested at higher metal levels. In Fig. 2.20b, a unique DUT is placed at each metal level, significantly reducing the silicon area per DUT. The disadvantage is that each DUT is accessible only at the corresponding metal level and cannot be used for further diagnostics after a higher metal layer in the stack is deposited. In Fig. 2.20c, the DUT is placed partially under the M2 layer

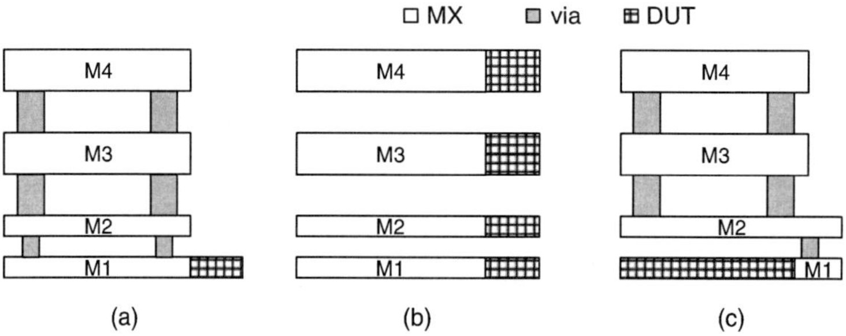

Fig. 2.20 Example cross sections of I/O pads in metals M1–M4 and relative placement of DUTs or parts of test structures: **a** test at any metal level, **b** test at only one metal level, and **c** test at M2 or above

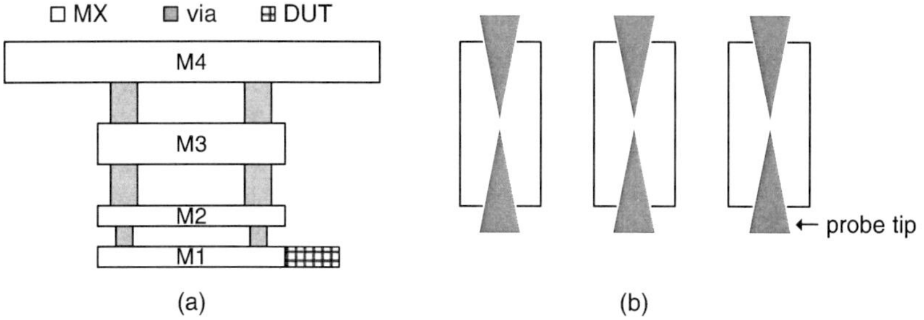

Fig. 2.21 **a** Schematic cross section of an I/O pad stack, showing a mushroom pad at the M4 metal layer. **b** Planar view of mushroom pads for landing two probes on each pad

to provide a larger surface area for the test structure design at M1. In this case, the tests are carried out at the M2 level or above. Other such schemes may be utilized for minimizing macro area. Complex DUTs and test structures designed with several metal layers are contacted only at the top layer.

Repeated testing may cause physical damage to the metal layers. As a result, the pad contact resistance may increase or there may be a failure to make an electrical contact altogether. One approach to overcome this problem is to increase the pad area and shift the probe contacting point in repeated touchdowns. The pad area may be increased with rectangular pads without affecting the pad pitch. This increases the macro area but the macro can still be tested with the standard probe card. Alternatively, the pad area may be increased only in the top layer, without impacting the test structure area allocation, as shown in Fig. 2.21a. These so-called mushroom pads are useful for repeated testing carried out after the top metal layer delineation on the wafer.

Increasing the I/O pad area provides an option of landing two probes on a single pad as shown in Fig. 2.21b. This arrangement facilitates a four-terminal Kelvin measurement as briefly discussed here and again in more detail in Section 3.2.3. Circuit schematics indicating parasitic resistances in series with the DUT for one and two probes per I/O pad are shown in Fig. 2.22. In Fig. 2.22a, with one probe per pad, the parasitic resistance is series with the DUT is $2\left(R_s + R_p\right)$. Here R_s is the contact resistance between the DUT and the probe tip, which includes the resistance of the pad itself and wire resistance between the pad and the DUT. R_p is the resistance of the metal probe and the cable connecting to the test equipment. With two probes per pad, the contribution of R_p is eliminated by forcing current through one pair of probes (contacted to AH2 and AL2) and measuring voltage across the second pair (AH1 and AL1). Mushroom pads are therefore useful for improving measurement accuracy. The probe card design is appropriately modified to enable four-terminal measurements with mushroom pads.

The physical layouts of the metal pads must follow the technology GRs. With the introduction of CMP of metal layers in standard CMOS processing, large solid areas of metal are forbidden because of dishing effects. Instead of a solid metal pad, a wire mesh design is used to reduce the effective metal area as illustrated in Fig. 2.23.

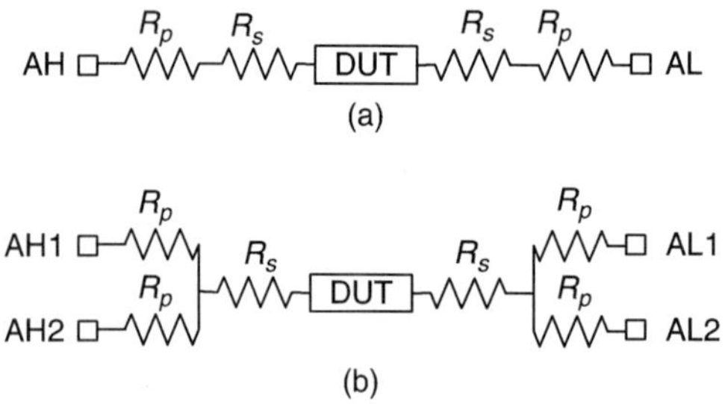

Fig. 2.22 Circuit schematics showing parasitic resistances in series with a DUT contacted with (**a**) two probes and (**b**) four probes. R_s is the resistance from the DUT to the pad and R_p is the probe contact resistance

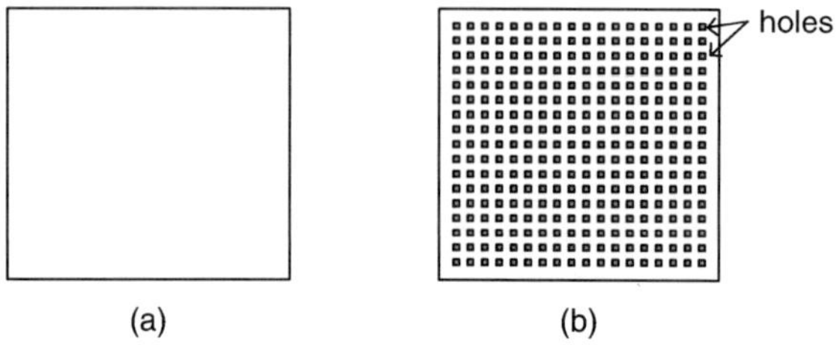

Fig. 2.23 Physical layout of I/O pads with (**a**) solid metal and (**b**) metal cheesing to form a mesh per technology GRs

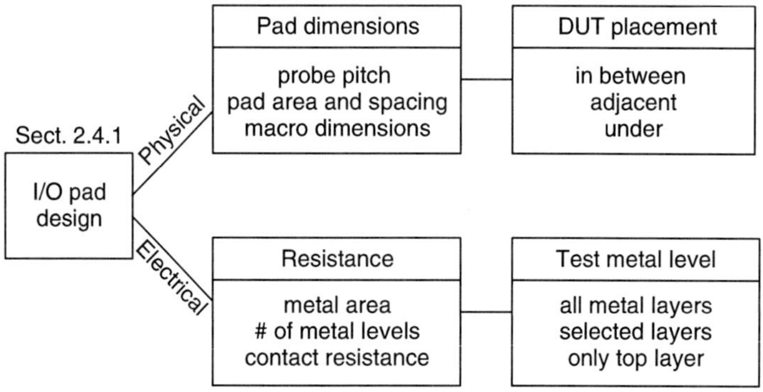

Fig. 2.24 Physical and electrical characteristics of I/O pads

The physical and electrical considerations for I/O pad design are summarized in Fig. 2.24.

2.4.2 Signal Propagation Delay of CMOS Logic Gates

The maximum operating frequency of a product is limited by the propagation delays of signals traversing circuit paths within a machine cycle time. Measurement of signal delays of logic circuit blocks and other high-speed characteristics is therefore an essential part of CMOS technology characterization. In addition, macro designs for DC and AC characterization may employ CMOS circuits for performing logic functions within a DUT or in peripheral control logic. For robust macro designs, the use of CMOS static logic gates is preferred.

Although sophisticated circuit simulators are available to carry out CMOS circuit designs, simple rules for estimating logic gate delays are very useful in sizing DUTs and their placement and in macro floorplanning early in the design phase. In a majority of macro design examples in this book, the sizing of logic gates and interconnect wires can be carried out with the delay rules described here. The *RC* equivalent circuit of logic gates described in this section also sets the foundation for extracting MOSFET and wire parameters from measurements of logic gate delays.

A pulse propagating through an inverting logic gate and associated timing diagram are shown in Fig. 2.25a, b. When the voltage level at the leading edge of the pulse at node A transitions from a "1" to a "0," the voltage at node Z rises from a "0" to "1" after a time delay τ_{pu}. This is a signal-falling-at-A to signal-rising-at-Z transition, also referred to as a fall-to-rise or pull-up (PU) transition. Similarly, the logic gate undergoes a rise-to-fall or pull-down (PD) transition for the trailing edge of the pulse with a propagation delay of τ_{pd}. In circuit simulations, the gate delays τ_{pu} and τ_{pd} are measured between 50% voltage levels ($V_{\mathrm{DD}}/2$) at circuit nodes A and Z. As indicated in Fig. 2.25c, the time for the signal level to transition from 10 to 90% V_{DD} is the rise time τ_{r} or the fall time τ_{f}. The signal propagation delays through a logic gate and rise/fall times depend on the *RC* load on the logic gate, in conjunction with its effective width, and current drive strength as explained below.

A MOSFET acts as a non-linear resistance whose value is a function of the gate-to-source and gate-to-drain voltages. This resistor, in parallel with the MOSFET capacitances, forms an *RC* network. The *RC* network representation of a logic gate is very helpful in relating logic gate characteristics to those of its constituent circuit elements.

The circuit schematic of an inverter and its simplified equivalent *RC* circuit is shown in Fig. 2.26a, b. The capacitance at input node A is the sum of the gate capacitances of the p-FET and n-FET which include gate-to-source, gate-to-drain, and gate-to-body capacitances (Section 4.1.3). This capacitance is expressed as

$$C_{\mathrm{inp}} W_{\mathrm{p}} + C_{\mathrm{inn}} W_{\mathrm{n}} = C_{\mathrm{in}} \left(W_{\mathrm{p}} + W_{\mathrm{n}} \right) = \mathrm{C}_{\mathrm{in}} W, \quad (2.1)$$

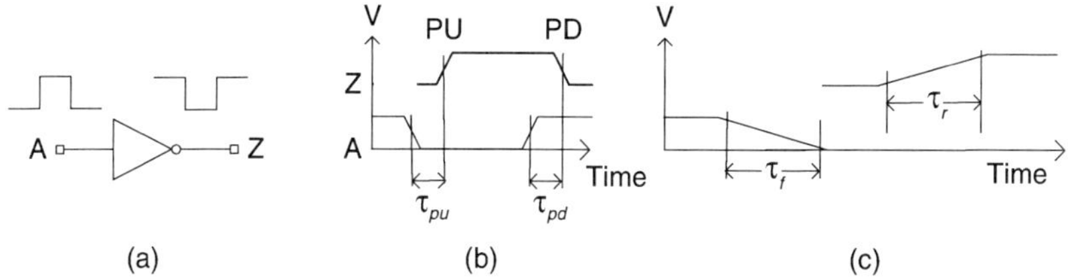

Fig. 2.25 **a** Inverter signal input and output. **b** Timing diagram for a PU and a PD transition. **c** Fall and rise times of signals on an expanded timescale

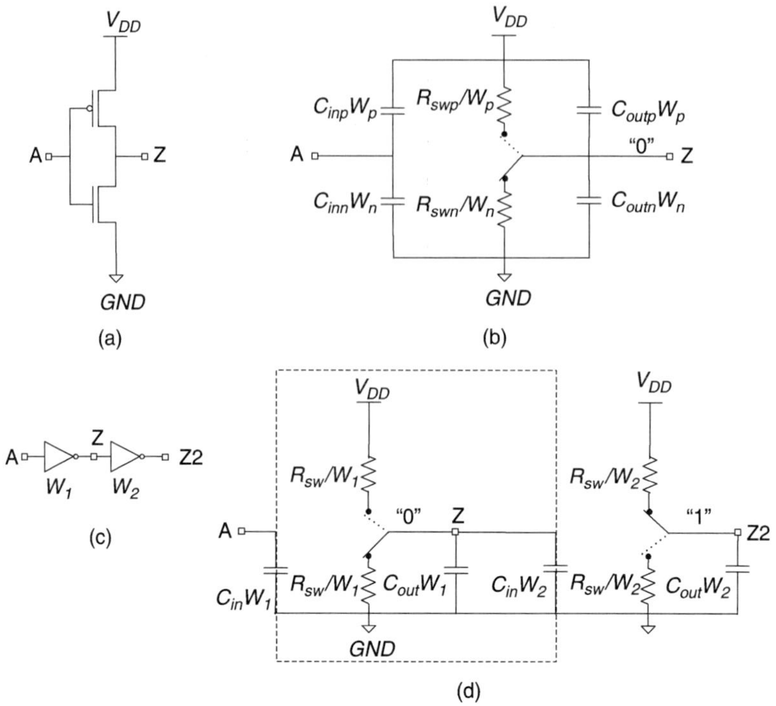

Fig. 2.26 Inverter: **a** circuit schematic and **b** equivalent *RC* model. **c** An inverter of width W_1 driving another inverter of width W_2 and **d** a simplified *RC* model of the circuit schematic in **c**

where C_{inp} and C_{inn} are the p-FET and n-FET gate capacitances per unit width, respectively, W_{p} and W_{n} are the widths of the p-FET and n-FET, respectively, and $W = W_{\text{p}} + W_{\text{n}}$. For simplification, it is assumed that $C_{\text{inp}} = C_{\text{inn}} = C_{\text{in}}$. Similarly, the net output drain-to-source capacitance per unit MOSFET width C_{out}, is expressed as

$$C_{\text{outp}} W_{\text{p}} + C_{\text{outn}} W_{\text{n}} = C_{\text{out}} \left(W_{\text{p}} + W_{\text{n}} \right) = C_{\text{out}} W, \tag{2.2}$$

where C_{outp} and C_{outn} are the p-FET and n-FET drain-to-source capacitances per unit width, respectively, and as with C_{in}, it is assumed that $C_{\text{outp}} = C_{\text{outn}} = C_{\text{out}}$.

In the *RC* model shown in Fig. 2.26b the inverter changes state as its internal switch is toggled between two possible positions in response to changes in voltage levels at input A. The effective resistances of the p-FET and the n-FET during switching are $R_{\text{swp}}/W_{\text{p}}\ (= r_{\text{swp}})$ and $R_{\text{swn}}/W_{\text{n}}\ (= r_{\text{swn}})$. Here, R_{swp} and R_{swn} are the effective resistances for a unit width of each MOSFET. Unlike the capacitance, r_{swp} and r_{swn} decrease as the MOSFET width is increased. With input A at "1," output Z is at "0," connected to GND through $R_{\text{swn}}/W_{\text{n}}$. When input A transitions from a "1" to a "0," the switch toggles to its upper position, and output Z is pulled up to a "1" with an *RC* time delay determined by $R_{\text{swp}}/W_{\text{p}}$ and the total capacitance being charged. When input A subsequently transitions from "0" to "1," the switch closes through $R_{\mathrm{swn}}/W_{\mathrm{n}}$ and the capacitance is discharged. Typically the current drive per unit width of a p-FET is smaller than that of an n-FET and $R_{\mathrm{swp}} > R_{\mathrm{swn}}$. This can be compensated for by making $W_{\mathrm{p}} > W_{\mathrm{n}}$ such that

$$\frac{R_{\mathrm{swp}}}{W_{\mathrm{p}}} = \frac{R_{\mathrm{swn}}}{W_{\mathrm{n}}}. \tag{2.3}$$

When Eq. (2.3) holds, the PU and PD delays are equal $\left(\tau_{\mathrm{pu}} = \tau_{\mathrm{pd}}\right)$. In circuits such as ring oscillators, or a chain of inverting logic gates, the measured gate delay is an average of PU and PD delays. In general, a switching resistance r_{sw} of a logic gate can be defined as the specific switching resistance R_{sw} divided by the relevant gate width W. For an inverter $W = W_{\mathrm{n}} + W_{\mathrm{p}}$ and $r_{\mathrm{sw}} = \left(r_{\mathrm{swp}} + r_{\mathrm{swn}}\right)/2$.

The signal propagation delay through a logic gate τ is $\sim RC$. In the case of a logic gate of width W_1 driving a logic gate of width W_2 as shown in Fig. 2.26c, d, the signal propagation delay τ from A to Z, averaged over PU and PD transitions, is given by

$$\tau = \frac{R_{\mathrm{sw}}}{W_1}\left(C_{\mathrm{out}} W_1 + C_{\mathrm{in}} W_2\right) = R_{\mathrm{sw}}\left(C_{\mathrm{out}} + \mathrm{FO} \times C_{\mathrm{in}}\right). \tag{2.4}$$

The ratio of the input capacitance of the load inverter to that of the driver inverter is the fanout (FO $\approx W_2/W_1$). If there is more than one gate connected to the output of the driver, W_2 includes the widths of all the gates. As an example, FO is equal to three when a logic gate drives three other identical gates. Parasitic R and C elements of interconnect wires also add to the delay. Estimation of equivalent FO for logic gate and wire loads is discussed in Section 2.4.4.

For a logic gate, all three delay parameters C_{in}, C_{out}, and R_{sw} vary with V_{DD}, temperature, and MOSFET properties such as gate oxide thickness and channel length. These delay parameters can be derived from circuit simulations or from ring oscillator-based test structures of the type described in Chapter 6. The delay parameters include the parasitic R and C elements associated with the MOSFET physical design. Although transient simulations are carried out to predict circuit behavior accurately, the simplified RC model can be used to estimate propagation delays within ~10% of the true values. These estimates are adequate for designing test structures where precise relative timing of signals is not required. Normalization of R and C parameters to MOSFET widths is helpful in sizing MOSFETs in CMOS circuits and in migrating these circuits from one technology node to another.

2.4.3 Wire R, C, and L

Electrical connections between DUTs and I/O pads, between circuit elements within DUTs and circuit blocks, and between DUTs and peripheral circuits are made with metal interconnect layers. Metal wires are also used for power supply distribution

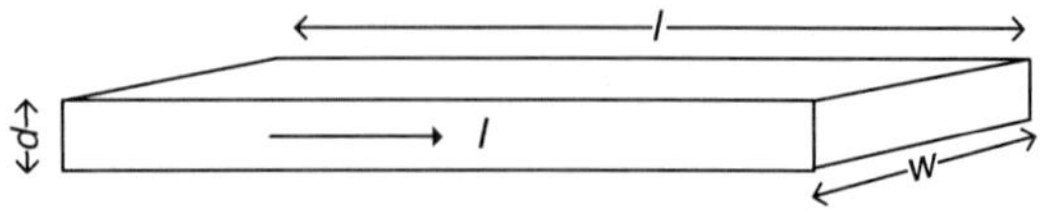

Fig. 2.27 A metal segment of width w, thickness d, and length l along the direction of current flow I

within a DUT and across a macro. The resistances and the capacitances of these metal wires form parasitic circuit elements which degrade voltage levels and signal waveforms applied to the CMOS logic gates and increase signal propagation delays.

A metal conductor segment of width w, thickness d, and length l along the direction of the current I is shown in Fig. 2.27. The resistance of this segment R is expressed as

$$R = \rho \frac{l}{wd}, \tag{2.5}$$

where ρ is the resistivity of the metal. The resistance can be expressed in terms of the sheet resistance ρ_{sh} of the metal layer:

$$R = \rho_{sh} \frac{l}{w} = \rho_{sh} n_{sq}, \tag{2.6}$$

where

$$\rho_{sh} = \frac{\rho}{d}. \tag{2.7}$$

The ratio l/w in Eq. (2.6) is the number of squares n_{sq}. Sheet resistance is expressed in units of $\Omega/\square$. The ρ_{sh} of a metal layer is calculated from the known resistivity and layer thickness or obtained from electrical measurements on test structures of the type described in Chapter 3. For CMOS circuit designs, ρ_{sh} of different conducting layers in a technology is supplied by the silicon foundry as part of technology definition.

The resistance of a wire can be roughly computed from the wire geometry by dividing it in squares and then counting n_{sq}. In Fig. 2.28, three different wire geometries are shown with dashed lines indicating a square. Note that the direction of current flow in the two identical wire geometries shown in Fig. 2.28a, b being orthogonal, their resistances differ by a factor of 25. In more complex wire geometries of the type shown in Fig. 2.28c, a simple method for estimating effective n_{sq} is to calculate n_{sq} for a solid metal of this shape and divide it by the fractional area occupied by the metal. Here, the solid block has two squares in the direction of current flow, and the fraction covered by the metal is 0.63 which gives a resistance of $3.17\rho_{sh}$ compared with $3.5\rho_{sh}$ from a resistance network of four parallel resistors of 14 squares each.

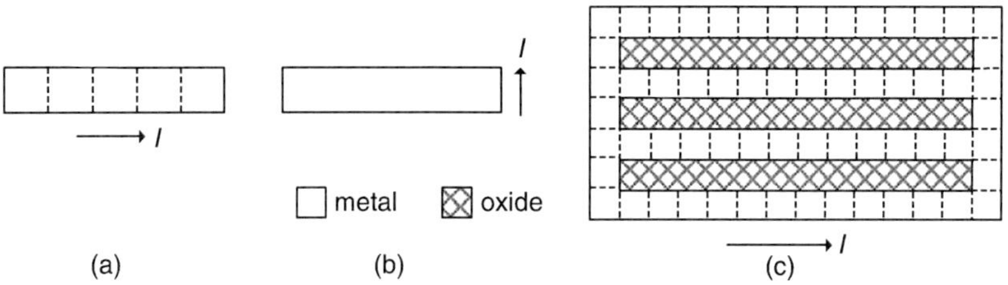

Fig. 2.28 Planar views of wire geometries with resistances of **a** $5\rho_{sh}$, **b** $0.2\rho_{sh}$, and **c** $\sim 3.17\rho_{sh}$. *Dashed lines* outline square shapes in the metal regions

Equation (2.6) is valid only for long and wide rectangular wires and does not take current crowding in wire bends and edges into account. A more accurate determination of resistance values of complex geometries can be obtained from the parasitic extraction models available for CMOS circuit design.

In DC circuits and wires connecting a DUT to an I/O pad, the voltage drop across the resistive wires ($= IR$) lowers the voltage level at the DUT or the circuit element. Wire connections to the gates of MOSFETs can have resistances of a few kΩ for carrying DC currents in the pA to nA range. Metal wires for power supply distribution carry current of a few mA and have more stringent requirements to minimize the *IR* drop as discussed in Section 2.4.6.

Wire capacitance comes into play in AC circuits in the form of power consumption during switching, time delay in signal propagation, and cross talk between adjacent wires. This parasitic capacitance is charged and discharged each time the voltage on a node transitions between a "0" and a "1" or any intermediate voltage level. Just as in the case of wire resistance, a rough estimate of capacitance can be made for sizing logic gates and wires in macro designs.

The capacitance of two parallel conducting plates shown in Fig. 2.29, each of length l and width w, separated by a dielectric material of thickness h and relative dielectric constant ε is given by

$$C = \frac{\varepsilon\varepsilon_0 lw}{h} = \frac{\varepsilon\varepsilon_0 A}{h}, \tag{2.8}$$

where A is the area of the plate and ε_0 $\left(= 8.854 \times 10^{-12}\ \text{F/m}\right)$ is the permittivity of free space. Capacitances of conducting layers in a CMOS circuit with more complex

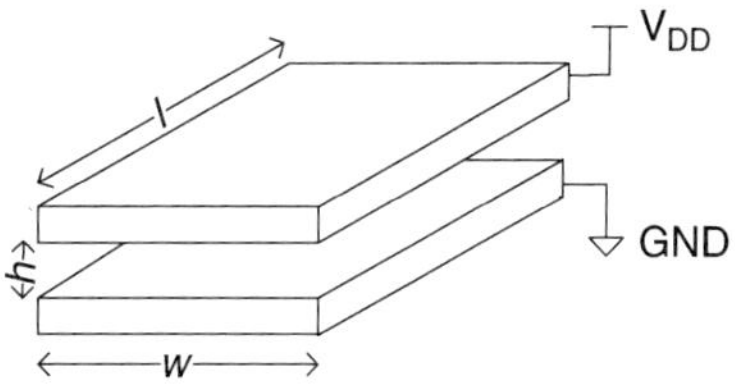

Fig. 2.29 A parallel plate capacitor of length l, width w, and plate separation h

geometries are dependent on the geometry of the entire conducting network and the dielectric properties of the non-conducting materials in the network. A rough estimation of wire capacitances can be made from the capacitance per unit length values provided by the silicon foundry as part of the technology definition.

For ease of wiring in high-density circuits, the metal layers are generally designed with long dimensions in orthogonal orientations on vertically adjacent layers. Cross sections of the first two or three metal layers, M1, M2, and M3, are shown in Fig. 2.30 for three different geometrical arrangements. The M2 capacitance to M1 and M3 layers are C_{down} and C_{up}, respectively. The lateral capacitances to the adjacent M2 wires are C_{left} and C_{right}. For these narrow wires, the parallel plate model is not valid as the electric field lines extend beyond the area between the metal plates. In long narrow wires, $l >> w$, it is therefore convenient to specify a capacitance per unit length. This wire capacitance per unit length C_{w} is the sum of all four capacitances:

$$C_{\text{w}} = C_{\text{up}} + C_{\text{down}} + C_{\text{left}} + C_{\text{right}}. \tag{2.9}$$

There are additional but smaller capacitance contributions from second nearest neighboring metal wires. However, the rules of thumb are that the vertical components C_{down} and C_{up} increase with w and the lateral components C_{left} and C_{right} increase as the separation s is reduced. In thin, dense wires, the lateral components dominate over the vertical components.

The capacitance per unit length C_{w} of different metal layers embedded in the same ILD material is nearly unchanged as long as w, s, and h scale by about the same factor. In order to reduce wire capacitance, SiO_2, the dielectric material of choice in earlier technology generations, is now being replaced with different lower dielectric constant materials. Hence, C_{w} of metal wires is dependent on the surrounding dielectric material properties in the stack. A capacitance calculator based on 2D or 3D electric field modeling for metal wires is usually available as an aid to circuit design.

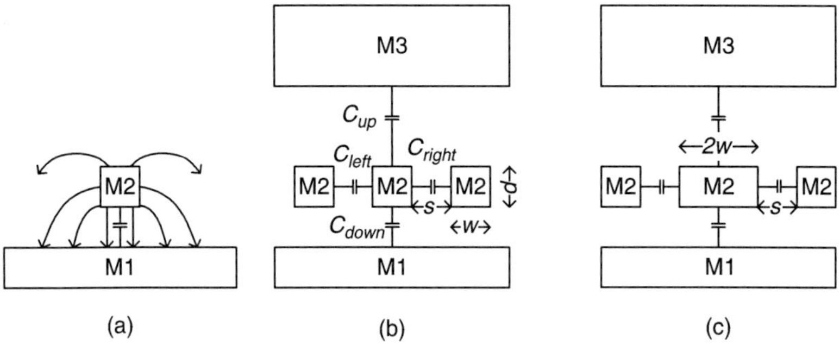

Fig. 2.30 Schematic cross sections of wire geometries for metals M1, M2, and M3, with M2 length orthogonal to M1 and M3, **a** for an isolated M2 wire, **b** for fully populated M2 wires of minimum width w, and **c** for one of the M2 wires of width $2w$

The inductance of an on-chip wire of length l can be expressed as

$$L = g\mu_0 l = L_w l. \tag{2.10}$$

where $\mu_0 = 1.26 \times 10^{-6}$ H/m ($= 1.26$ pH/μm) is the permeability of free space, g is a geometrical factor, and L_w is the inductance per unit length. The value of the function g varies slowly with the cross-sectional geometry and relative positions of the signal wire and current return paths (Section 4.5). In most on-chip wiring, g has a value in the range of 0.5–1.5. As will be clarified in a subsequent example, inductance does not play a significant role in on-chip signal paths for most test structures. Inductance of probes and cables may be significant and result in instantaneous changes in voltage (spikes) in response to rapid changes in current. The effect of inductance on power supply distribution in AC circuits is discussed in Section 2.4.6.

A distributed *RLC* model of a long wire is shown in Fig. 2.31a. The conductance of the dielectric material is assumed to be negligible and not shown in this transmission line representation of a wire. Typically, the wire lengths within a macro are limited by buffer insertion to restore signal waveforms as discussed in Section 2.4.4. In this case, a simplified lumped element *RC* model shown in Fig. 2.31b can be used for sizing wire dimensions. If R_w and C_w are the resistance and capacitance per unit length of a wire of length l, respectively, the signal propagation delay across the wire, to a first order, is expressed as

$$\tau = \frac{R_w C_w l^2}{2}. \tag{2.11}$$

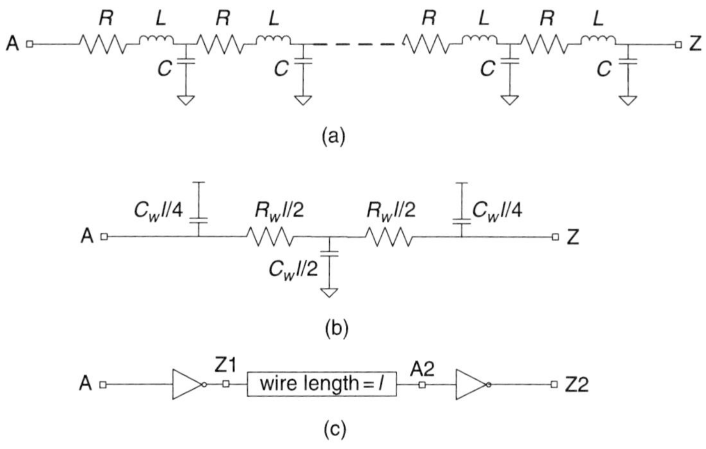

Fig. 2.31 **a** Distributed transmission line RLC model of a long wire with $l = 1$ in each section. **b** Lumped *RC* model of wire of length l. **c** An inverter driving a second inverter through a wire of length l

In Fig. 2.31c, an inverter drives another identical inverter through a wire of length l. The wire has a source R and C, in this case the switching resistance R_{sw}/W and output capacitance $C_{out}W$ of the driving inverter, and is terminated with the gate capacitance of the load inverter $C_{in}W$. Let us look at a practical example using the parameters in Appendix A to get a feel for what is and what is not important. Consider a case where $R_w = 2.0\ \Omega/\mu m$, $C_w = 0.2$ fF/μm, and $L_w = 0.5$ pH/μm. If a wire of length $l = 100\ \mu m$ connects the two inverters with $W = 2\ \mu m$ each ($R_{sw}/W = 1{,}000\ \Omega$, $C_{in}W = C_{out}W = 2.0$fF), the relevant characteristic times include the RC time constant, the L/R time constant, and the electromagnetic wave propagation delay through the wire $l\sqrt{(L_w C_w)}$. Neglecting the 200 Ω wire resistance, the lower limit on the RC time constant is 1,000 Ω × 24 fF = 24 ps, whereas L/R = 50 pH/1,000 Ω = 0.05 ps and $l\sqrt{(L_w C_w)} = 1.0$ ps. The RC time constant of the circuit dominates, as it is 480× the L/R time constant and 24× the electromagnetic wave propagation delay.

DC inputs for digital or analog controls to the gates of MOSFETs in circuits such as decoders or bias circuits generally carry very low currents – the standby current flowing through these wires is the gate oxide tunneling current which is typically in the pA to nA range. These DC input signals can be carried over high resistance wires without any penalty and the location of the I/O pads with respect to the DUT is not important. The wires may have high resistance segments of PS or DF layers, with ρ_{sh} of 10–30 Ω/□, for crossing under metal wires in M1 testable macros. Note that any additional current and associated voltage drop during a change in the DC input voltage level has no impact on measurements made on a DUT at a later time.

2.4.4 *Buffer (Driver) Sizing and Noise Reduction*

A logic gate may drive other logic gates and interconnect wires. When signals are carried over distances of the order of a few mm or more across a macro, the interconnect wire delay may become significant. A circuit schematic of a logic gate D1 of total width W_1 driving a wire of length l connecting it to the input of a logic gate D2 of width W_2 is shown in Fig. 2.32. The signal delay from A1 to A2 is given by

$$\tau = R_{sw1}\left(C_{in1}\frac{W_2}{W_1} + C_{out1}\right) + \left(\frac{R_{sw1}C_w}{W_1} + R_w C_{in2} W_2\right) l + \frac{R_w C_w l^2}{2}. \quad (2.12)$$

In short wires, $R_{sw1} >> R_w l$, and Eq. (2.12) reduces to

$$\tau = R_{sw1}\left(C_{in1}\frac{W_2}{W_1} + C_{out1}\right) + \frac{R_{sw1}C_w l}{W_1}. \quad (2.13)$$

In this case, for estimating signal path delay, only wire capacitance needs to be considered and the wire RC delay can be ignored. As an example, again using the

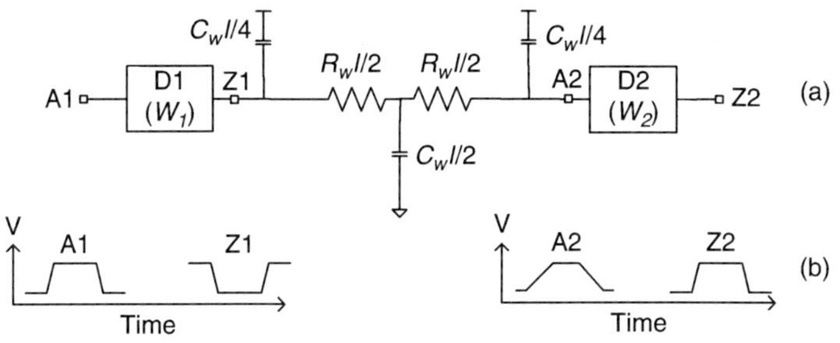

Fig. 2.32 **a** A logic gate D1 of width W_1 driving another logic gate D2 of width W_2, through a wire of length l. **b** Voltage pulse waveforms at nodes A1, Z1, A2, and Z2

parameters in Appendix A, a 40 μm long M1 wire of width 0.1 μm has a resistance of 80 Ω and a capacitance of 8 fF. Hence, its RC time constant is 0.32 ps. When driven by a standard inverter with $W = 2$ μm and $R_{\text{sw}} = 2{,}000$ Ω μm, the $R_{\text{sw1}} C_{\text{w}} l / W_1$ delay is 8 ps which is much larger than the RC delay of the wire itself. Such wire lengths are considered "short." Thus for estimating delays, wires within a DUT, placed in the 40 μm space between I/O pads, are considered as pure capacitive loads.

Characterization of a complex DUT often requires that the output signal from the DUT be sent through various on-chip circuits to an off-chip driver. Similarly, in the case of high-speed bench tests, fast input signals may have to travel through active circuitry before reaching the DUT. It is important that the necessary signal integrity be maintained as inputs and outputs pass through this circuitry. In one common situation, the output signal from a DUT is fed to a frequency divide-by circuit or an off-chip driver that is hundreds of micrometers or more away. In these long wires $\left(R_{\text{sw1}} \gtrsim R_{\text{w}} l\right)$, the rise/fall time at the input of the receiver gate is significantly longer. The shape of a pulse travelling from A1 to Z2 for a long wire is shown in Fig. 2.32b. The pulse amplitude at A2 is lowered, and as the wire length is increased, the pulse may ultimately disappear. The maximum length of a wire is therefore restricted and the rule of thumb is to maintain the rise/fall time at A2 to be < one-third of input signal pulse width T_{w}. The rise/fall time to reach 90% of V_{DD} at A2 is $\sim 2.2\tau$. The maximum length of wire $l_{\max}$ is expressed in terms of the delay parameters [12] as

$$l_{\max} = \left[\left(S^2 + \frac{2}{R_{\text{w}} C_{\text{w}}} \left(\frac{T_{\text{w}}}{3.3} - R_{\text{sw1}} \left(C_{\text{out}} + \frac{C_{\text{in}} W_2}{W_1} \right) \right) \right)^{0.5} - S \right], \qquad (2.14)$$

where,

$$S = \frac{C_{\text{in}} W_2}{C_{\text{w}}} + \frac{R_{\text{sw1}}}{R_{\text{w}} W_1}.$$

If signals have to travel over distances longer than the $l_{\max}$, additional logic gate drivers are inserted in the wires to maintain signal integrity. These drivers or buffers

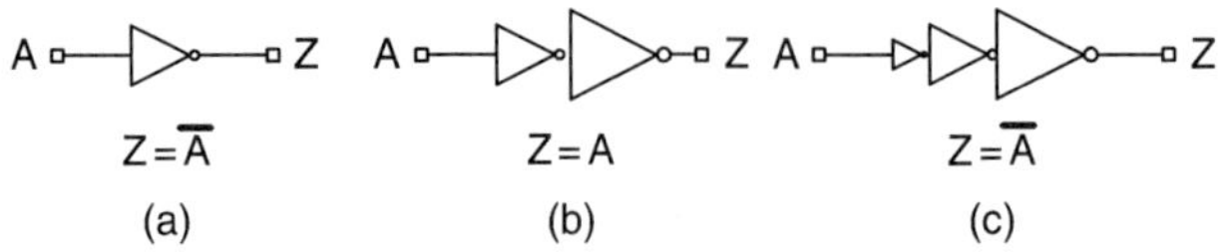

Fig. 2.33 Buffers with **a** one, **b** two, and **c** three inverter stages

may be a single-stage, two-stage, or three-stage inverters as shown in Fig. 2.33. The second and third stages are wider than the first stage so that the buffer can drive a longer wire at its output but with less loading to the previous stage. The device width ratios in the buffer stages (FO) are ~3–4. A one-stage buffer is useful in designs implemented at the M1 metal level and wherever the available space is limited. The two-stage buffer outputs a non-inverting signal. This is an advantage in designs sensitive to the logic-level input. The three-stage buffer is suitable in I/O drivers as discussed in Section 2.4.5.

In test structure macros, precise calculation of l_{max} is not required. Generally there is enough physical space to add buffers and a rule of thumb is to limit the equivalent FO to ~10. Consider a 1,000 μm long signal path. From Appendix A, a 0.2 μm wide M1 wire, 1,000 μm in length, has a total capacitance of 200 fF and a resistance of 1,000 Ω. Our standard two-finger inverter has $(C_{in} + C_{out})\ W =$ 4 fF and $R_{sw}/W = 1{,}000\ \Omega$. An FO = 10 for this inverter is a 20 fF capacitive load equivalent to a 100 μm segment of 0.2 μm wide wire with a resistance of 100 Ω. Standard inverter buffers may be distributed at a spacing of 100 μm along the wire path. A buffer spacing of 100 μm is especially convenient as it matches the I/O pad pitch in a standard macro. This strategy reduces buffer design complexity, standardizes DUT loading, and facilitates polarity adjustment by the addition of a single inverter if necessary.

An alternative is to use fewer buffers of a larger size. In the extreme case, one could drive the 1,000 μm segment of wire with a single large 20-finger inverter. With $R_{sw}/W = 100\ \Omega$, the 1,000 Ω wire resistance is now dominant. This buffer will actually deliver the signal to the far end of the line >2× faster but will draw a > 5× higher instantaneous current with approximately the same total charge transfer.

Signals travelling on long wires are subject to coupled noise from neighboring wires. In test structure macros, it is highly advisable to keep the space between nearest neighbors to > 4× the minimum spacing and avoid cross talk in signal wires.

2.4.5 I/O Drivers and ESD Circuits

DC outputs from a DUT or a circuit are connected to voltmeters, current meters, or SMUs. The I/O pads dedicated to these outputs are connected to the test equipment through probe needles or bonding wires and cables. The additional series resistance of these connections is taken into account in determining measurement accuracy.

An I/O driver circuit is included in the DUT or macro when high-frequency outputs are fed to AC test equipment such as a frequency counter, a standard oscilloscope, a sampling scope, or a spectrum analyzer. This is to ensure that the AC signal is not degraded by *RC* loading of the connectors and cables. A two- or a three-stage buffer shown in Fig. 2.33b, c also serves as an I/O driver.

When measuring frequency output of a ring oscillator and counting the number of transitions between a high- and a low-voltage level in a specified period of time, a voltage swing of $\gtrsim V_{DD}/2$ at the test equipment is desirable. The output stage is generally designed to have a total impedance of ~50 Ω. As there is some series resistance in the output line, we can design an I/O inverter driver with an impedance of 40 Ω. For R_{sw} of 2,000 Ω μm from Appendix A, we get W_p~30 μm and W_n~20 μm. The inverter can be drawn with 30 PS fingers in parallel so that each finger width is $\lesssim 1$ μm. The driver is placed close to the I/O pad and the resistance of the wire connecting the output of the driver to the I/O pad is minimized.

High-frequency circuit delay measurements using a differential technique are made using a sampling oscilloscope as described in Chapter 7. The I/O driver in this case has higher impedance, a lower output current, and a low output voltage swing to minimize cross talk. The output stage of the buffer is typically designed with W_p~12 μm and W_n~8 μm, to give an output impedance of ~100 Ω.

DC inputs from power supplies or SMUs are directly connected to a DUT or a circuit. When the input is connected to the gate of a MOSFET, the thin gate oxide is susceptible to damage from electrostatic discharge (ESD) from momentary high current spikes. Human handling of the silicon may also generate large discharge currents as the human body can build a static potential of up to 25,000 V which can discharge to ground through an electronic device. This effect is more pronounced in a low-humidity (<40%) environment. When handling test structures on the bench or when handling packaged chips, use of grounding wrist straps by operators is recommended to prevent the body from building up static charge. Antistatic-treated garments and floor mats also provide protection from ESD damage. Automated wafer probers prevent direct human contact with silicon and further reduce the circuit failure from ESD.

Current spikes may be generated when power supplies are turned on or off and damage the thin gate oxide in MOSFETs. Power supply output transients can be suppressed by incorporating a low-pass filter. Standard ATE may be equipped with low-noise power supplies to minimize such undesirable spikes.

Damage from ESD and current spikes generated by test equipment can be prevented by a protection circuit connected to the input I/O pad. One version of ESD circuit is shown in Fig. 2.34. It comprises a dual-diode structure which acts as a current sink and a voltage clamp. The p^+/n diode suppresses positive transients and n^+/p diode suppresses negative transients. Two or more p^+/n diodes in series may be used on the V_{DD} side to increase the voltage-handling capability. As the current flows mainly in the periphery of the diode, the current sinking capability is increased with a striped design to maximize the perimeter to area ratio. The diode area is optimized for current-handling capacity as well as for capacitance in high-frequency I/Os in RF circuits. A resistor of a few hundred ohms is connected in series with

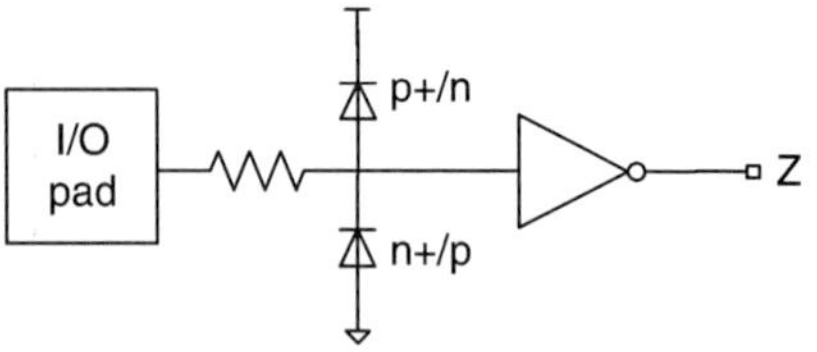

Fig. 2.34 An ESD protection circuit with dual diodes and a resistor to limit the peak current

the input pad to serve as a current limiter. This circuit occupies a large area and recommended only if a macro is handled in the lab for repeated testing.

2.4.6 Power Supply Distribution

Electrical measurements on test structures require external power supplies to deliver either constant voltage or constant current to the DUT. In the vast majority of test structures for measurements of resistance, capacitance, MOSFET parameters, ring oscillators, and signal propagation delays, a constant voltage is applied and the output voltage or current is measured. The accuracy of measurements is dependent on the actual voltage delivered to the DUT which is given by

$$V = V_{\mathrm{DD}} - I_{\mathrm{DD}}R, \tag{2.15}$$

where V_{DD} and I_{DD} are power supply voltage and current, respectively, and R is the resistance in series with the DUT. This series resistance has several components. Resistance of the probe, I/O pad, and probe-to-pad contact is part of the probe card and I/O pad design. Resistance from the I/O pad to the DUT is dependent on the wire connections within the macro. The third component is the resistance of the power distribution within the DUT itself. Voltage drops in the wires comprising the DUT may introduce a variable voltage distribution within the DUT. The effect of voltage reduction on a linear property of the device is proportional to the percentage change in applied voltage. In case of a non-linear device, such as a MOSFET in the subthreshold region, the percentage error in the measurement may be much larger.

The total series resistance, comprised of all the components mentioned above, is typically of the order of 1–5 Ω. If the measurements are made at a power supply voltage of 1.0 V and the test circuit draws 1.0 mA of current, the error in measurement for a linear circuit element will be well under 1%. The error for a non-linear DUT such as a diode or a MOSFET will vary with the region in which the DUT is operating and can be considerably larger as discussed in Chapter 5. In the case of precision measurements where measurement accuracy of better than 0.5% for linear circuit elements is desired, the rule of thumb is that the circuit should draw a current of <1 mA per I/O pad. If the circuit draws more current, a number of I/O pads can be connected in parallel to achieve the desired measurement accuracy. Alternatively, a three- or a four-terminal measurement configuration can be used.

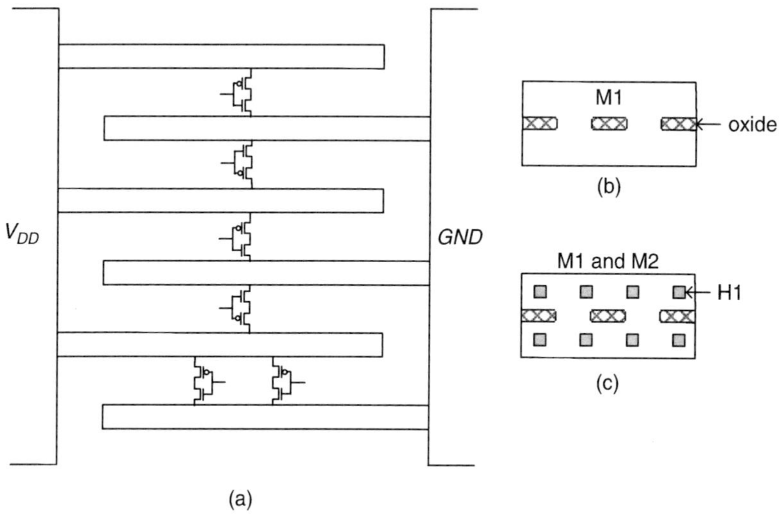

Fig. 2.35 **a** Interdigitated M1 power grid for placement of circuits in the space between the I/O pads. Planar views of metal layer patterns designed to meet GR maximum width requirements: **b** M1 metal and **c** metals M1 and M2

Power distribution schemes within the macro and DUTs vary with the physical size of the test structure, power requirements, and I/O signal frequencies. A common configuration for DUTs testable at the M1 metal level is shown in Fig. 2.35a, where the DUT receives its power through interdigitated fingers emanating from adjacent V_{DD} and GND pads. The power grid resistance can be lowered by effectively increasing the M1 wire width beyond the GR maximum by cheesing the metal as shown in Fig. 2.35b. The V_{DD} and GND busses can be strapped with additional layers of metal to further lower the across-DUT voltage variations, when tests are conducted beyond the M1 test stop, as shown in Fig. 2.35c. Results from tests at higher levels of metal can then also be compared with those of M1 tests to assess any impact of internal voltage drops.

In DUTs testable only at upper levels of metal, the power supply distribution system can be configured to be a product-like grid design with multiple metal layers. The layers alternate in orthogonal directions, with via interconnects between ILD layers as shown in Fig. 2.36. In Fig. 2.36a, vertically adjacent logic gates share the V_{DD} and GND wires and their physical layouts are alternately flipped. In Fig. 2.36b, with independent V_{DD} and GND wires on M1, all logic gates have the same orientation. Both of these arrangements may span many pad pitches and employ multiple V_{DD} and GND pads for test structures drawing current of more than a few mA. M3 and M4 and higher metal layers may be added parallel to M1 or M2 to reduce power grid resistance. The effective power grid sheet resistance for two metal layers of the same thickness, and equal linewidth and spacing is $\sim 0.5\times$ the sheet resistance of a single solid metal layer, plus a correction term for via resistance. The power grid resistance for the arrangements described here may be estimated in the initial design phase and for floorplanning, as discussed in Section 2.4.3.

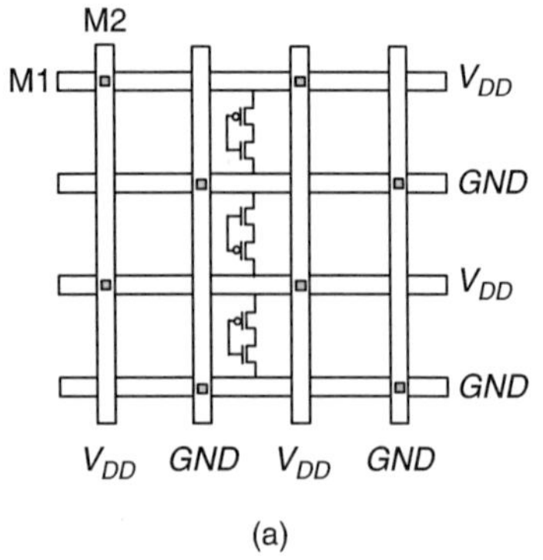

Fig. 2.36 Power grid with **a** shared V_{DD} and GND wires, and **b** independent V_{DD} and GND wires for adjacent cells

Design and analysis of the power delivery and distribution system becomes more complex for DUTs in which the power supply current I_{DD} varies with time. In the ideal case, the DUT is powered by a perfect voltage source in conjunction with a perfect GND. In practice, the situation is similar to that shown in Fig. 2.15, where the power supply is located remotely in the ATE or is a stand-alone unit in a test rack. The voltage source is connected to the circuit to be tested through coaxial cables, probes, and on-chip wires, which can be approximately represented by an *RLC* network, a version of which is shown in Fig. 2.37. If the current drawn by the circuit is constant, the reactive components in the circuit play no role and only the series resistances act to modify the voltage applied to the circuit from that delivered at the source, as previously described.

When the current drawn by a DUT varies with time due to internal switching of circuits, the reactive components in the power supply system can become very

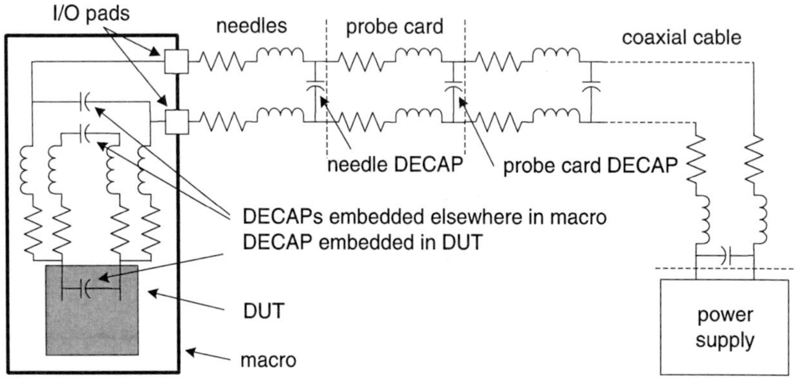

Fig. 2.37 RLC network of wires connecting the power supply in the test equipment to the DUT. DECAPs may be added as indicated

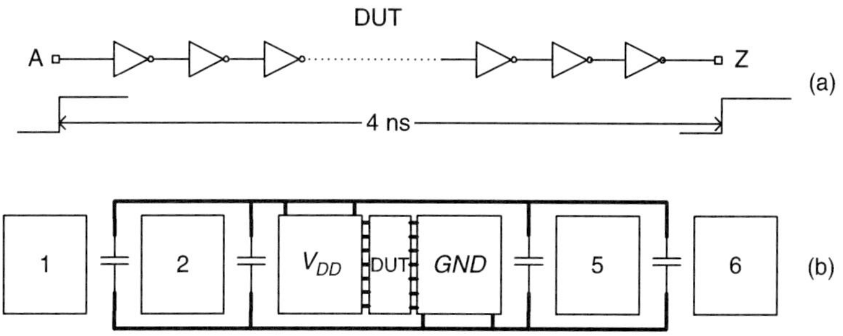

Fig. 2.38 **a** A DUT comprising a delay chain with 1,000 inverters. **b** DECAP placement relative to a DUT and I/O pads for reducing V_{DD} droop in a transient measurement

important. An extreme but not uncommon case is one in which a circuit is operated in an isolated pulse mode. As an example, consider a delay chain of 1,000 inverters with FO = 1 shown in Fig. 2.38a. With one PS finger per MOSFET and $W_p + W_n = 1.0\,\mu$m, this delay chain can be placed entirely within the 40 μm × 60 μm space between two pads, with power supply connections as shown in Fig. 2.38b. The chain is inactive for a long period, biased at $V_{DD} = 1.0$ V, and drawing a leakage current of a few μA. At some point in time, a voltage step input is applied to the chain and a single edge propagates from one end to the other, over a time period of approximately 4.0 ns (4.0 ps delay/stage). When one inverter goes through a PU transition following a PD transition in the previous stage, a net charge of

$$Q_c = C_{total} V_{DD} = \delta I_{DD} \tau \tag{2.16}$$

is transferred from V_{DD} to ground. Here δI_{DD} is the current drawn during the combined PU and PD transitions occurring over time τ. With C_{sw} $(= C_{in} + C_{out})$ W = 2.0 fF and $V_{DD} = 1.0$ V, this amounts to about 2.0 fC of charge. With 500 such pairs of inverters in the chain, a total of 1.0 pC of charge is transferred from V_{DD} to GND, corresponding to a current of about 0.25 mA flowing during the 4 ns transient. With an ideal power supply connected directly to the chain, this charge would come directly from the supply with a voltage of V_{DD} remaining across all of the inverters during the event. With all the parasitic elements in series with the power supply shown in Fig. 2.37, the true V_{DD} and GND are in fact isolated somewhat from the delay chain. Any charge coming directly from the power supply has to pass through a significant inductance, of the order of 1 μH, to get to the circuit. In addition, the power supply itself is likely to have an internal time constant larger than the entire event duration. It is thus inevitable that the power supply voltage across the circuit will droop significantly during the event and that the performance of the circuit will be strongly influenced by this droop. Designing a robust power supply distribution for AC applications requires sophisticated tools and impedance models for probe cards and other accessories. In almost all test structures, simpler design techniques using the guidelines described below are sufficient to produce the desired measurement accuracy of <1%.

One way to minimize such a transient V_{DD} droop is to provide significant ballast or decoupling capacitance in very close proximity to the circuit so that inductance between the DUT and the external power supply plays a negligible role. This can be accomplished by placing a bank of DECAPs within the macro itself. Questions that immediately arise are just how much DECAP area is sufficient and where should it be placed? The total area between two I/O pads is 40 μm $\times$ 60 μm = 2,400 μm^2. If this area is packed with DECAPs with a capacitance density of 5 fF/μm^2, the resultant total capacitance is ~ 12 pF and can store ~12 pC of charge at 1.0 V. If this capacitance is tied directly across inverter delay chain experiment, the 1 pC of charge required to power the transient event can be provided by the DECAPs, but with an associated voltage droop of ~8%. With additional DECAPs to add another 12 pF in the available space, either between I/O pads or elsewhere within the macro, the droop can be decreased to ~4%.

The physical configuration for limiting the voltage droop to 2% during a 4 ns transient with four *RLC* loops is shown in Fig. 2.38b. The inductance of each loop of $L_w \times$ distance around the loop from Eq. (2.10) is <0.6 nH for the largest loop. For the four loops in parallel, the total inductance reduces to <0.15 nH. The effective resistance of the active circuit during the transient for an ~0.25 mA current draw at a V_{DD} of 1.0 V is ~4 kΩ. Thus, the *L*/*R* time is <1 ps, while the *RC* time for a *C* value of 48 pF is ~200 ns. This implies that the charge in the DECAP is available to sustain the circuit V_{DD} ($RC >> L/R$).

The approach described above works well when a standard low-frequency probe card is used, which is typically the case for in-line test. For smaller circuits it is often possible to include sufficient decoupling capacitance immediately adjacent to or imbedded within the DUT itself.

Even in a circuit switching periodically and drawing a nearly constant steady-state current, such as a ring oscillator (RO), there can be dips and spikes in the current draw that will at most involve a disturbance with an integrated area of $< Q_c$. Furthermore, averaged over the cycle of the ring oscillator, the sum of the charge involved in the dips and spikes is zero against the background of the average constant current. If the RO stage delay varies linearly with V_{DD}, the net effect on RO frequency from these variations will be negligible. As a specific example, for a 101-stage ring oscillator the minimum charged capacitance at any time is of order 200 fF corresponding to a stored charge of about 200 fC at 1.0 V. Assuming the transient charge is provided by other nearby devices in the ring oscillator, the resultant local V_{DD} variation would be less than 1%. This can be further mitigated (by as much as 10$\times$) by placing explicit DECAP cells in close proximity, for example, a 200 μm^2 bank of DECAPs with an associated charge of about a 1 pC. In this case, adding DECAPs to further reduce the V_{DD} variations is optional; however, such additional capacitance also helps to provide immunity from other sources of power supply noise. The rule of thumb here is to add DECAPs if there is otherwise unutilized space within or near the DUT.

Custom probe cards can provide other decoupling options. As an example, a manufacturer may provide a probe card with a combination of very wide bandwidth I/Os of 50 Ω impedance having built-in decoupling capacitors along with a set of DC I/Os. Such cards are routinely used for high-speed bench tests where both I/O impedance of 50 Ω and good decoupling are required. Referring to Fig. 2.37, a manufacturer may be able to place ~120 pF decoupling capacitors within 300 μm of the needle tips. This large decoupling capacitance is no more than 1 nH away from DUTs in the macro. In addition, DECAPs of ~100 nF can be located further back in the probe card. This can substantially ease the requirements for DECAPs embedded within the macro itself.

There is one other power supply decoupling option that can be used in a high-speed probe card with low, but variable, current draw. With this option, the DUT is powered directly through a very wide bandwidth high-speed line driven by a 50 Ω source. When the current draw changes from say 0.0 to 0.25 mA, there will be an immediate 1.25% drop in V_{DD} corresponding to the 12.5 mV, 0.25 mA signal edge that will be sent back to the 50 Ω source. This arrangement is equivalent to driving the experiment directly with an ideal voltage source in series with a 50 Ω resistor. It sacrifices a high-speed line that could otherwise be used to deliver or observe a fast signal and it is really useful only when the current drawn is low.

In a macro comprising multiple complex DUTs, it is frequently of advantage to have several power supply sectors. This scheme has an additional advantage of enabling accurate determination of the off-current (*IDDQ*) of individual or small groups of DUTs. In the case of parallel test, it can also enable one to obtain the on-current of individual DUTs. This partitioning of power delivery is especially important in situations where the output of a complex DUT is sent off-chip at high or moderate speed via an off-chip driver, concurrently with operation of the DUT. The off-chip driver and associated buffers and other I/O devices should always be placed on a dedicated power supply to provide isolation between the DUT(s) and the I/O. The off-chip driver itself may draw a current 10–20 times larger than that of an active DUT. A driver delivering 1.0 V into a 50 Ω line requires 20 mA! Since in-line probe cards typically have no DECAPs, there is a large droop in the power supply of, for example, an off-chip driver providing a 1 MHz square wave output from a ring oscillator macro to the input of a frequency counter. The frequency counter can be set to have an input impedance of either 50 Ω or 1 MΩ. In either case, there is a large transient every 500 ns, while a non-ideal yet robust 1 MHz signal is received by the frequency counter. It is important to prevent these transients from impacting the measurement accuracy.

Multiple power supply sectors require additional power supply I/O pads. A workable solution is to have a common GND for all sectors and a dedicated V_{DD} pad for each sector. As will be shown in subsequent sections, there may still be several GND pads, but they are electrically connected internally within the macro and also sometimes at the probe card or external electronics. It should also be noted that while multiple power supply sectors do cost I/O pads, this can be mitigated somewhat by using power supply voltage inputs as control signals for the I/O circuitry.

Important cross-checks can and should be performed to verify adequate isolation between different power supply sectors. A voltage sense line may be connected to the power supply grid in a sector to measure the voltage levels for experimentally validating the macro designs.

2.4.7 Differential Measurement Schemes

Absolute measurement of many device and circuit properties is complicated by end effects from wire parasitics and signal distortion. A majority of these spurious effects can be eliminated with differential design and measurement techniques used in many of the examples in this book.

In Fig. 2.39a, two sets of nominally identical DUTs are connected in series with the number of DUTs in one set being greater by at least one than in the second set. The difference in parameter values for the two sets gives a measure of the parameter value of a single DUT, eliminating any end effects. The total number of DUTs in a series-connected set may be increased to correctly account for end effects that penetrate beyond a single DUT. In Fig. 2.39b, the differential method is applied to circuit elements connected in parallel. This is suitable for capacitance and MOSFET current measurements. In both of the methods described here, additional I/Os or peripheral circuits to select one or the other set of DUTs are required. It is also assumed that the difference in properties of nominally identical DUTs from random variations does not compromise the differential measurements.

The configuration shown in Fig. 2.39c is suitable for AC test structures. An input signal is steered through circuit block EA or EB by setting control switch “S” to a “1” or a “0,” respectively. All other peripheral circuits and I/Os for the two measurements are identical and the difference in delay between the two blocks is determined. The two circuit blocks may comprise different number of circuits in series or circuits differing in a property of interest. Applications of this technique are covered in Chapters 7 and 8.

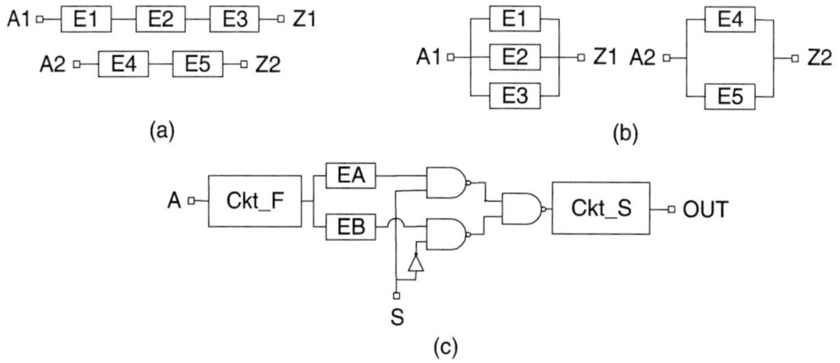

Fig. 2.39 Differential measurement schemes; DUTs connected **a** in series and **b** in parallel. **c** Input S sets the signal path through either circuit block EA or EB

2.4.8 Commonly Used Circuit Blocks

In characterizing circuit behavior, the DUTs may comprise one or more types of logic gates described in Section 2.1.5. Logic gates are also used in the peripheral circuitry of test structures. Other circuit blocks commonly employed in peripheral circuits of both DC and AC test structures are buffers, latches, decoders, multiplexers, and shift registers.

Decoders and multiplexers are used for reducing the number of I/Os and hence the area of macros comprising, for example, arrays of DUTs. In a decoder, there are N inputs and 2^N outputs. By setting different combinations of input voltage levels, the voltage level of any one and only one of the 2^N outputs is set to a "1" or a "0," while all other outputs remain at a complementary level of "0" or "1." A decoder can also be configured to set none or all the outputs to the same level. A two-input decoder (also called a 2-bit decoder) circuit schematic implemented with NAND2 and inverter logic is shown in Fig. 2.40a. The symbol and the logical truth table for the decoder are shown in Fig. 2.40b, c respectively.

A multiplexer is used for passing only the selected input level to the output. One implementation of this is an "OR" function with NAND2 and inverter logic as shown in Fig. 2.41, along with its symbol and truth table. This scheme is well suited for macros with large aspect ratios placed in the scribe line. The circuit schematic in Fig. 2.42 is of a two-way multiplexer or demultiplexer. It is implemented with n-passgates for passing "1." A complementary circuit with p-passgates is used for passing "0" and another version with transmission gates works well for passing

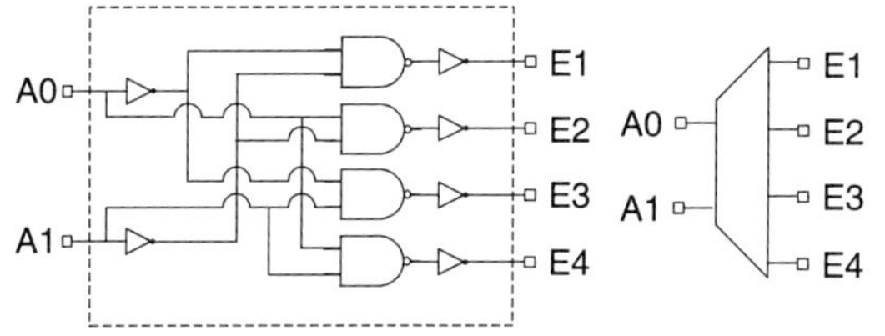

(a) (b)

A1	A0	E1	E2	E3	E4
0	0	1	0	0	0
0	1	0	1	0	0
1	0	0	0	1	0
1	1	0	0	0	1

(c)

Fig. 2.40 A 2-bit decoder: **a** circuit schematic, **b** symbol, and **c** logic truth table

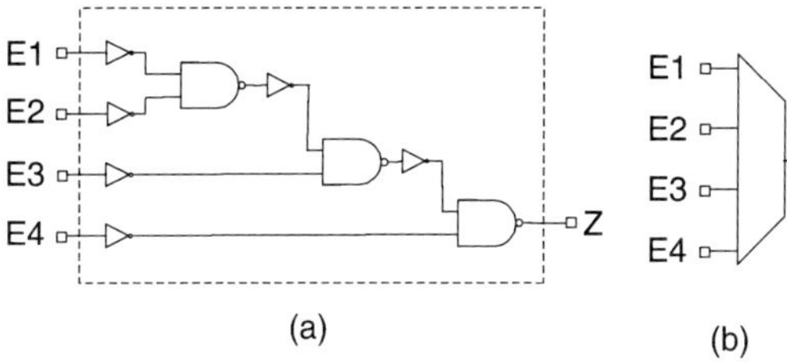

(a) (b)

E1	E2	E3	E4	Z
E1	0	0	0	E1
0	E2	0	0	E2
0	0	E3	0	E3
0	0	0	E4	E4

(c)

Fig. 2.41 An "OR" circuit implemented with NAND2 and inverter logic gates for multiplexing: **a** circuit schematic, **b** symbol, and **c** logic truth table

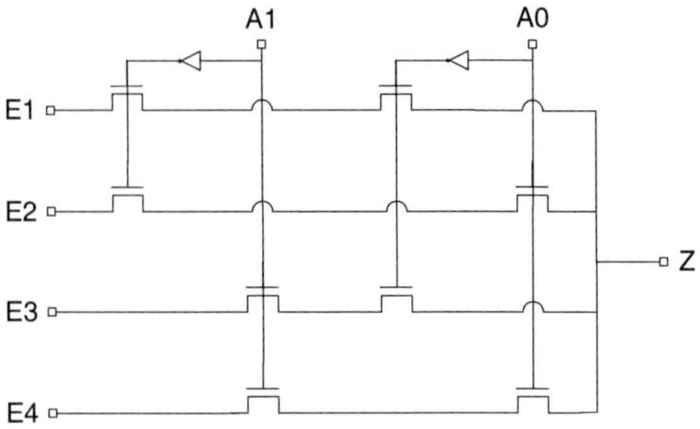

(a)

A1	A0	Z
0	0	E1
0	1	E2
1	0	E3
1	1	E4

(b)

Fig. 2.42 A two-way multiplexer/demultiplexer implemented with n-passgates: **a** circuit schematic and **b** logic truth table

either "1" or "0." The decoder and multiplexer circuits in Figs. 2.40, 2.41, and 2.42 can be operated with DC inputs.

The decoders and multiplexers in peripheral circuits may utilize low-leakage, slow switching devices to minimize background leakage power. The circuit blocks draw a small current during switching and the switching speeds and output voltage levels are not critical for macro functionality. Hence, a higher power grid resistance can be tolerated allowing greater flexibility in the placement of I/O pads with respect to the circuits.

Memory elements in CMOS logic circuits are built with latches to hold a "1" or a "0". Static CMOS latches are also useful in test structures for controlled initialization of an event, for generating a sharp rising signal edge or pulses, and in counters

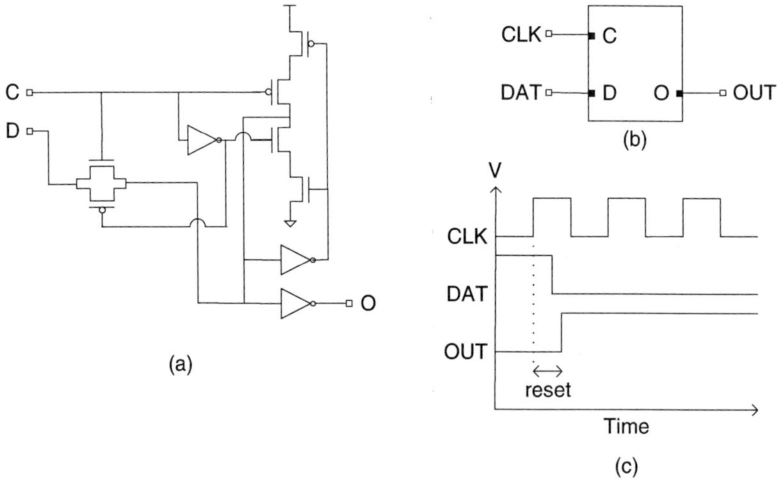

Fig. 2.43 A level-sensitive latch: **a** circuit schematic, **b** symbol, and **c** timing diagram showing CLK, DAT, and OUT signals

and shift registers. A commonly used robust design is the level-sensitive latch also called a Level-sensitive scan design (LSSD) latch. One implementation of an LSSD latch is shown in Fig. 2.43. Input signal DAT is captured in the latch and appears at OUT as long as it arrives before the falling edge of the clock (CLK) signal. The

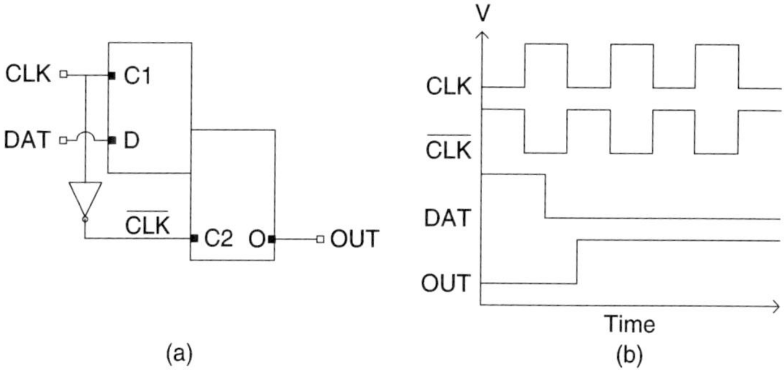

Fig. 2.44 Master–slave (MS) latch: **a** symbol and **b** timing diagram

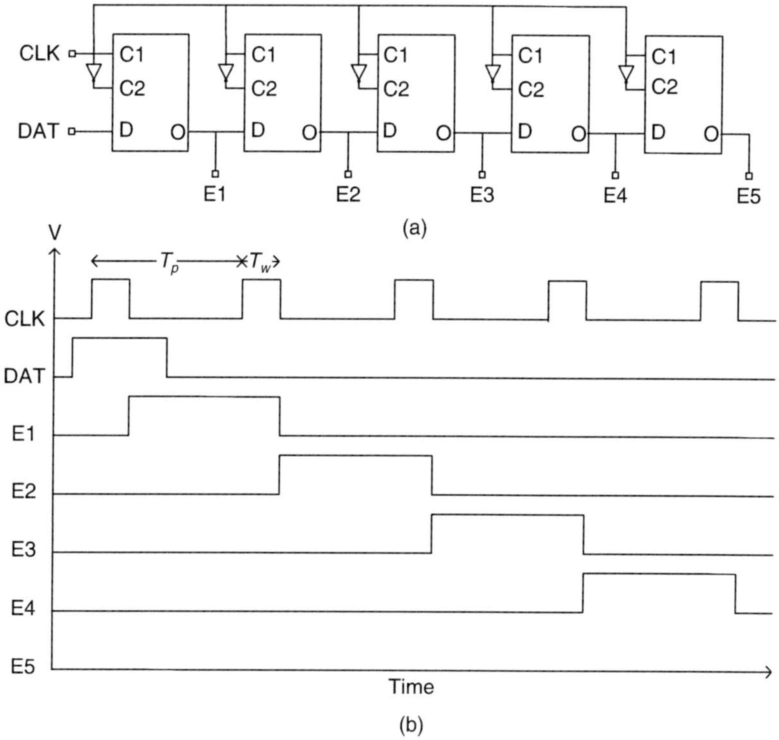

Fig. 2.45 **a** Circuit schematic of a scan chain comprising five MS latches. **b** Timing diagram showing sequential activation of latch outputs for an input clock pulse width T_w and time period T_p

same latch design is also used to store the output of a high-speed circuit which can then be read out as a DC signal at any time after the completion of a switching event.

A master–slave (MS) latch is constructed with two LSSD latches with complementary CLK inputs (Fig. 2.44). A shift register built with five MS latches is shown in Fig. 2.45. This circuit block can be used for sequentially selecting one DUT in each CLK cycle. Combining a shift register with a decoder gives the greatest flexibility in selecting one out of a large number of DUTs with very few I/Os. In a 2D array, row and column decoders are used to select a DUT in the array. By combining a shift register with a decoder, as shown in Fig. 2.46, with only two inputs, any one of 16 DUTs may be selected. This type of scheme is used for reducing I/O count and packing a large number of DUTs in a compact macro.

Digital circuits and latches require clock signals which may be provided by the test equipment or generated inside a macro (on-chip). The circuit schematic of a ring oscillator for generating on-chip clock waveforms is shown in Fig. 2.47. The ring oscillator comprises an odd number of inverting stages and the clock can be turned on or off by toggling the EBL input between "1" and "0." The frequency of the output signal is lowered with a frequency divider circuit. This circuit is covered in detail in Chapter 6.

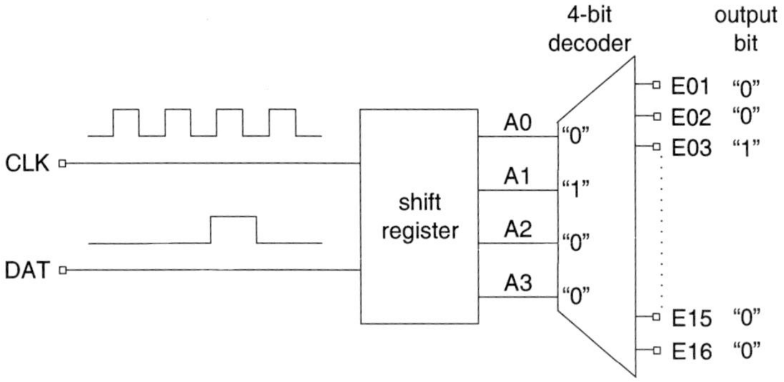

Fig. 2.46 A circuit schematic with two inputs for selecting any one of 16 experiments. This circuit comprises a shift register and a 4-bit decoder

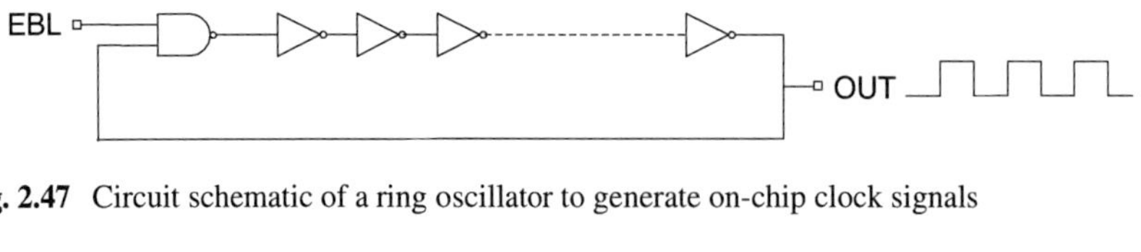

Fig. 2.47 Circuit schematic of a ring oscillator to generate on-chip clock signals

2.5 Macro Templates and Design Methodology

A macro is devoted to a single test structure or DUT or a set of independent DUTs. In technology characterization, similar measurements are carried out on a large number of DUTs. For example, MOSFET and ring oscillator measurements are made on many flavors of device types, sizes, and physical layouts, and resistances and capacitances are measured for all metal layers. Design and test efficiency are significantly improved by creating a macro template for a set of similar DUTs. A macro template comprises standard DUT designs and the required peripheral circuitry. The template design is optimized for area, number of I/O pads, and ease of test and data analysis while maximizing the total number of DUTs. New macros are created by substituting different DUTs into replicas of the macro template without altering the peripheral circuits and I/O pad assignments. In this section, we discuss the design considerations of macro templates, followed by examples of discrete element macros and macros designs with 1D arrays and 2D arrays of DUTs.

In a macro template, a shell is created into which DUTs with similar characterization requirements are inserted. The template is assembled using the guidelines covered in the previous sections. A standard cell library may be created for the circuit blocks in common use following the practice in product designs. Circuit block and DUT designs may also be shared amongst different templates for improving design efficiency and for consistency. It is generally a good practice to keep the form factor of the macro (width/height), number of available I/O pads, probe card options, and wiring constraints in mind while creating a circuit diagram and physical layout of a macro template. When macros are tested in a manufacturing line, total test time is reduced by maintaining a standard I/O pad footprint for all the macro templates which can then be tested with the same probe card.

Another advantage of the macro template concept is that only the template needs to be fully checked for technology compliance and design verification. Some of the commonly used checking tools are design rule checker (DRC) and layout vs. schematic (LVS). Complex macro templates may require logic verification, along with noise and power analysis. Once the physical layout is completed, parasitic extraction and circuit simulation tools are used for checking macro functionality. After the checking process is successfully completed, different DUTs can be substituted in the replicas of the template to create new macros. Macros generated from a common template may also share the test procedures and the software used for data analysis. This facilitates generation of standardized charts for displaying and reviewing the data in an automated fashion. The test and data analysis methodology for macros sharing the same template is discussed in more detail in Chapters 9 and 10.

2.5.1 DUT Designs and P cells

DUT designs may be generated with a parameterized cell (P cell) approach followed in standard cell libraries in product designs. A P cell is defined in SKILL

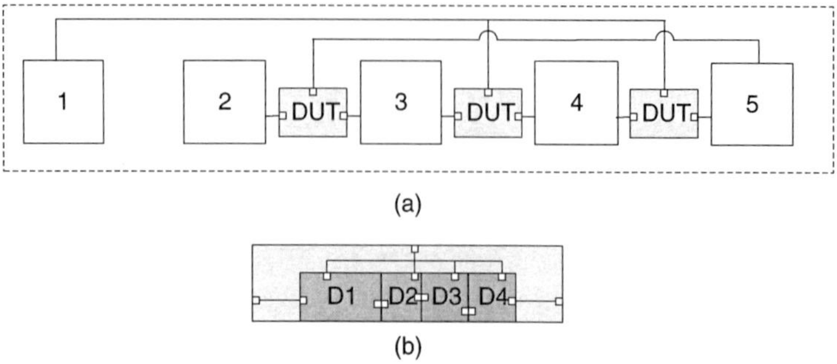

Fig. 2.48 **a** Example macro template section with three DUTs. **b** Expanded view of a DUT with replaceable blocks D1–D4 having fixed I/O pin locations

programming language to create physical layouts of a family of DUTs within user-defined and technology GR constraints. The DUT specifications (dimensions, spacing, and locations for each layer) are entered as variables in a spreadsheet as inputs to the SKILL code. The spreadsheet also serves to document the DUT and macro content. With this approach, the effort to create documentation is reduced and any errors in characterization originating from incorrect documentation are minimized.

The concept of using standardized DUT and macro designs is illustrated in Fig. 2.48. The area of each DUT type and its placement in the macro template are specified. The positions of wires or contacts (displayed as "pins" in the schematic) from the DUT to the I/O pads or other circuits in the macro are fixed to allow replacement of the DUTs in a straightforward manner. The DUTs can be replaced without altering the macro wiring and I/O pad assignments. Similarly, circuit blocks within a DUT are arranged in a hierarchical fashion. These can also be easily replaced with blocks of different design or physical layout as long as the position of the I/O connections within each block is unchanged.

2.5.2 Discrete Element Macros

In discrete element macros, the DUTs are directly connected to the I/O pads and there are no active peripheral circuits. In early technology development and in short-loop test vehicles with a limited number of metal layers and no active elements, there are a large number of discrete element DUT macros. A number of different schemes are used to improve silicon area utilization by I/O pad sharing. Two examples are shown in Fig. 2.49. In Fig. 2.49a, each terminal of a MOSFET is connected to an I/O pad and only six MOSFETs can be accommodated in our standard 1×25 padset macro. In Fig. 2.49b with pad sharing between the S and D terminals of adjacent MOSFETs, and common G and B terminals for all, the number of MOSFETs in a macro is increased to 22. The economy in space utilization is usually at a cost of reduced measurement accuracy for some parameters and a higher vulnerability to defects.

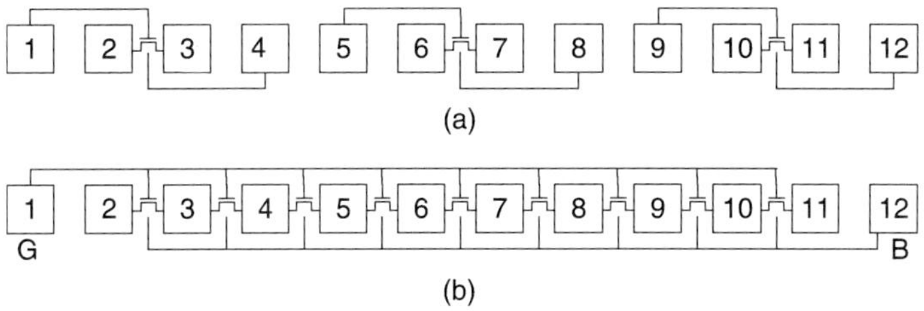

Fig. 2.49 Discrete element macros for MOSFET characterization: **a** with isolated I/O pads and **b** with shared I/O pads

Test time is reduced if a number of DUTs are tested in parallel. In Chapters 3, 4, and 5, we provide examples of discrete element macros for resistance, capacitance, and MOSFET characterization with different degrees of parallel test capability, respectively, and present the advantages and drawbacks of such designs.

2.5.3 One-Dimensional Array Macros

The area and test efficiency of discrete element macros is improved with a moderate level of design complexity in one-dimensional (1D) arrays. A number of such 1D array designs with active and passive elements can be implemented at the M1 metal level. These macro designs are very useful in the technology development phase and for process monitoring early in the manufacturing cycle.

A 1D array comprises DUTs connected either in series or in parallel. In the example of parallel-connected DUTs shown in Fig. 2.50, a decoder is used to select any one of the DUTs. A voltage is applied to I/O pad AV1 and the measured current flows only through the selected DUT. This design functions correctly as long as none of the DUTs is shorted and the total leakage current of the unselected DUTs is negligible compared to the current through the selected DUT. Different schemes to reduce the leakage current by applying negative bias or adding resistance in series to the unselected DUTs may be used to allow a larger number of DUTs to be placed in parallel. A number of 1D arrays are placed within a macro and the arrays share the decoder input pads A0, A1, A2, and A3. A pair of arrays may also share a common GND pad to improve area efficiency. The arrays can be tested in parallel to reduce test time.

In an array of series-connected DUTs, a switch of the type shown in Fig. 2.10 is inserted in series with each output node and a decoder connects the selected node to the I/O pad AV1 as shown in Fig. 2.51. A constant current flows through all the DUTs, and the voltage at each node is measured sequentially. This design is ideally suited for measuring resistors. The presence of an electrical open in any DUT is easily detected by a single measurement of these series-connected DUTs, a useful feature in yield test structures. As in the case of parallel-connected 1D arrays, the series-connected arrays can also be tested in parallel.

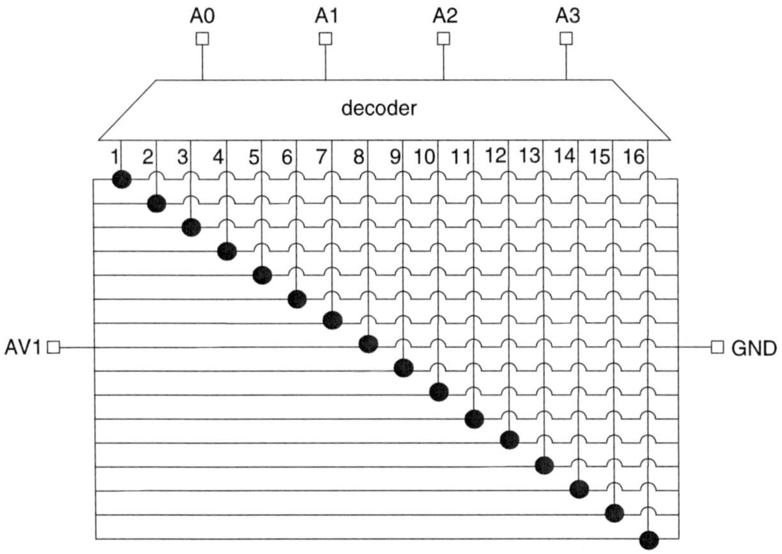

Fig. 2.50 A 1D array of DUTs connected in parallel with a 4-bit decoder to select any one of 16 DUTs represented by a *solid circle*

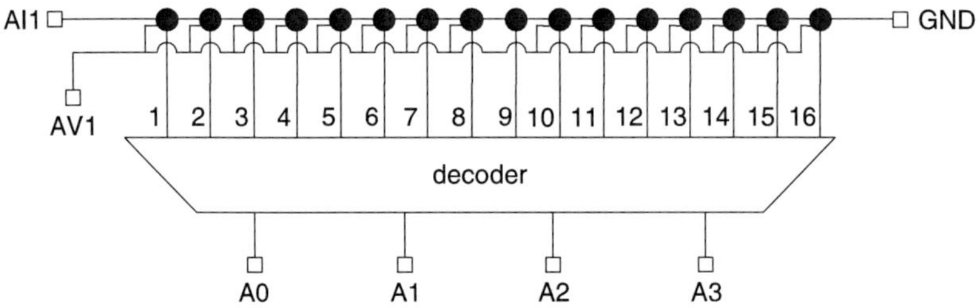

Fig. 2.51 A 1D array of series-connected DUTs shown as *solid circles*. A multiplexer connects the selected output node to the I/O pad AV1 through a switch placed in the DUT

Arrays with identical DUTs give a measure of statistical variations in the DUT parameters. All DUTs can be measured together to get the average parameter value and the standard deviation is obtained from the individual measurements. This feature is useful in DFM and process development applications for collecting moderate statistics while conserving silicon area and test time.

2.5.4 Two-Dimensional Array Macros

Macros comprising one or more two-dimensional (2D) arrays are commonly used when measurements on a large number of DUTs are required. These array designs are complex and the form factor is usually not suitable for placement in the scribe line. The wiring of DUTs and peripheral circuits requires several metal layers, design tools for checking, and careful consideration of measurement errors due to parasitic elements and leakage currents of circuits used for steering signals in the array. The area occupied by the I/O pads is minimized with the use of decoders, scan chains, and other circuits. Such 2D arrays are therefore very useful in yield macros in which sufficient critical area is necessary as discussed in Section 3.3. A wide variety of 2D array designs have been demonstrated for resistors, capacitors, MOSFETs, ring oscillators, and various combinations of them. Here, we present some of the basic ideas of 2D array designs. Specific examples and references are included in other chapters.

In Fig. 2.52, the concept of a 2D array is shown with 4-bit row and column address decoders to test 256 DUTs. The row and column decoders uniquely select one DUT in the array to connect it to an SMU for making the desired measurements. Each DUT circuit includes four switches to steer inputs applied via AVF1 and AVF2 to the DUT and outputs from the DUT to ZVM1 and ZVM2. A representative switch design to select a circuit is shown in Fig. 2.53a. Decoder inputs CAD and RAD from the column and row decoders must both be "1" to turn on the transmission gate and connect input A to Z. The switch connections to a resistor DUT are shown in Fig. 2.53c, using the circuit symbol in Fig. 2.53b. This design has full flexibility in isolating the resistor DUT for making a four-terminal measurement but also a large overhead of 32 MOSFETs per DUT. This design can be simplified by replacing

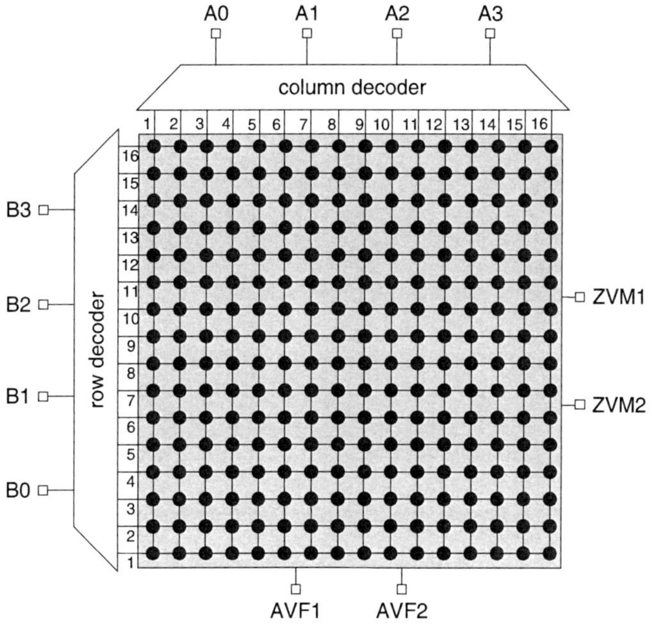

Fig. 2.52 A 2D array with 4-bit row and column decoders and a 16 × 16 matrix of 256 DUTs

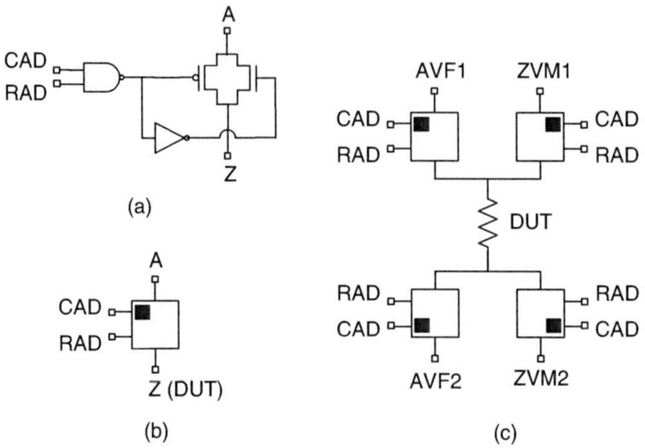

Fig. 2.53 **a** Circuit schematic and **b** symbol of a switch. **c** Resistor DUT with four switches to connect its inputs and outputs to the respective I/O pads

the transmission gates with n-passgates for the GND voltage connections and p-passgates for V_{DD} connections, thereby reducing the number of MOSFETs to 20. The I/O pad count in a 2D array is further reduced by using a shift register in place of or in combination with decoders.

A macro and DUT template for 2D arrays is very useful in simplifying the designs, and sharing probe cards and test codes among different macros. Macro design migration to other technology nodes, while maintaining the same I/O pad footprint, allows the use of a common custom probe card across different technology generations in a manufacturing line.

2.5.5 *High-Speed Macros*

High-speed measurements are classified by high-frequency operation (10 MHz to several GHz) or by single-shot pulse events with rise/fall times of <100 ps. Macro designs for these types of measurements can be classified by the test equipment requirements. In a situation where parametric ATEs are most commonly used, as in a manufacturing line at the M1 test stop, circuit delay and some transient measurements can be accommodated in a standard 1×25 padset macro and tested using the same DC probe card. With the high-speed actions taking place within the macro and with only DC input and outputs, the macro is immune to noisy manufacturing test environment. Inputs and outputs for such a macro are shown in Fig. 2.54 [13]. The DC inputs control latches, decoders, analog tuning of delay, and rise/fall times for generating pulses, selecting experiments, and tuning signal waveform shapes. The output is either a DC voltage output of a latch or a low-frequency (<10 MHz) signal from a ring oscillator.

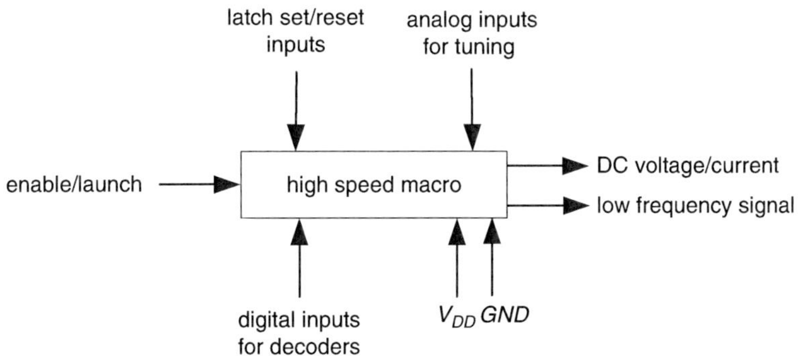

Fig. 2.54 High-speed macro with DC inputs and DC or low-frequency outputs

The second type of high-speed macro may have either high-frequency (~10 MHz to several GHz) inputs and outputs or signals with <100 ps rise and fall times or both. In this type of macro, pads for high-speed I/Os are best arranged in a GND–signal–GND fashion and a probe card matching the macro footprint of this pad arrangement is required. Macro templates for high-speed measurements are very useful and allow the use of a common probe card for a large variety of experiments. With a differential time measurement scheme, sub-ps time resolution can be achieved in bench measurements. High-speed macro designs are covered in Chapters 7 and 8.

2.5.6 Scaling of Macro Designs

Semiconductor packaging and probe card manufacturers have been making advances to follow CMOS scaling. However, although the feature size on silicon has been scaled by ~0.7× in each technology generation, the probe pitch has not been shrinking at the same rate. For example, the I/O pad area has been reduced from 90 μm × 60 μm with 60 μm pad spacing at the 180 nm technology node to 40 μm × 40 μm with 40 μm spacing at the 45 nm technology node. This is a factor of 3.3× reduction in pad area and 1.5× reduction in space between the pads compared to a 16× reduction in circuit area. Because of this difference in scaling factors for I/O pads and CMOS circuits, not all aspects of macro designs can be faithfully scaled in migrating designs from one technology node to the next.

Three different scenarios in migrating macro designs are illustrated in Fig. 2.55. In Fig. 2.55a, the pad size and the pitch remain the same but as the DUT size shrinks, longer wires are required to connect the DUT to the pads. In order to maintain a constant parasitic resistance of these wires, the wire widths should be increased in a scaled design. In the second case shown in Fig. 2.55b, a DUT is accommodated between the pads after scaling, thereby reducing the parasitic R and C values. The height of the macro is also reduced and the entire macro is compacted. In the third case in Fig. 2.55c, the pad size is reduced from 60 μm × 60 μm to 40 μm × 40 μm

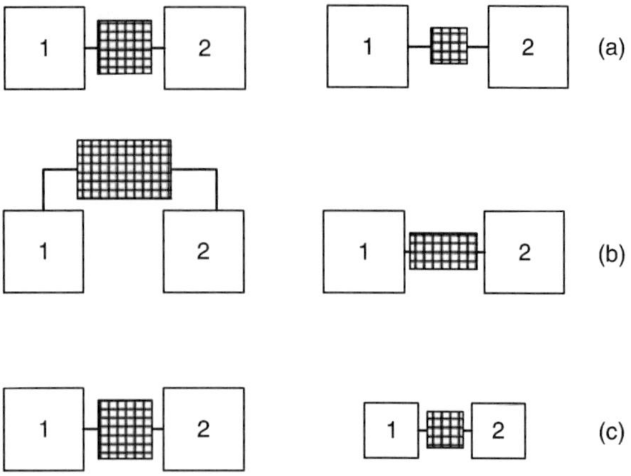

Fig. 2.55 Migration of macro designs with a DUT scaling factor of 0.7× **a** for the same pad size and pitch with DUT between the pads and **b** for the same pad pitch with the DUT shifted from outside pads to between pads. **c** Pad dimensions scaled by ~0.7×

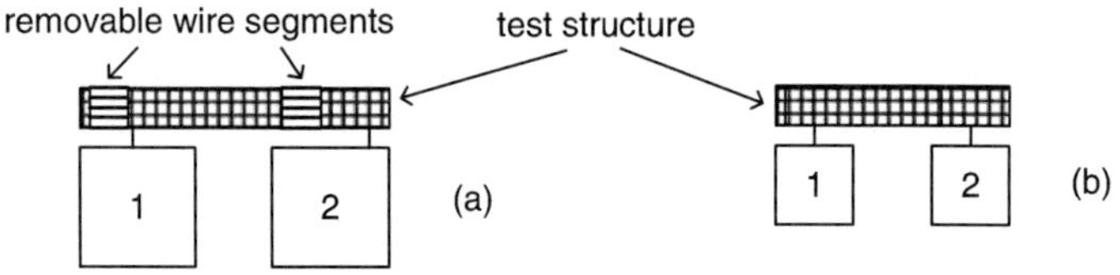

Fig. 2.56 Physical placement of test structure showing **a** wire segments to be removed to accommodate a scaled pad pitch and **b** placement with scaled pad pitch

and the pad pitch from 120 to 80 μm. The design can now be nearly faithfully scaled by ~0.7×.

In some cases, a change in pad pitch may be introduced during the technology development or the manufacturing phase. In this situation, modified designs need to be revalidated. If pad pitch changes are anticipated, the templates can be created with wire segments which are removed and the design collapsed to fit a different pad pitch as shown in Fig. 2.56, thereby maintaining the integrity of the design with minimum effort.

References

1. Weste NHE, Eshraghian K, Smith MJS (2000) Principles of CMOS VLSI design, 2nd edn. Addison Wesley
2. Baker RJ (2010) CMOS circuit design, layout and simulation, 3rd edn. Wiley, Hoboken, NJ
3. Uyemura JP (2001) CMOS logic circuit design. Kluwer Academic
4. Taur Y, Ning TH (2009) Fundamentals of modern VLSI devices, 2nd edn. Cambridge University Press, New York, NY
5. Sze SM (2002) Semiconductor devices: physics and technology, 2nd edn. Wiley, New York, NY
6. Jaeger RC (2001) Introduction to microelectronic fabrication, vol 5, modular series on solid state devices, 2nd edn. Prentice Hall, Englewood-Cliffs, NJ

7. Campbell SA (2001) The science and engineering of microelectronic fabrication, 2nd edn. Oxford University Press, New York, NY
8. Schroder DK (2006) Semiconductor material and device characterization, 3rd edn. Wiley, Hoboken, NJ
9. Moore GE (1965) Cramming more components onto integrated circuits. Electronics 38: 114–117
10. Natarajan S, Armstrong M, Bost M, Brain R, Brazier M, Chang C-H (2008) A 32 nm logic technology featuring 2nd generation high K+ metal-gate transistors, enhanced channel strain and 0.171 μm^2 SRAM cell size in 291 Mb array. IEEE international electron device meeting, IEDM-2008: 1–3
11. Weeden O (2003) Probe card tutorial. http://www.accuprobe.com/Downloads/Probe%20Card%20Tutorial.pdf. Accessed 15 Mar 2011
12. Bakoglu HB, Meindl JD (1985) Optimal interconnection circuits for VLSI. IEEE Trans. Electron Dev ED-32:903–909
13. Ketchen MB, Bhushan M (2006) Product-representative "at speed" test structures for CMOS characterization. IBM J Res Dev 50:451–468

Chapter 3
Resistors

Contents

DC resistance measurements of precision and parasitic resistors form an important part of CMOS technology characterization. Sheet resistances are measured to monitor and control material properties such as film thickness and doping levels in silicon layers. Resistance measurements are utilized in monitoring product yield loss caused by electrical opens in metal wires and inter-level vias, or shorts between neighboring wires. In metrology and for process tuning applications, linewidths are extracted from resistance measurements. For CMOS circuit simulations, electrical models of resistors are constructed based on data collected from test structures.

In this chapter, properties of resistor elements are described in Section 3.1. Resistance measurement techniques are covered in Section 3.2. DUT designs for resistance and yield characterization are described in Section 3.3. In Section 3.4,

M. Bhushan, M.B. Ketchen, *Microelectronic Test Structures for CMOS Technology*,
DOI 10.1007/978-1-4419-9377-9_3, © Springer Science+Business Media, LLC 2011

five examples of test structure designs, ranging in complexity from discrete resistors to 2D arrays, are provided. Metrology applications and extraction of process variations from resistance measurements are discussed in Section 3.5.

A comprehensive treatment of resistance measurement techniques for semiconductor materials and devices can be found in reference [1]. A list of selected publications on film resistance measurements and resistor macro designs is provided in the reference section.

3.1 DC Resistance

The electrical resistance of a metal wire is determined by connecting its two terminals to a constant voltage source and measuring the current flowing through it. This method of measuring resistance was first reported by George Simon Ohm in 1827 and the well-known Ohm's law bears his name. The application of resistance as a measure of quality control of commercial copper for underwater telegraph cables was established in 1857 [2]. In 1958, Van der Pauw reported a method for measuring specific resistivity of semiconductor films of any arbitrary shape [3]. Over 50 years after his publication, resistance measurements in the semiconductor industry still continue to challenge engineers as sub-μm geometries, parallel test methods for rapid data collection, and defect detection and localization techniques are incorporated in test structure designs.

Resistance test structures are employed for (1) measurement of resistances of circuit elements, (2) measurement of sheet resistances of conducting layers, and (3) detection and localization of electrical opens and shorts. DUT designs to measure contact, parasitic, and wire-end resistances are included to correct for measurement errors and to build accurate resistor models. Variability and matching characteristics of resistors are measured to determine systematic and random process variations. Temperature coefficient of resistance is measured to cover the operating range of CMOS products, which is typically –40°C to 120°C. Test structures for metrology applications such as linewidth extraction exploit the dependence of resistance on layer geometry. Macro designs and measurement methods are dependent on the resistance range, desired measurement accuracy, number of DUTs to be accommodated, allocated silicon area, and test time constraints.

3.1.1 Properties of Resistors

The resistance R of a uniformly conducting film is a function of its sheet resistance ρ_{sh} and its physical dimensions as discussed in Section 2.4.3. In the case of a straight long planar wire of length l, width w, and thickness d, the relationships are restated here:

$$R = \rho_{sh} \frac{l}{w}, \tag{3.1}$$

and

$$\rho_{sh} = \frac{\rho}{d}, \tag{3.2}$$

where ρ is the resistivity of the film. The term l/w is the number of squares (□s) n_{sq} in the film and ρ_{sh} is expressed in units of Ω/□. If the dimensions of the film are precisely known, ρ_{sh} can be calculated from the measured value of R using Eq. (3.1). In resistor elements with complex geometries or very small dimensions, current crowding at the edges, end effects, and shape distortion during processing, such as corner rounding, are taken into account in modeling the resistance.

Metal resistivity ρ and in turn R vary approximately linearly with temperature:

$$R_2 = R_1 \left\{1 + \mathrm{TCR}\,(T_2 - T_1)\right\}, \tag{3.3}$$

where R_1 and R_2 are resistances at temperatures T_1 and T_2, respectively, and TCR is the temperature coefficient of resistance. The TCR of Cu and Al metal layers and of heavily doped (n+ and p+) silicon layers is ~0.35%/°C and slightly smaller for precision resistors. A 5°C increase in the temperature of a copper wire produces an ~1.65% increase in its resistance.

The temperature dependence of precision resistors in silicon can be more accurately approximated by adding a quadratic term:

$$R_2 = R_1 \left\{1 + \mathrm{TCR}_1\,(T_2 - T_1) + \mathrm{TCR}_2\,(T_2 - T_1)^2\right\}, \tag{3.4}$$

where the coefficients TCR_1 and TCR_2 are determined empirically. If the temperature dependence is more complex, other terms may be added to get a best fit to the experimental data. The temperature dependence of resistance is exploited in temperature sensors to measure silicon chip temperature and to detect local temperature variations in CMOS products.

The voltage drop V across a conductor is a function of the current I flowing through it and its resistance R as stated by Ohm's law:

$$V = IR. \tag{3.5}$$

The power dissipation P in the resistor is

$$P = I^2R. \tag{3.6}$$

The heat generated by the current flow results in an increase in the resistor temperature, thereby changing its resistance. Measurement error arising from internal heating in a resistor can be minimized as discussed in Section 3.3.

The measurements of interest for conducting layers are as follows: (1) ρ_{sh}, (2) resistance per unit length (R/l) for minimum linewidth, and (3) R/l for specific linewidths, line spaces, and pattern densities. Precision resistors are defined in either silicon diffusion or polysilicon (unsilicided) layers with ρ_{sh} of 50–1,000 Ω/□. The

Table 3.1 Resistance ranges for conducting layers, vias, and defects in the 65 to 45 nm CMOS technology nodes

Conducting element	Properties
Precision resistor	$\rho_{sh} = 50-1{,}000\ \Omega/\square$
Silicided silicon diffusion/polysilicon	$\rho_{sh} = 10-20\ \Omega/\square$
Metal wire	$\rho_{sh} = 0.01-0.20\ \Omega/\square$
Interconnect via, H0	$R = 1-300\ \Omega$
Interconnect via H1 and higher	$R = 0.001-2\ \Omega$
Electrical open	$\geqslant 10^9\ \Omega$
Electrical short	$\leqslant 10^{-3}\ \Omega$

minimum allowed linewidths for the metal layers are restricted by technology physical ground rules (GRs). Thick metal layers with smaller values of ρ_{sh} have wider minimum linewidths compared to thin layers. Hence, R/l values of metal layers may vary over a wide range, from ~5 Ω/μm for the M1 metal layer to ~0.001 Ω/μm for thick metal layers.

Single via resistances for H0 (silicon to first metal M1 contact) have been steadily increasing with technology scaling, from ~1 Ω in earlier technology generations to as high as 300 Ω in the 32 and 22 nm technology nodes. Resistances of vias between thick metal layers may be as low as 0.001 Ω. The resistance limits for detecting electrical opens and shorts in metal wires and vias may be set $>10^9\ \Omega$ for opens and $<10^{-3}$ for shorts. The resistance ranges of conducting layers, vias, and defects are summarized in Table 3.1.

3.2 Resistance Measurements

The resistance range of conducting layers, vias, and defects spans over 12 decades. Standardization of test equipment and test structure macro templates is simplified if all resistor DUTs have resistance values in the range of ~1–1,000 Ω. This can be accomplished by appropriately sizing the resistor DUTs as discussed in Section 3.3. The resistance measurement accuracy requirement is generally < 1% for precision resistors and metal wires and more relaxed for opens and shorts. Higher precision is obtained by accounting for parasitic resistance contributions of I/O pads, probe contacts, and cable connections to the test equipment. Test equipment requirements and methods of improving measurement accuracy are described in the following sections.

3.2.1 Resistance Range and Test Equipment

A selection of standard off-the-shelf resistance meters, with measurement accuracies of 0.1–1%, cover resistance ranges from very low $\left(10^{-5}-10^{-3}\ \Omega\right)$ to very high

Table 3.2 Test current and voltage ranges of off-the-shelf resistance meters

Resistance (Ω)	Test current	Test voltage
10^{-5}	100 A	1 mV
10^{-3}	10 A	10 mV
1.0	100 mA	100 mV
10^{3}	1 mA	1 V
10^{6}	1 μA	1 V
10^{9}	1 nA	1 V
10^{12}	1 pA	1 V
10^{15}	1 pA	1,000 V

$(10^3-10^{15}\ \Omega)$. The test current and voltage ranges for resistance measurements are shown in Table 3.2. At the low resistance end ($\leqslant 1\Omega$), the current level is increased to keep the voltage drop across the resistor to $\geqslant$ 1mV. At the high resistance end $(\geqslant 10^{12}\ \Omega)$, voltage level is increased to maintain the current in a measurable (pA) range. The optimum test levels and measurement accuracies are dependent on the specification of the test equipment which should be carefully considered prior to designing resistor DUTs.

In silicon manufacturing, resistance measurements are carried out with parametric ATE mentioned briefly in Chapter 2 and covered in more detail in Chapter 9. The SMUs in these testers are set to either voltage force current measure (VFIM) or current force voltage measure (IFVM) configurations. A suitable voltage force range for parametric ATE is ~10 mV to 10 V and the current measure range is ~10 nA to 100 mA. An upper limit on the current is set by the current handling capability of probe needles, which is typically $\lesssim$100 mA. The lower limit on both current and voltage is set by the measurement accuracy of the ATE. Typically, the range over which resistance measurements may be made with a precision of 1% is thus ~0.1 to ~10^9 Ω.

The optimum resistance measurement range is further restricted at the low end by parasitic series resistances and at the high end by test time. Wire connections from the resistor DUT to the I/O pads and probes may add a parasitic series resistance of a few Ω at each DUT terminal. Their effect is minimized by optimizing the physical layout of the DUT and by designing the DUT resistance to be at least a few hundred Ω. Measurement methods to reduce the parasitic contact resistances are described in Section 3.2.3.

At the high resistance end, because of the large *RC* time constant of the DUT, the time to reach full applied voltage at the DUT terminals gets longer. In the VFIM mode, the internal measurement time of the tester in the low current (nA to pA range) is larger than for currents in the mA range. Hence, the measurement time is longer, decreasing the test throughput. As an example, the total measurement time for a 100 Ω resistor is ~10 ms increasing to ~200 ms for a 10^{12} Ω resistor.

3.2.2 Two-Terminal Measurements

In a two-terminal resistance measurement, a voltage V_f is applied across the two I/O pads connected to the resistor element, and the current I_m flowing through the resistor is measured. Alternatively, a current I_f is forced into the resistor and the voltage across its terminals V_m is measured. The resistance R is then

$$R = \frac{V_f}{I_m} = \frac{V_m}{I_f}. \tag{3.7}$$

An advantage of a two-terminal measurement, also known as a two-wire or a two-probe measurement, is that only two I/O pads per resistor DUT are required as shown in Fig. 3.1. In practice, however, the measured resistance R_m includes the resistance of the on-chip wire connections from the I/O pads to the DUT terminals, R_{s1} and R_{s2}, and the sum of probe-to-pad contact, probe wire, and cable resistances R_p:

$$R_m = R + R_{s1} + R_{s2} + 2R_p. \tag{3.8}$$

Probe contact resistance is dependent on the pressure applied to the probe tips, extent of skidding, and electrical properties of the pad material. In an ideal test setup, R_p is $\lesssim 1\ \Omega$ for all pad connections and $R_s\ (= R_{s1} = R_{s2})$ can be designed to be $\lesssim 1\ \Omega$. The DUT resistance limits are set by the desired measurement accuracy. For example, assuming $2\left(R_s + R_p\right) \approx 4\ \Omega$, R must be >400 Ω to limit the measurement error to <1%.

3.2.3 Four-Terminal (Kelvin) Measurements

The accuracy of a resistance measurement is improved with a four-terminal measurement scheme shown in Fig. 3.2a. This method, first developed in 1861 by Lord Kelvin, to measure small resistances is also known as Kelvin sensing. A current is

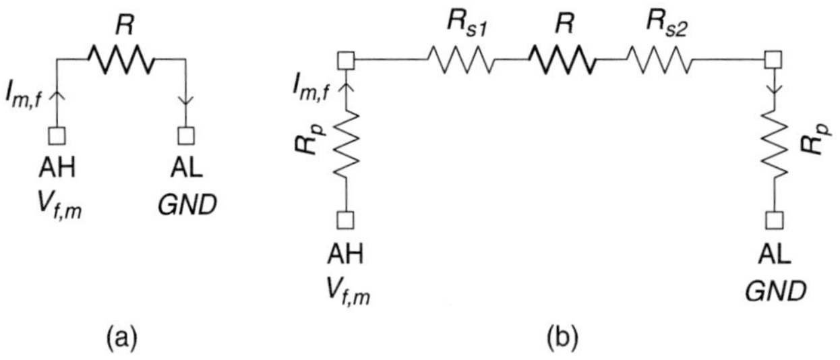

Fig. 3.1 Two-terminal resistance measurement of **a** an ideal resistor R and **b** a resistor R with parasitic wire resistances R_{s1} and R_{s2}, and contact/probe/cable resistances R_p in series

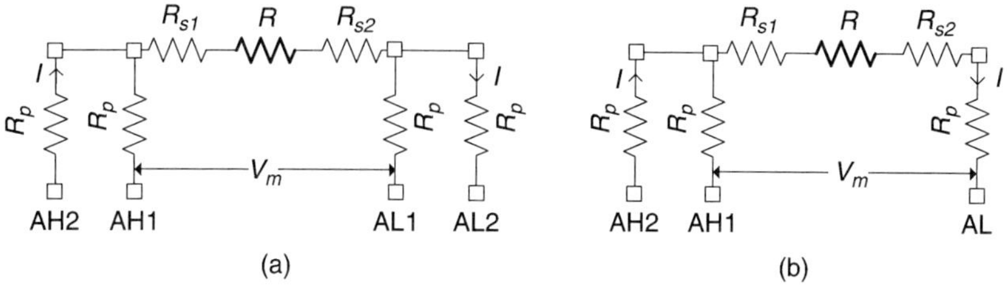

Fig. 3.2 Resistance DUT R and its parasitic resistances R_{s1}, R_{s2}, and R_p are shown for measurement configurations: **a** four-terminal and **b** three-terminal

forced through the outer terminals AH2 and AL2, and the voltage across the inner terminals AH1 and AL1 is measured. Alternatively, a voltage is forced at the outer terminals and the resultant current I and the voltage across the inner terminals V_m are measured. As the current flowing through the inner voltage sense leads is negligibly small, any voltage drop across the series resistor R_p is ignored.

In a third arrangement shown in Fig. 3.2b, only three terminals are used by combining the functions of AL1 and AL2 pads. In this configuration, more DUTs can be accommodated with the same number of I/O pads compared to a four-terminal measurement scheme. A correction for resistance R_p in the common AL node can be determined from the difference between two-terminal and three-terminal measurements or by dedicating a DUT with $R = 0$. These schemes are discussed in more detail in Section 3.4.1. When only two I/O pads are available per DUT, a four-terminal measurement can be made by landing two probes on each pad as described in Section 2.4.1 and thereby reduce R_p. To exclude the resistance of cable connection from the probe to the test equipment, the probe card may be designed with two cables per probe tip.

3.2.4 Contact Resistance Measurements

Probe-to-pad contact resistance, which is included in R_p, is typically $<1.0\ \Omega$ but may increase to $5\ \Omega$ or more with misalignment and mechanical wear of the probe tips and I/O pads. As this resistance is in series with the DUT resistance, its impact may become significant in a two-terminal measurement.

In silicon manufacturing, a test for mechanical and electrical integrity of probe contacts may be carried out on each reticle field after a certain number of probe touchdowns. A test structure macro for monitoring probe contact resistance described here also serves for checking probe alignment and for probe contact pressure adjustment. The basic design of the macro is shown in Fig. 3.3. Each pair of I/O pads is electrically shorted together with a low resistance metal ($\ll 1\ \Omega$) wire. The resistances of these short metal segments, dominated by their respective probe contact resistances, are measured and the data evaluated to ascertain the quality of probe contacts. A correction term for the DUT resistance, derived

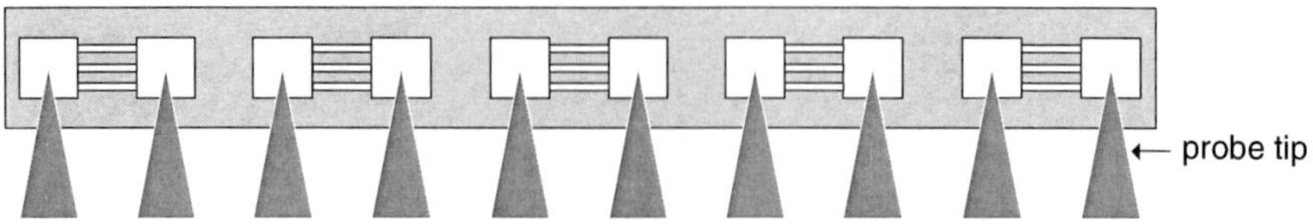

Fig. 3.3 Top view of a macro for measuring probe contact resistances

from these measurements, may be used for improving the accuracy of two-terminal measurements.

An upper resistance limit for R_p, typically ~2–5 Ω, is defined for test quality control. If the measured resistance values exceed this limit, data collected on a reticle field or wafer may be rejected and the probe card replaced. In order to prevent loss of data, probe cards are periodically replaced after a specified number of touchdowns.

3.2.5 Sheet Resistance Measurements

The sheet resistance of a film of uniform thickness and any arbitrary shape can be determined by the Van der Pauw method [3]. The DUT is contacted at four points along the periphery of the film as shown in Fig. 3.4a, keeping the contact area small compared to the area of the film. A current is forced through two adjacent contacts and the voltage measured across the other two contacts. This procedure is repeated by starting with two other adjacent contacts. For example, resistance $R_{12,34}$ is determined by passing current through A1 and A2 and measuring the voltage across A3 and A4. Similarly, resistance $R_{23,41}$ is determined by passing current through A2 and A3 and measuring the voltage across A4 and A1. The sheet resistance of the film is given by

$$\rho_{sh} = \frac{\pi}{\ln 2} \frac{(R_{12,34} + R_{23,41})}{2} F, \tag{3.9}$$

where F is a correction factor whose value depends on the ratio $R_{12,34}/R_{23,41}$. If the shape of the film is symmetric as shown in Fig. 3.4b, $R_{12,34} = R_{23,41} = R,\ F = 1$, and Eq. (3.9) reduces to

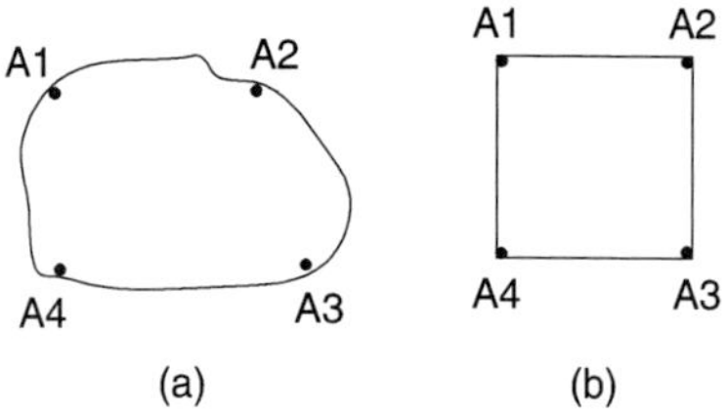

Fig. 3.4 Thin film with four contacts A1–A4 to facilitate ρ_{sh} measurements by the Van der Pauw method of **a** an arbitrary shape and **b** a symmetric shape

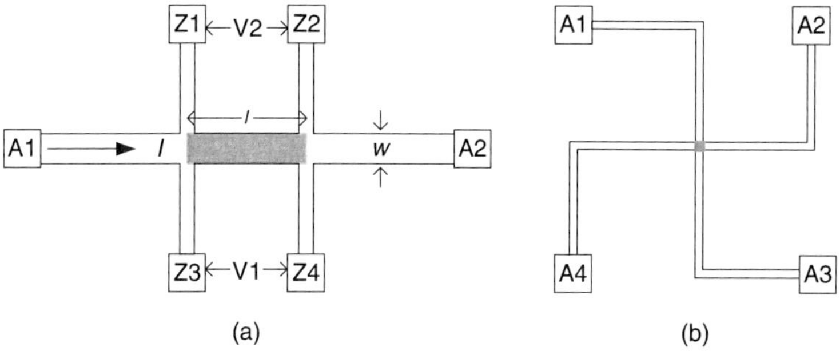

Fig. 3.5 Physical layout of a DUT for measurement of ρ_{sh}: **a** resistance bridge and **b** Greek cross

$$\rho_{sh} = \frac{\pi R}{\ln 2} = 4.532R. \tag{3.10}$$

A bridge structure for measuring resistance of a metal wire is shown in Fig. 3.5a. A four-terminal Kelvin resistance measurement is made with a constant current I flowing from A1 to A2. The accuracy of voltage measurement is improved by taking the average of V_1 and V_2. The ρ_{sh} of the metal layer is estimated from known l and w of the wire using Eq. (3.1).

As feature sizes are reduced, the linewidths of narrow wires may depend on the lithography and etching processes and may deviate from the dimensions defined in the DUT designs. The linewidth and film thickness of a metal layer may also be influenced by chemical mechanical polishing (CMP) process and its sensitivity to local pattern density variations. An independent test structure for sheet resistance measurement coupled with the bridge structure for resistance measurement described above is therefore very useful for characterizing material layer properties.

A Greek cross, shown in Fig. 3.5b, is commonly used for measuring ρ_{sh} of narrow wires with an accuracy of <1% [4]. In this application of the Van der Pauw technique, first the voltage $V(43)$ is sensed at terminals A4 and A3 with the current $I(12)$ flowing from terminal A1 to A2. The direction of current is reversed to correct for any voltage offsets arising from, for example, thermoelectric effect, and $I(21)$ and the voltage $V(34)$ are measured. This procedure is repeated by forcing the current between A2 and A3 and measuring the voltage between A1 and A4. The resistances with the orthogonal measurements are as follows:

$$R_{12,34} = \frac{V(43) - V(34)}{I(12) - I(21)} \quad \text{and} \quad R_{23,14} = \frac{V(14) - V(41)}{I(23) - I(32)},$$

where $I(21)$, $I(32)$, $V(34)$, and $V(41)$ are negative quantities. Assuming the resistor material to be homogeneous and of uniform thickness, ρ_{sh} can be calculated from Eq. (3.9) with $F = 1$. Accurate measurements of linewidths may be made with a

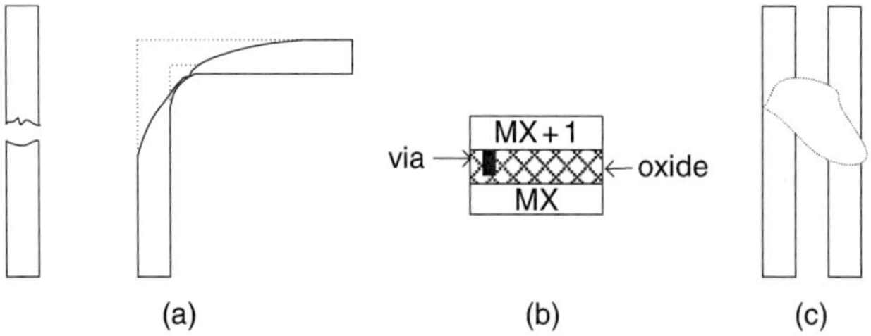

Fig. 3.6 **a** Planar views of electrical opens in wires. **b** Cross section of a defective via causing an electrical open. **c** Planar view of a metal short between two wires

combination of a bridge structure and a Greek cross. This application is discussed in more detail in Section 3.5.

3.2.6 Electrical Opens and Short Detection

Process or particulate-related defects may result in electrical opens in metal wires and vias, and electrical shorts between neighboring wires in the same layer and in vertically adjacent layers. An open is defined as a resistance greater than a specified value and similarly a short is defined as a resistance smaller than a specified value. Some examples of defects in metal wires and vias are shown in Fig. 3.6. In Fig. 3.6a, an electrical open in a metal wire occurs due to missing metal, and linewidth narrowing in a bend in a wire results in a high resistance. In Fig. 3.6b, incomplete metal filling in a metal interconnect via causes an electrical open. In Fig. 3.6c, two metal wires are shorted together because of incomplete metal etching caused by a particulate or a defect.

3.3 Resistor DUT Designs

In this section, examples of DUT designs for resistance measurements of conducting layers and interconnect vias, and for defect monitors are provided. The lower limit for the DUT resistance is based on desired measurement accuracy. The upper limit is based on available area, minimum current limit, and test time. DUT placement in the macro is based on available area and generally compact designs are preferred. For determination of resistor matching characteristics, or extraction of linewidth and end resistance using a differential pair of resistors, DUTs are placed in close physical proximity and in identical local environments. The physical layout of DUTs must meet the technology GRs except where these rules are deliberately violated to determine a process window.

In the case of a two-terminal measurement, it is preferable to keep the DUT resistance >400 Ω as discussed in Section 3.2.2. A rough estimate of resistor DUT

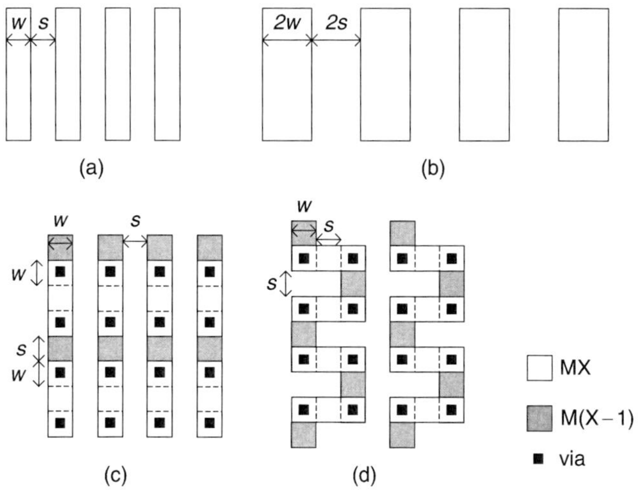

Fig. 3.7 Planar views of metal wires with **a** minimum width and spacing, and **b** double width and spacing. Via chain segments **c** of metal M(X−1) and MX travelling in the same direction, and **d** for orthogonal M(X−1) and MX

area A for specified values of R, ρ_{sh}, linewidth w, and spacing s shown in Fig. 3.7a, b is given by

$$A = \frac{Rw\,(w+s)}{\rho_{sh}}. \tag{3.11}$$

A few examples of calculated A, for a wide range of ρ_{sh} and geometries for 400 Ω resistor DUTs, are listed in Table 3.3. Such DUTs can be placed in the space (60 μm × 40 μm) between our standard I/O pads (Appendix A) for silicon and polysilicon resistors and thin metal layers with narrow linewidths but generally not for thick metal layers.

Resistance per interconnect via, from H0 to vias connecting top metal layers, may vary by as much as five orders of magnitude. The DUT for measuring average via resistance comprises a chain of vias. The length of the via chain is dependent on the desired total resistance and available area. The allowed via dimensions, spacing between vias, and metal overlap past vias, as governed by technology GRs, should

Table 3.3 Estimated area of a 400 Ω resistor DUT for conducting layers

Layer	ρ_{sh} (Ω/□)	w (μm)	s (μm)	~Area (μm^2)
Precision resistor	100	0.20	0.20	0.32
Polysilicon	10	0.04	0.16	0.32
Thin metal	0.2	0.10	0.10	40
Thick metal	0.01	0.35	0.35	9,800

be taken into account in calculating the area per via. Two example layouts of via chains with parallel and orthogonal metal layers are shown in Fig. 3.7c, d.

The approximate area of a DUT with N vias, for the geometries shown in Fig. 3.7c, d, is given by

$$A = N\,(w + s)\,(w + s)\,. \tag{3.12}$$

The DUT areas for via chains, for the case $w = s$ and a total resistance of 400 Ω, are listed in Table 3.4. The resistance of the metal lines connecting the vias must be taken into account when measuring low resistance vias. A reference DUT with the same geometrical outline as the via chain with only one of the metal layers is shown in Fig. 3.8. The resistance of the reference DUT is subtracted from the via chain resistance. If the two metal wires in the via chain have significantly different resistances per unit length, two reference DUTs may be included.

Four-terminal DUTs are designed to accurately measure resistance values of <400 Ω. The lower resistance limit is set by the desired accuracy of measurement and the voltage accuracy of the tester. Let us assume a desired measurement accuracy of ±1% and a tester voltage accuracy of ±1 mV. For a 10 Ω resistor, this sets $V_f = 100$ mV and $I_m = 10$ mA.

Temperature rise from Joule heating in the DUT adds to the measurement error and an upper current limit needs to be considered. The increase in temperature ΔT of the DUT during the measurement time of a few ms or more is dependent on the power dissipation, thermal properties of the DUT and the surrounding insulators,

Table 3.4 Estimated areas of a 400 Ω via chain DUT

Via	R/via (Ω)	No. of vias	$w = s$ (μm)	~Area (μm^2)
H0 (Si to M1)	50	8	0.1	0.32
H1 (M1 to M2)	1.0	400	0.1	16
H3 (M3 to M4)	0.1	4,000	0.5	4,000
HX (to MT)	0.005	80,000	2.0	1,280,000

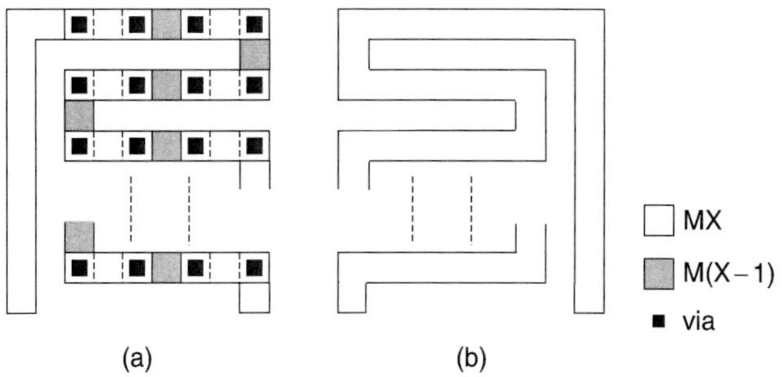

Fig. 3.8 Planar views of **a** a via chain and **b** a reference metal serpentine

DUT area, insulator thicknesses, and cooling effect at the edges. In general, ΔT can be expressed as

$$\Delta T = R_{\text{th}} P, \quad (3.13)$$

where R_{th} is the thermal resistance and P is the power dissipated in the DUT [5]. An exact calculation of R_{th} may be complex but often an approximation is sufficient. Consider, for example, a DUT consisting of a conducting film of length l and width w, dissipating energy uniformly at a rate P. If this film is vertically separated from the underlying silicon substrate by an insulating SiO_2 layer of thickness h, then using a parallel plate approximation, R_{th} can be calculated as

$$R_{\text{th}} = \frac{h}{wlK}. \quad (3.14)$$

Here K is the thermal conductivity of SiO_2 $\left(\sim 1.4 \times 10^{-6}\ \text{W}/\mu\text{m}^{\circ}\text{C}\right)$ and the silicon substrate is assumed to be a thermal "ground" at a constant temperature T. Equation (3.14) holds only if $h << w,\ l$, and the heat flows uniformly in the vertical direction with no lateral flow at the DUT edges. It sets an upper limit on ΔT.

Consider, for example, a precision resistor element with $l = 20\,\mu\text{m}$ and $w = 2\,\mu\text{m}$ placed on a SiO_2 film of thickness $h = 1\ \mu\text{m}$. With a ρ_{sh} of 100 Ω/□, it has a resistance of 1,000 Ω. If the resistance measurement is carried out with a current of 1.0 mA, with a heat dissipation of 10^{-3} W, the increase in resistor temperature calculated from Eqs. (3.13) and (3.14) is 18°C. For a TCR of 0.35%/°C, the measurement error from this self-heating is ~6%. Edge and end effects will lower this but a substantial error is nevertheless introduced. The error is reduced to <0.1% if the current is lowered to 100 μA. Note that cooling effects from the connecting metal wires produce non-uniform heating across the resistor, limiting the temperature rise, especially in small-area resistors. For a specified ΔT, an upper limit of the current flowing through a DUT can be estimated from the relationship

$$I_{\text{m}} = \sqrt{\frac{w\Delta T}{lR_{\text{th}}\rho_{\text{sh}}}}. \quad (3.15)$$

Although the measurement error from heat dissipation in the DUT can always be reduced by lowering the current flowing through it, a lower current limit is set by the accuracy of the tester. In low resistance DUTs, the minimum current is set by the accuracy of voltage resolution, typically in the range of 1.0–10 mV for ATE. In very high resistance DUTs, although the heating effect may be small, the measurement accuracy is worse as the current is lowered to the nA range. The test time for high resistance DUTs is also longer as the time to reach the full applied voltage across the DUT is governed by the resistance and self-capacitance of the DUT.

Another consideration in sizing DUTs is the effect of statistical variations. Linewidth variations in narrow wires are averaged over the length of the wires. In a chain of vias, large deviations in a single or a few vias may be undetectable. As an example, consider a series chain of 10,000 vias with a mean resistance of 10,000 Ω and a standard deviation σ of 1% or 100 Ω. If, because of a local defect or an incomplete processing, one of the vias has a resistance of 100 Ω (100 times its nominal value of 1.0 Ω), the via chain resistance will increase by 1%. This increase is well within 3σ (± 3%) of the distribution and hence not detectable. If on the other hand, the number of vias in the chain is reduced to 1,000, the increase in resistance by 100 Ω (10%) due to one defective via can be easily detected.

Metal and silicon precision resistor DUTs, delineated by rectangles or serpentines, may be placed in the space between two I/O pads. Physical layout examples for a short DUT and a long narrow wire serpentine DUT are shown in Fig. 3.9. Pad connections may be on the same metal layer as the resistor or a different layer, in which case a number of vias between the metal layer connection to the pads and the wide contact metal are included to reduce the parasitic series resistances. This resistance at the ends, which may include via contact resistance and spreading resistance, may be eliminated by designing two DUTs of different lengths and taking the difference in their measured resistance values. DUTs in thick metal layers, where the minimum area for desired accuracy is larger than the area between the pads, comprise long serpentines placed in the area above the pads.

A number of DUT layouts are used to detect opens and shorts between horizontally or vertically adjacent metal layers. Comb and serpentine layouts are used for detecting opens and shorts between metal lines in a single layer and also for measuring wire resistances. Line drawings of a serpentine and a comb are shown in Fig. 3.10. A very high resistance between AH and AL indicates an electrical open in the line in Fig. 3.10a. A short is detected as a very low resistance measured between pads AH and AL in Fig. 3.10b. A maize and a multiple serpentine design (MPS) shown in Fig. 3.11 are very useful for monitoring both opens and shorts as well as for measuring wire resistances and capacitances. A more complex DUT with two entangled via chains for detecting opens and shorts between two metal layers is shown in Fig. 3.12.

For resistance measurements, it is desirable to keep the DUT area small to achieve a compact layout and to minimize spatial variations. DUT area for defect monitors, on the other hand, must be large to detect the occurrence of relatively rare random defects. The critical area for a certain type of defect is defined as the effective DUT area vulnerable to the occurrence of that defect type and can be expressed

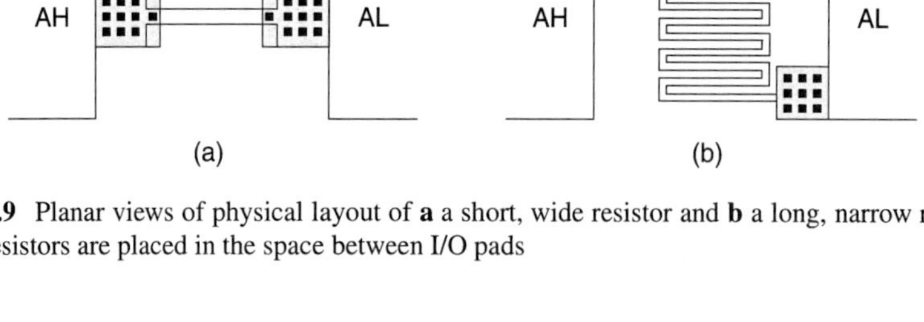

Fig. 3.9 Planar views of physical layout of **a** a short, wide resistor and **b** a long, narrow resistor. The resistors are placed in the space between I/O pads

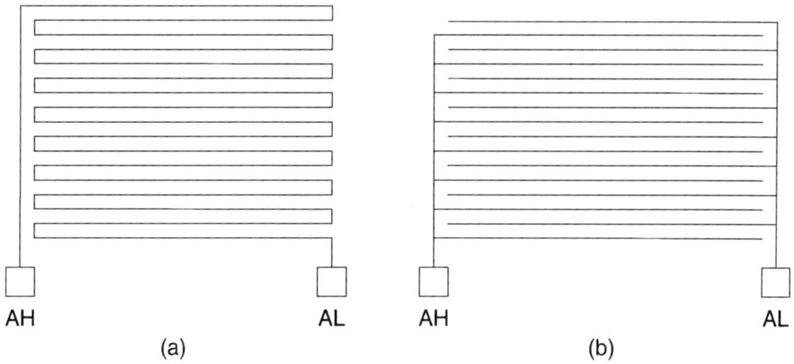

Fig. 3.10 Line drawing of defect monitors: **a** a serpentine for resistance measurements and **b** a comb structure

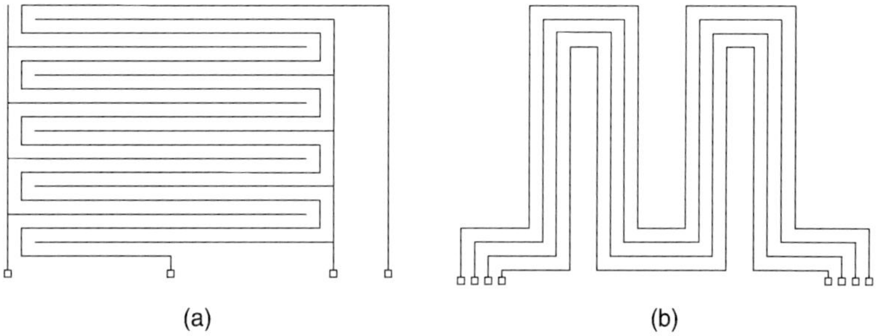

Fig. 3.11 Line drawings of defect monitors: **a** a maize and **b** an MPS structure, both used for resistance measurements and for monitoring opens and shorts

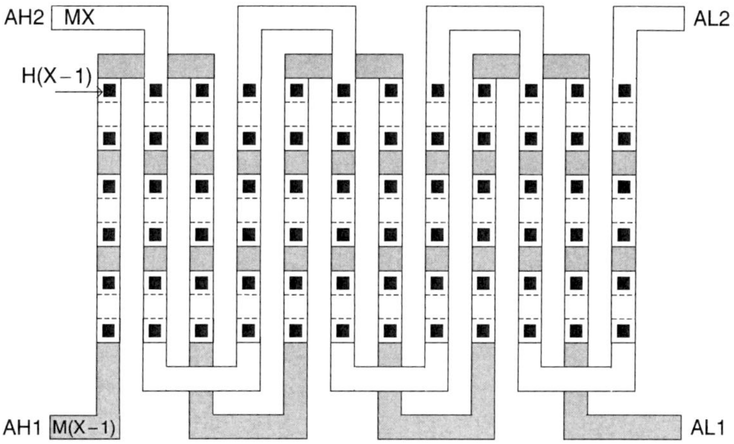

Fig. 3.12 Top view of two entangled via chains in two metal layers to detect via opens and leakage paths between vertically adjacent layers

as γA, where A is the area of the DUT and the area efficiency γ is generally <1. The critical area necessary to detect a given defect type increases as defect density is reduced in the manufacturing line. DUT layouts may be optimized to maximize γ and thus improve DUT efficiency for defect detection [6]. As an example, γ of defect test structures is increased by interleaving and entangling the serpentines and chains.

Early in technology development, the defects are typically process limited and relatively smaller area coverage but more varied layout styles are needed for systematic and random defect detection. As the technology matures, the defects are limited by particulates and the necessary critical area increases. However, a larger area is more readily available in test vehicles used in early technology development and in short-loop test vehicles than on a product wafer with limited scribe-line area. A large number of passive test structures may be placed on test vehicles, while complex high γ designs with control circuitry are placed on product wafers.

The defect density DD is calculated from the measured yield and the DUT critical area as

$$DD = \frac{1 - Y}{\gamma A}, \tag{3.16}$$

where Y is the yield of the DUT [7]. The relationship between DUT yield and area is shown in Fig. 3.13. A DUT is most effective as a defect density monitor when its yield is well away from both 0 and 1.

Electromigration in wires carrying current causes metal agglomeration in the direction of current flow, creating metal-deficit areas. The wire resistance increases with time and ultimately the wire becomes non-conducting, thereby resulting in an electrical open. Electromigration is enhanced at high current densities and at high chip temperatures. The wire resistance may also increase with both self-heating and local heating from surrounding circuit activity which can further accelerate electromigration-induced failures. Resistance test structures are employed to determine electromigration failure rates. Careful consideration is given to the DUT

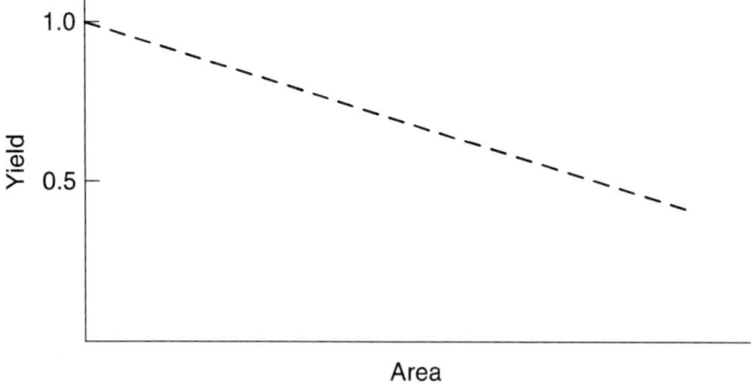

Fig. 3.13 Conceptual plot of yield as a function of chip area for a CMOS product

designs designated for accelerated stress tests for electromigration. Typical wire lengths are of the order of a few hundred μm and widened at the two ends to reduce contact resistance. Standard DUT designs for electromigration have been developed by the National Institute of Standards and Technology [1].

3.4 Resistor Macro Designs

There are six major considerations in the design of resistor macros: (1) availability of MOSFETs, (2) number of metal levels, (3) number of resistor DUTs per macro and the DUT area, (4) resistance range and desired accuracy of measurements, (5) parallel test capability of the test equipment, and (6) test time budget.

Resistor macros placed on short-loop test vehicles for technology development or for routine process monitoring may not undergo MOSFET processing. The DUTs in these macros are built with one or more conducting layers but without any active circuit elements or peripheral circuitry. The number of DUTs within a macro is severely limited by the number of available pads. Silicon utilization is improved with the availability of CMOS active circuit elements in macros comprising 1D or 2D arrays of DUTs. These complex designs are area efficient and suitable for yield monitoring and for characterization of variability in resistor elements.

In this section, five examples of macro designs for resistance measurements and defect monitoring are described. The macro configurations with two-, three-, or four-terminal measurements covered in Example 1 can be implemented with only a single conducting layer. In Example 2, with multiple metal levels, more complex pad sharing schemes to increase the number of DUTs per macro are feasible. Designs of 1D arrays with and without a decoder circuit for DUT selection are described in Example 3. These 1D array designs, using MOSFET control circuits, can be implemented at the M1 metal level and multiple arrays are placed within one macro. In Examples 1, 2, and 3, DUTs can be accommodated in a minimum size macro with our standard 1×25 linear array of pads shown in Appendix A. Complex 2D arrays in Examples 4 and 5 are utilized for large-area yield structures including defect localization for failure analysis. In each example, requirements for parallel test implementation and schemes for test time reduction are included.

3.4.1 Example 1: Discrete Resistor Macros

Macro designs for discrete resistors are implemented with a minimum of one conducting layer. These types of designs are particularly suitable for both resistance measurements and defect detection in short-loop test vehicles or in test vehicles early in the technology development cycle. I/O pads may be configured for two-terminal, three-terminal, or four-terminal measurements with different degrees of I/O pad sharing. The relative merits of I/O pad configurations and test methods are examined for each macro design type.

In Fig. 3.14, different I/O configuration options for two-terminal, three-terminal, and four-terminal measurements with isolated or shared pads are shown. In Fig. 3.14a, b, all DUTs are electrically isolated. In Fig. 3.14c, d, DUTs are connected

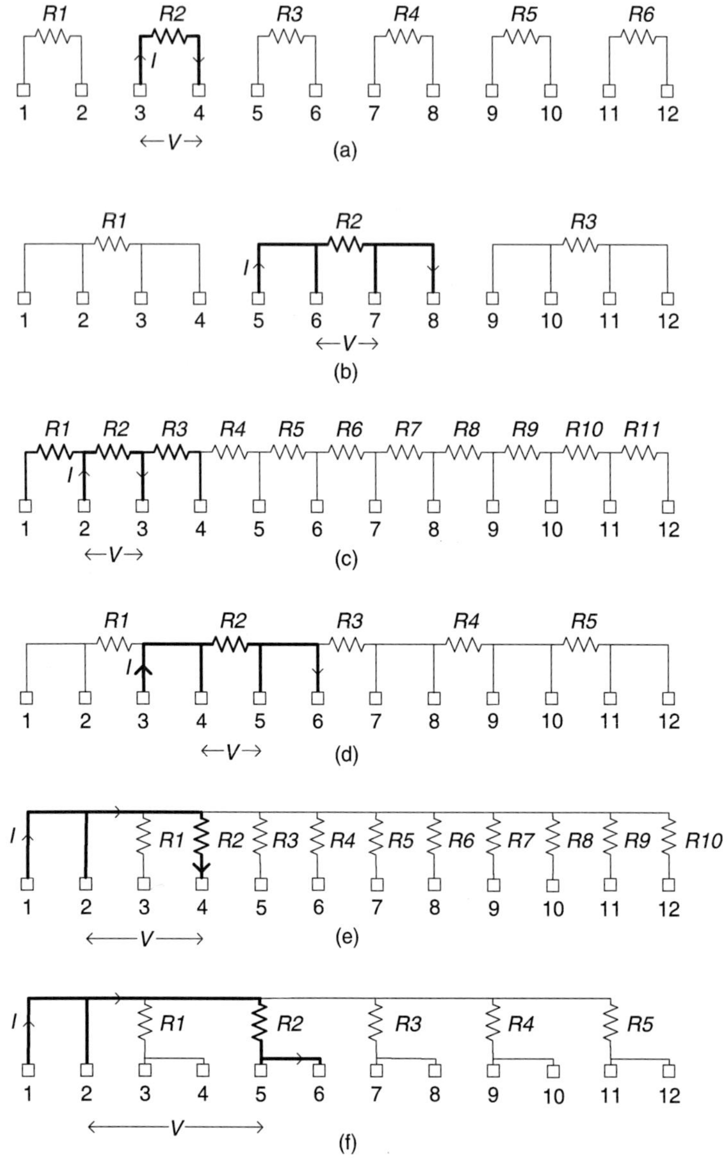

Fig. 3.14 I/O pad assignments for resistor measurements: **a** for isolated two-terminal DUTs, **b** for isolated four-terminal DUTs, **c** for series-connected two-terminal DUTs, **d** for series-connected four-terminal DUTs, **e** for star-connected three-terminal DUTs, and **f** for star-connected four-terminal DUTs. Current *I* and voltage *V* for measuring resistor DUT *R2* are indicated

in series and pads are shared between adjacent DUTs. In Fig. 3.14e, f, the DUTs are connected in a star configuration, with at least one common pad for injecting current in all the DUTs. Only two-terminal measurements are possible for DUTs in Fig. 3.14a, four-terminal measurements in Fig. 3.14b–d, f, and three-terminal measurements in Fig. 3.14e. More DUTs can be accommodated in the arrangement shown in Fig. 3.14c compared with the arrangement in Fig. 3.14d. However, for four-terminal measurements in series-connected DUTs shown in Fig. 3.14c, current to the I/O pads flows through adjacent DUTs and this scheme works only if none of the neighboring DUTs are electrically open.

For the different configurations discussed above, the total number of DUTs in 1×25 padset macros is listed in Table 3.5. Macro templates may be created for the selected configurations. Macro designs based on a common template can share test automation software.

When a large number of macros are present, a significant reduction in test time is obtained by measuring the DUTs in parallel [8]. In standard parametric ATE, the number of SMUs is generally $\lesssim 8$ and only a few DUTs can be tested in parallel. ATE designed with an independently controlled SMU for each I/O pin may be equipped with as many as 200 SMUs and in this case, a large number of DUTs (50–100) can be tested in parallel.

Isolated DUTs may be measured in parallel in either VFIM or IFVM modes. In series-connected DUTs, IFVM mode is preferred if no DUTs are open. A current is forced through the DUTs and the voltages across the DUTs are measured in parallel. The difference in the measured voltages between two adjacent pads gives the voltage across the DUT. A shorted DUT has no impact on the measurement and the short can be easily detected as the voltage drop across it is extremely small.

If open mode failures are expected in series-connected DUTs, the VFIM method may be used for parallel testing. The test method is illustrated in Fig. 3.15 for four series-connected resistors. A voltage is forced at each node and the current in the SMUs, I_{m1}, I_{m2}, I_{m3}, I_{m4}, and I_{m5}, recorded. The voltage across a DUT is the difference in its node voltages. The currents through the DUTs, I_{r1}, I_{r2}, I_{r3}, and I_{r4}, are computed as shown in Fig. 3.15a, taking the sign of the currents into consideration. The resistance of a DUT ($R2$) is given by

Table 3.5 Total number of DUTs in a 1×25 padset macro for the configurations shown in Fig. 3.14

Macro type	Total DUTs	Measurement	Comments
a	12	Two-terminal	Isolated
b	6	Four-terminal	Isolated
c	24	Two-terminal	Four-terminal if no opens
d	11	Four-terminal	Four-terminal
e	21	Three-terminal	Three-terminal
f	11	Four-terminal	Four-terminal

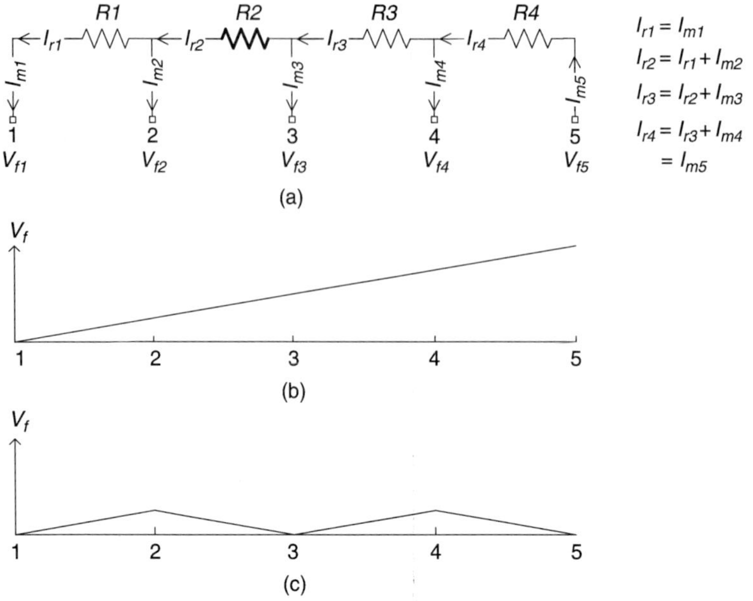

Fig. 3.15 **a** Series-connected DUTs configured for parallel test. Force voltages at the nodes: **b** monotonically increasing and **c** alternating

$$R2 = \frac{(V_{f3} - V_{f2})}{I_{r2}}. \tag{3.17}$$

If *R2* is open, it can be easily detected as $I_{r2} = 0$ (or $|I_{m1}| = -|I_{m2}|$). The force voltage is incremented at each node monotonically as shown in Fig. 3.15b. Note that if the DUT resistances are exactly the same, $|I_{m1}| = -|I_{m5}|$, and the current drawn by the intermediate SMUs is zero. The force voltage can be alternated, as shown in Fig. 3.15c, to limit the absolute maximum applied voltage on a node.

In case of star-connected DUTs, the VFIM mode is used for parallel test. A voltage is forced on the common node and the currents through all the DUTs are measured in parallel. If a DUT is open, it does not influence the measurement, but its presence is detected as no current flows through it. If one or more DUTs are shorted, the currents in the corresponding SMUs will hit compliance. A second measurement pass is made after disconnecting the failing DUTs.

A drawback of star-connected DUTs is the measurement error from *IR* drop in the common line. Care should be taken to ensure that the common line resistance is low and any voltage drop in it is insignificant. Additional I/O pads may be used to feed into the common line to reduce *IR* drops. Alternatively, the common line may be held at GND and by alternating the force voltage on each node, the current in the common line is balanced as long as the resistor values of the DUTs are similar.

In the macro designs discussed here, test time is substantially reduced by parallel testing. When testing a large number of such macros, the probe index time to step from one macro to the next can become a substantial fraction of the total test time and ultimately limit the test efficiency as discussed in Section 2.3.3.

3.4.2 Example 2: Passive Array Macros

The types of macro designs in this example are implemented with more than one conducting layer in the metal stack to facilitate low resistance multiple metal wire connections. The number of DUTs in a macro is substantially increased with more complex I/O pad sharing schemes. However, parasitic wire resistances may be higher and these resistance values, relative to the DUT resistance, must be carefully considered in the physical layout of the macro.

In one example shown in Fig. 3.16, a group of DUTs is formed by connecting one terminal of all the DUTs to a single I/O pad. The other terminals of the DUTs

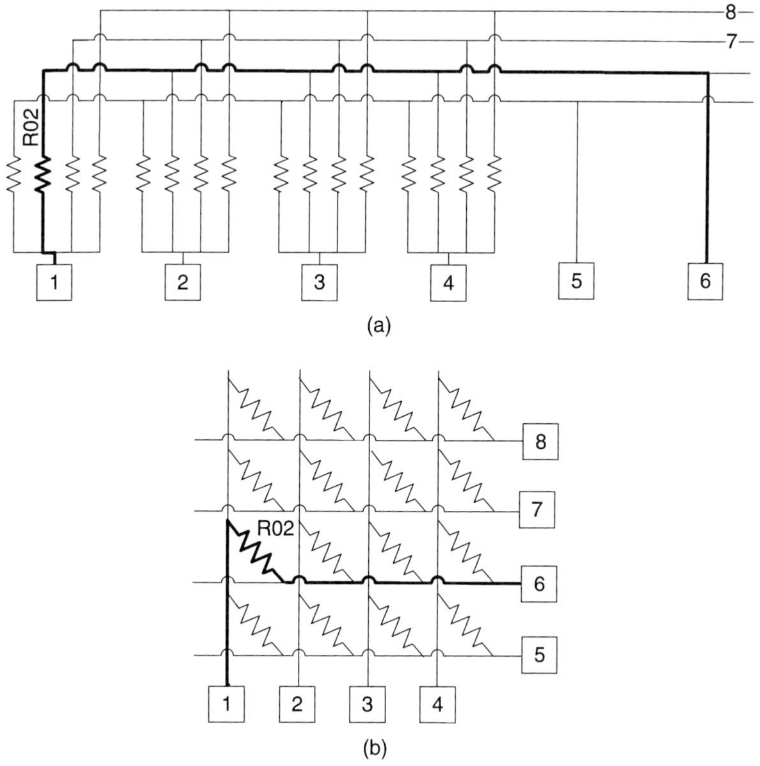

Fig. 3.16 **a** Circuit schematic of a section of a passive resistor array macro with four DUTs/group, and **b** its 2D array representation. Wire connections for resistor R02 to the I/O pads are highlighted

are connected to a set of pads in a way that no two DUTs share the same two pads [9]. The implementation of this scheme in a standard 1D linear pad array macro is shown in Fig. 3.16a, and an equivalent 2D array representation in Fig. 3.16b. A group comprises four DUTs with one common terminal. The second access terminal of each DUT is shared with DUTs in the other three groups. For convenience, the wires from the access I/O pads 5, 6, 7, and 8 travel the length of the macro so that connections may be tapped for all the groups in the macro. Resistance of these wires should be much smaller than the resistance of the DUTs. This design is well suited for applications in which measurement precision is not critical such as a yield monitor for open and short detection.

Four DUTs can be measured in parallel with a dedicated SMU for each of the common terminals (pads 1, 2, 3, and 4). The current flowing through the DUTs is measured with one of the second terminals (pad 6 for R02) held at GND. The other three access pads (5, 7, and 8) are left floating to prevent parasitic currents in the other DUTs from interfering with the measurement. Hence, the number of parallel measurements is equal to the number of groups. An alternative approach for reducing the parasitic current flow is to add a diode in series with each DUT [10].

The total number of DUTs in a 1×25 padset macro as a function of the number of DUTs in a group is shown in Fig. 3.17. The maximum number of DUTs in a macro is achieved with the number of DUTs per group equal to the number of groups ($N = 12$), and the total number of DUTs is $\sim N^2$. This optimization is much the same as the area of a rectangle of a given perimeter being maximized for equal length and width, and scaling as the square of the perimeter.

3.4.3 Example 3: 1D Addressable Array Macros

In an integrated process, with the availability of at least one n-FET type and one p-FET type, I/O pad sharing schemes with the aid of CMOS peripheral circuits

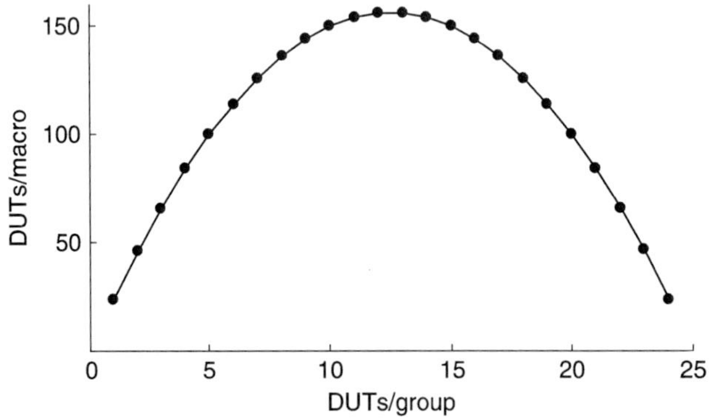

Fig. 3.17 Total number of DUTs in a macro with 25 I/O pads as a function of number of DUTs per group

are possible. The macro designs become more complex, but the area efficiency is improved with an increase in the total number of DUTs per macro. Test efficiency is also improved with parallel testing.

Here, we describe macro designs which incorporate a number of active 1D arrays of DUTs in which a single DUT in an array is addressed at a time [11]. In one design using a decoder for DUT selection, each array may contain as many as 32 DUTs and seven arrays may be accommodated in a macro with a 1×25 padset. We show how this design can be implemented for test at the M1 metal level. The small resistor arrays can be placed in the space between pads to minimize macro area. This design is particularly suitable for precision resistors, MOSFET parasitics such as PS and DF resistors, H0 vias, and may also be used for monitoring random variation in linewidths for all metal levels. Two related designs without the use of a decoder but with MOSFET switches to select a DUT are also described.

The basic building blocks of the macro template are an electrical 1D array comprising 16 or 32 DUTs, a 4- or a 5-bit decoder for DUT selection, and a control circuit for each DUT. The control circuit is used for steering a current through the selected DUT and measuring the voltage across it. Macro area efficiency is improved if a three-terminal measurement is made and a correction applied for the common GND wire resistance.

A DUT circuit configured for a four-terminal measurement is shown in Fig. 3.18a. The DUT design is simplified for a three-terminal measurement shown in Fig. 3.18b along with its symbol in Fig. 3.18c. The DUT is selected when the CAD input is high or "1." A voltage applied to the AVF node generates a current through the resistor. The voltage across the resistor is measured between the AVS and GND nodes. If the CAD input is low or "0," only the leakage current of the n-FETs flows through R. By placing two n-FETs in series, this background leakage current of the n-FETs is reduced substantially. Alternatively, low leakage n-FETs may be used, if available.

A number of DUT circuits are placed in parallel and a decoder selects the DUT to be measured. The array placement in the space between two I/O pads along with a 4-bit decoder is illustrated in Fig. 3.18d, using the symbol for a three-terminal DUT shown in Fig. 3.18c. The 16 DUTs are placed in three rows between the AVF and GND pads. The AVS wire is connected to a nearby I/O pad. Since this is the voltage sense wire, a higher series resistance to the pad is acceptable. A number of arrays are placed within a macro as shown in Fig. 3.19. The decoder input pads for A0, A1, A2, and A3 in all the arrays are shared. Two adjacent arrays share a common GND connection to conserve I/O pads. Two I/O pads, one at each end of the macro, supply the decoder V_{DD} (V_{DDC}). The decoder GND is supplied by the array GND pad. All the GND pads are connected inside the macro to form a common GND. The decoder draws current only during switching and the switching speed of the decoder is not critical. A large transient voltage droop in the long V_{DDC} wires is therefore acceptable.

The circuit schematic of a 4-bit decoder is shown in Fig. 3.20. The decoder function is implemented with NAND2 and inverter logic. The schematic is arranged to illustrate its physical implementation at the M1 metal level. The V_{DDC}, GND, and

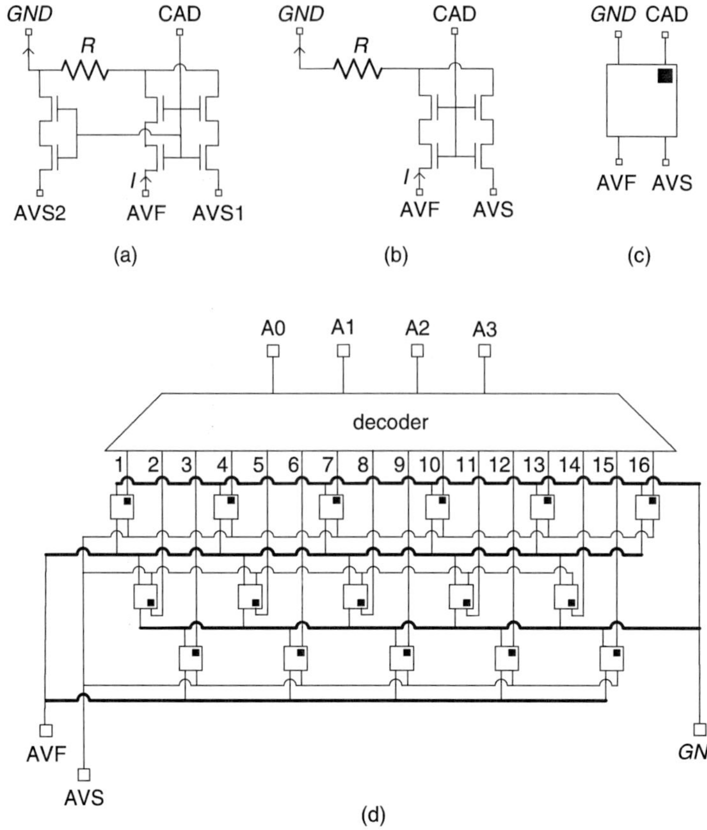

Fig. 3.18 Resistor DUT circuit schematic for **a** four-terminal measurements and **b** three-terminal measurements. **c** Symbol for the circuit in **b**. **d** A 1D array with 16 DUTs and a 4-bit decoder for DUT selection, configured for three-terminal resistance measurements

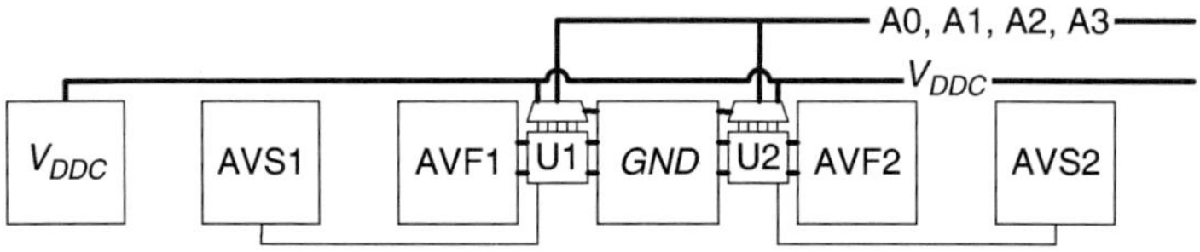

Fig. 3.19 Floorplan of a macro section showing I/O pad sharing between arrays

other horizontal wires are drawn in M1. The V_{DDC} and GND busses are split to allow the placement of horizontal M1 wires. The underpasses are in short sections of PS layer. In the 65 nm technology node and beyond, the direction of travel of long PS wires may be restricted. In this case, macro orientation on silicon is considered in floorplanning the test vehicle or the scribe line early in the design phase. The

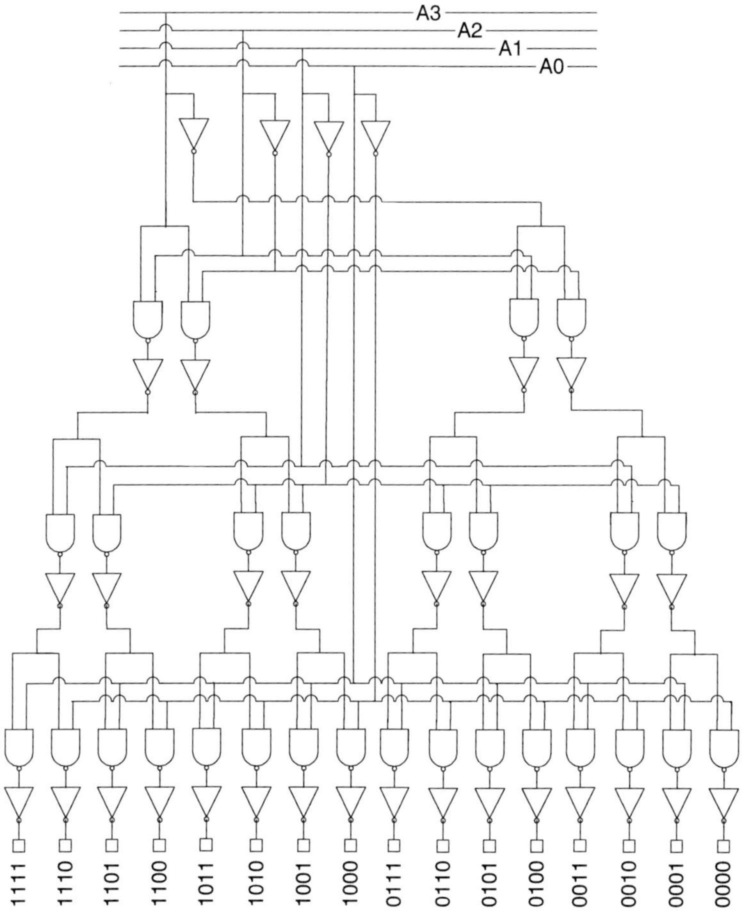

Fig. 3.20 Circuit schematic of a 4-bit decoder for a 1D resistor array, oriented to illustrate wiring at the M1 metal level

detailed physical layout of a similar decoder in a 1D MOSFET array is described in Section 5.4.3.

The physical layout of the DUT circuit is shown in Fig. 3.21 with the circuit schematic reproduced from Fig. 3.18b. The AVF and GND wires travelling in the horizontal direction in each row are directly connected to the I/O pads. The AVS wire crosses the AVF bus in DF layer, runs parallel to the power supply wires in each row, and then connects to a nearby I/O pad. The resistor element is positioned within this DUT template which is carefully sized to accommodate 16 or 32 DUTs in an array.

In a three-terminal measurement, the series resistance of the common GND wire is included in the measured resistance. For a precise resistance measurement, this parasitic resistance must be subtracted from the measured resistance. There are two

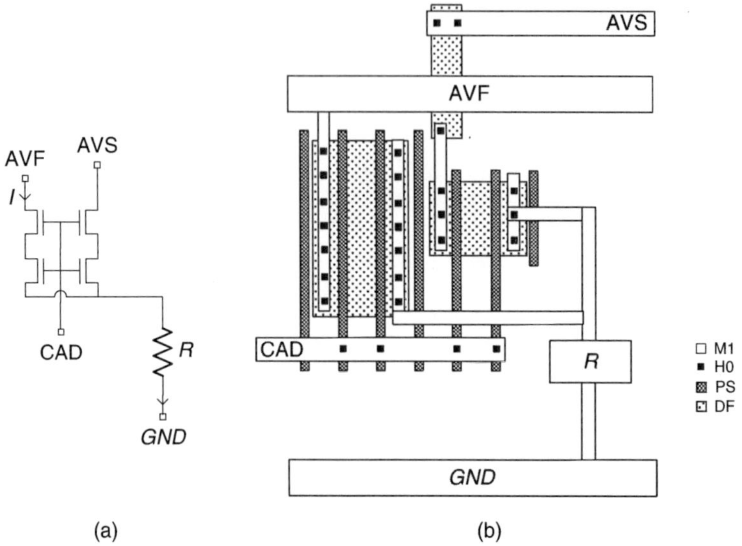

Fig. 3.21 Resistor DUT circuit for a 1D array: **a** schematic and **b** physical layout

possible ways to determine a correction. In one option, all resistor elements comprise a short metal wire with negligible resistance in one array in the macro. The resistance of all DUTs in this case is simply the GND wire resistance as shown in Fig. 3.22. These reference offset values are subtracted from the corresponding measured resistances in the other arrays. In a second option, metal shorts are placed as the resistor elements in the first and last DUT of each array. The correction factor for each DUT location in the array is obtained from a linear fit between the resistances of the DUTs at the two ends.

With a shared GND pad in two adjacent arrays, a total of seven arrays are accommodated in a standard 1×25 padset macro. The I/O pad assignment for such an

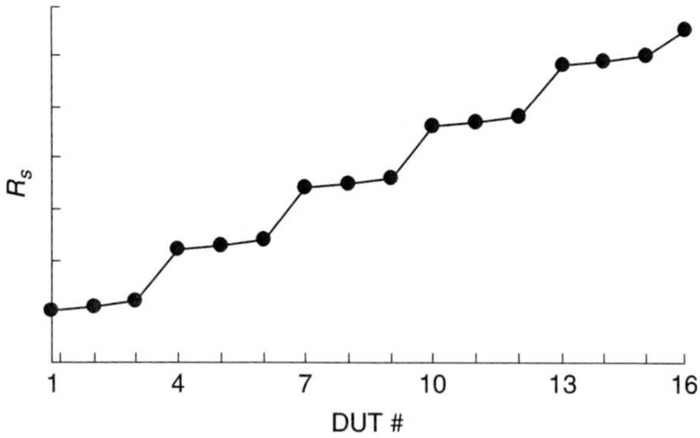

Fig. 3.22 Parasitic GND resistance R_S in series with the DUT as a function of DUT # for the circuit schematic shown in Fig. 3.18d

1	2	3	4	5	6	7	8	9	10	11	12	13	14	15	16	17	18	19	20	21	22	23	24	25
V	A	A	G	A	A	A	A	G	A	A	A	A	A	A	A	A	A	G	A	A	A	A	G	V
D	V	V	N	V	V	V	V	N	V	V	0	1	2	3	4	V	V	N	V	V	V	V	N	D
D	S	F	D	F	S	S	F	D	F	S						S	F	D	F	S	S	F	D	D
C	1	1		2	2	3	3		4	4						5	5		6	6	7	7		C

Fig. 3.23 I/O pad assignment for a macro with seven 32 DUT 1D arrays, implemented at the M1 metal level

arrangement is shown in Fig. 3.23. The decoder inputs are placed in the center of the macro and the two I/O pad numbers for the decoder V_{DDC} are 1 and 24. In the 45 nm technology node and beyond, 32 DUTs can be placed in 1D array situated between two I/O pads with a spacing of 40 μm. Hence the total number of DUTs in a minimum size macro is 224. Only six arrays may be placed in this macro if substrate or n-well and p-well contacts are needed.

One limitation of the M1 implementation with 32 DUTs is that the resistor area is small ($\sim 1\ \mu m^2$). This is acceptable for H0, precision resistors, and short lengths of PS and DF and thin metal wires. For thicker metal wires, the DUT and array size is typically larger and the array is placed above the I/O pads. With the availability of more metal layers, the use of PS and DF layers for underpasses is eliminated.

A drawback of this array design is that in early technology development the decoder employing ~100 logic gates may have a low yield. In such applications, a simplified 1D array implementation may be carried out without a decoder, but with fewer DUTs per macro. A number of resistor DUT circuits are connected in parallel between two I/O pads. In this multi-DUT unit (md-unit), an independent voltage bias (CAD) line is dedicated to each DUT circuit (Fig. 3.18a–c). This concept is illustrated in Fig. 3.24 with six DUT circuits connected in parallel between AVF and GND. A single DUT is selected for resistance measurements by setting the corresponding CAD (AVg) input to "1" and keeping all other CAD inputs at "0." Measurement precision is improved with the GND resistance correction scheme.

When the seven 1D arrays in Fig. 3.23 are replaced by md-units, each with seven DUTs, a standard macro can accommodate 49 DUTs. The CAD inputs for selecting

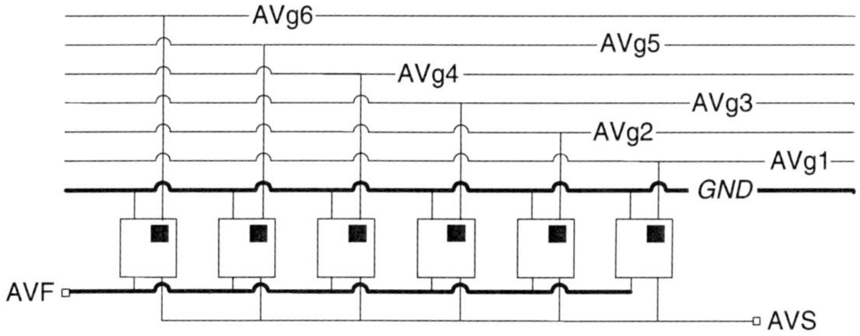

Fig. 3.24 A resistor md-unit with six DUT circuits in parallel. The AVg pads are shared with other DUT circuits in the macro. The DUT circuit is identical to the circuit and symbol shown in Fig. 3.18b, c

DUTs are shared by all md-units and all seven md-units can be tested in parallel. A more favorable design has six md-units. With 10 DUTs in each md-unit, the total number of DUTs in the macro is 60. The macro design is compact and either three-terminal or four-terminal measurements can be made. Each md-unit is functional with any of the DUTs shorted or open.

In another simplified version of a 1D array macro, DUTs within md-units are connected in series. This design is preferable if none of the resistors are expected to fail in an open-circuit mode. A series-connected md-unit with five resistor elements, configured for four-terminal measurements on all the DUTs, is shown in Fig. 3.25. Two series-connected n-FETs act as a switch to select the node at which voltage is measured at AVS. The switch is turned on with a voltage applied to the gates of the n-FETs. All md-units in a macro share the pads for the switch inputs (AVg1–AVg6) and can be tested in parallel. The DUTs and the n-FETs in the switches should be carefully sized to minimize the leakage current contribution of the n-FETs to the DUT resistance measurement.

In Fig. 3.26, the physical layout of the center section of an md-unit macro with 25 pads is shown. Each md-unit has independent AVS and AVF pads but shares a common GND pad with an adjacent md-unit. The current or voltage force terminals of the md-unit have short, low resistance wires connecting to the I/O pads. The control wires, AVg1–AVg6, travel across the width of the macro and underpasses from the md-unit are in PS or DF layers for implementation at the M1 metal level. As these control wires connect the n-FET gates to the I/O pad, series resistances of a few kΩ are acceptable.

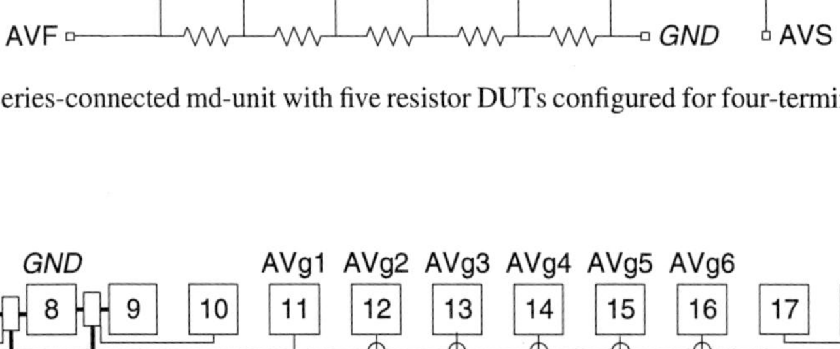

Fig. 3.25 Series-connected md-unit with five resistor DUTs configured for four-terminal measurements

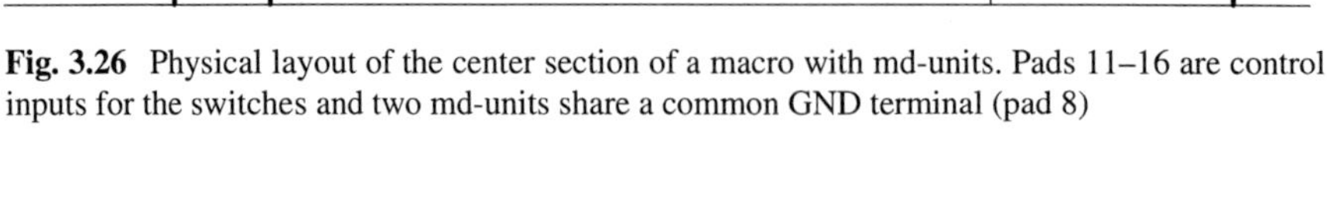

Fig. 3.26 Physical layout of the center section of a macro with md-units. Pads 11–16 are control inputs for the switches and two md-units share a common GND terminal (pad 8)

Table 3.6 Total number of DUTs for four-terminal measurements in a 1D array macro with 25 I/O pads

DUT select	DUT connection	DUTs/unit	No. of parallel tests	Total DUTs
Decoder	Parallel	30	7	210
Switch	Parallel	8	6	48
Switch	Series	9	6	54

Series-connected md-units can also serve as efficient yield monitors for electrical opens as, for example, with a chain of vias. By first measuring the total resistance of the series-connected DUTs, an open in any one of the DUTs can be detected in a single measurement. If the DUTs are nominally identical, any significant deviation from the expected mean resistance value indicates the presence of one or more deviates. Individual DUT measurements may then be made to locate these deviates.

In Table 3.6, the total number of DUTs in a 1×25 padset macro for all three array designs is summarized. In each design, four-terminal measurements for all the DUTs (with a parasitic resistance correction for the common GND terminal in parallel-connected DUTs) are possible. With 210 DUTs per macro, the 1D array design having decoder selection scheme has a clear advantage. The md-unit designs, on the other hand, have less design complexity.

The efficiency of md-unit macro designs is improved if two 1×25 padset macros are placed adjacent to each other and the control I/O pads are shared. The number of DUTs in a 1×50 padset macro, configured for four-terminal measurements, is listed in Table 3.7. For md-unit designs, there is a fourfold increase in the number of DUTs, and 12 such md-units can be tested in parallel. The number of DUTs in a macro can be increased if a three-terminal measurement configuration is used for some or all the DUTs instead. Note that a series-connected md-unit is functional only if there are no open DUTs.

3.4.4 Example 4: 2D Array Macros Implemented at M1

The number of DUTs per macro is significantly increased with the 2D array design concept introduced in Section 2.5.4. The DUTs are arranged in an $N_c \times N_r$, matrix where N_c and N_r are the number of DUTs in a column and a row, respectively.

Table 3.7 Total number of DUTs for four-terminal measurements in a 1D array macro with 50 I/O pads

DUT select	DUT connection	DUTs/unit	No. of parallel tests	Total DUTs
Decoder	Parallel	30	16	480
Switch	Parallel	18	12	216
Switch	Series	19	12	228

Column and row decoders set the address of the DUT to be measured and connect the voltage or current force and sense wires to the selected DUT.

One feature of the 2D array design described in Section 2.5.4 is that any single DUT may exhibit an electrical open or short without affecting the measurement of other DUTs in the matrix. The location of defective DUTs with an open, a short, or a resistance value outside a set limit is known from the decoded address bits of the DUT. By first establishing the physical locations of a defective DUT in a large array from electrical measurements, failure analysis techniques, such as optical and electron microscopy, can be used more efficiently to determine the root cause of failure.

Large 2D array designs do have some drawbacks. The designs are complex and often require the aid of more sophisticated design and checking tools. They do not, in general, fit in a standard 1×25 padset macro. Because of the number of wires across the macros and crossovers, four or more metal layers are needed for wiring the DUTs. In this case, the macros can only be tested at a test stop later than M1 in the manufacturing line.

Here, we begin with the discussion of 2D arrays with a simplified macro design implemented in a 1×25 padset at the M1 metal level. The 2D array comprises rows of series-connected DUTs and switches to connect the DUTs in a row to the voltage force and sense wires. A decoder is used to select a row and the same decoder output bit makes or breaks the connection of the terminals of all the DUTs in that row to the voltage sense I/O pads. A four-terminal resistance measurement can be made on all the DUTs in any row in parallel. This design is specially suited for precision resistors, H0 vias, and short metal lines as well as for some GR validation and design for manufacturability (DFM) applications.

In Fig. 3.27, the wiring scheme for a 2D array is illustrated with a 3×2 matrix of DUTs, a 1-bit decoder, and switches to select and measure all the DUTs in one row in parallel. A switch configuration with two n-FETs in series and its symbol are shown in Fig. 3.27a, b. An output bit of the decoder operates all the switches in a row. When this RAD input is high or "1," it connects one end of the selected row to the voltage force terminal AVF1. The switches connecting the DUT terminals to the voltage sense I/O pads (AVS1, AVS2, AVS3, and AVS4) are also turned on simultaneously. The voltage on all the sense pads can be read out in parallel for test time reduction. Low resistance wires must be used for connections between the DUTs, keeping the wire resistance much smaller than the DUT resistance. The vertical voltage sense wires can be resistive, with underpasses in PS or DF layers in this M1 metal implementation. The array remains functional if one or more DUTs are shorted and the location of the failing DUTs is known from voltage sense measurements. A row is disabled if a DUT has a very high resistance or is open. In this case, localization of the defective open DUT can be obtained from voltage sense measurements on the GND side.

The number of DUTs in a 2D array can be increased further with the use of a larger decoder. In a 1×25 padset macro, with a 5-bit decoder and 32 rows of 15 DUTs each, a total of 480 DUTs can be accommodated. This doubles the number of DUTs compared to the 1D arrays described in Example 3. The macro design can be

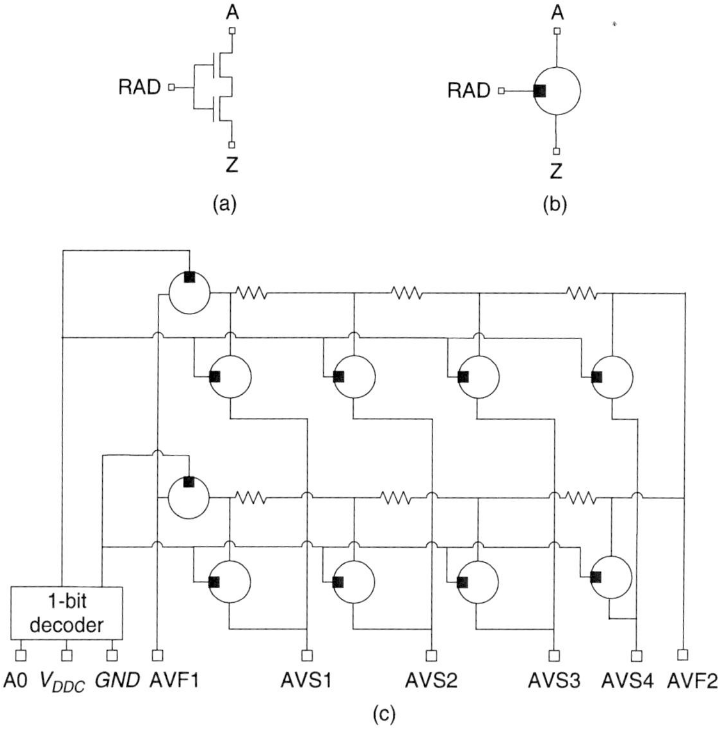

Fig. 3.27 **a** Circuit schematic of a switch and **b** switch symbol. **c** A 2D array (3×2) with a 1-bit decoder to select any one of the two rows

implemented at the M1 metal level by distributing the rows across the macro. The wiring within the macro using this scheme is illustrated in Fig. 3.28 with 3 of the 15 DUTs in each of the two rows situated between two I/O pads. The 32 decoded address bit wires and the 16 voltage sense M1 wires travel parallel to each other above and below the I/O pads, respectively. Wire connections from the decoder outputs to the DUTs and voltage sense wire connections to the I/O pads are vertical M1 wires with underpasses in PS or DF. When the DUT areas are large, the rows are placed above the pads and wired as shown in Fig. 3.27c. This increases the macro area beyond that of a minimum size macro and additional metal layers are required for wiring.

The I/O pad assignments of a compact 2D array macro with a 5-bit decoder and 32 rows comprising 15 DUTs each are shown in Fig. 3.29. The first seven pads are assigned to the decoder input bits A0–A4 and the decoder power supply. The decoder is situated between its power supply pads, V_{DDC} and GND, in a similar fashion to Example 3. The voltage force pads AVF1 and AVF2 (pads 15 and 16) are placed in the center of the group of voltage sense pads AVS1–AVS16. Alternatively, with a 1×50 padset, 32 pads can be used to drive 32 address lines directly and the decoder eliminated.

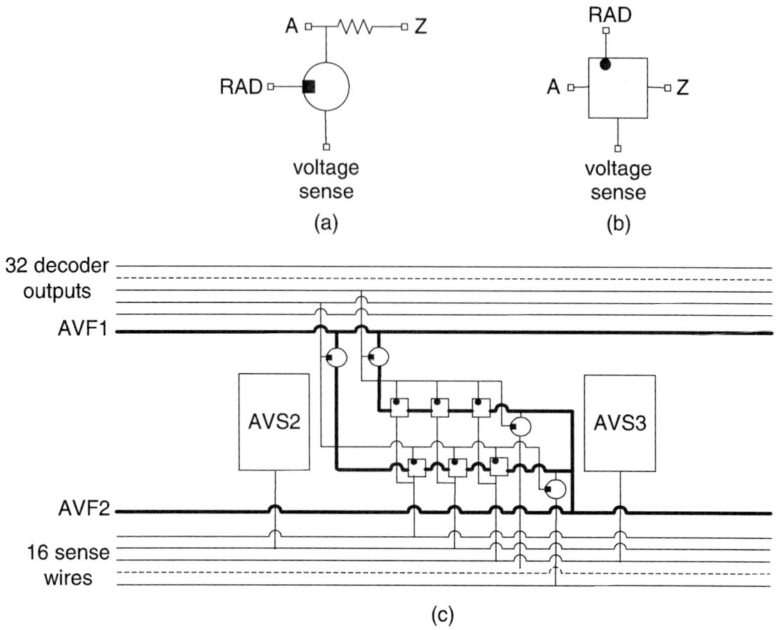

Fig. 3.28 **a** Circuit schematic of a DUT element with a switch and **b** corresponding symbol. **c** Macro wiring scheme illustrated with first three DUTs in 2 of 32 rows placed between AVS2 and AVS3 I/O pads

1	2	3	4	5	6	7	8	9	10	11	12	13	14	15	16	17	18	19	20	21	22	23	24	25
A0	A1	A2	A3	A4	VDDC	GND	AVS01	AVS02	AVS03	AVS04	AVS05	AVS06	AVS07	AVS08	AVF1	AVF2	AVS09	AVS10	AVS11	AVS12	AVS13	AVS14	AVS15	AVS16

Fig. 3.29 I/O pad assignment for a 2D array macro at M1. A0–A4 are inputs to a 5-bit decoder to select one of 32 rows, each with 15 DUTs. There are two voltage force and 16 voltage sense I/O pads

The decoder V_{DDC}, characteristic of the n-FETs in the switch, and parasitic wire resistances are carefully sized to meet the DUT resistance range and accuracy requirements. The voltage or current force wires travel a distance of ~ eight pad pitches (800 μm) and the maximum *IR* drop in a 4 μm wide M1 wire for a current of 1 mA would be 50 mV ($n_{sq} = 200$, $\rho_{sh} = 0.2\ \Omega/\square$, pattern density = 80%, and $R = 50\ \Omega$). The M1 wires for the decoder output bits and voltage sense may be narrower (0.5 μm pitch with $R < 1\ k\Omega$). If the DUT resistance including local wiring is 25 Ω, the total resistance of a row of 15 DUTs is 375 Ω. The decoder V_{DDC} is set to 1.5 V. With a voltage of 1.5 V applied at AVF1, a drop of ~1.0 V occurs across the n-FETs, and a current of 1 mA passes through the DUTs. It results in a voltage drop of ~25 mV across each DUT which can be measured with a reasonable degree of accuracy. The widths of the n-FETs are selected to support 1 mA current at a V_{gs} of 0.5 V. The leakage current of the switches in the unselected rows is reduced by a factor of ~100 by offsetting GND potential by ~ –0.2 V.

The design details mentioned above are valid for DUT resistances >25 Ω as well. For a DUT resistance of 1 kΩ each, the current flow is reduced to ~35 μA and the voltage drop across a DUT is still ~35 mV.

3.4.5 Example 5: Large 2D Array Macros

Complex 2D arrays are extensively used in CMOS characterization and for monitoring product yield. The I/O pad utilization for such arrays is more efficient and large critical area coverage can be achieved and scaled as the technology matures. With wide decoders and a large number of wires and switches, the I/O pads are generally in a square or a rectangular arrangement and may require a dedicated custom probe card. Measurements are carried out on a digital tester for rapid data acquisition. Localization of the failing DUTs is available from the address bits. Data analysis tools are similar to those used for memory arrays, providing a bit image of the arrays, highlighting the failing DUTs. The design challenges of a complex 2D array have led to a number of different approaches and trade-offs in design and test. Here, we describe one design in which each DUT can be controlled independently.

A 2D array schematic with 3-bit column and row decoders and 64 DUTs is shown in Fig. 3.30a. The DUT selection circuit is shown in Fig. 3.30b and it symbol in Fig. 3.30c. The transmission gate switch is activated by a NAND2 with CAD (column address decode bit) and RAD (row address decode bit) inputs. Each DUT, as shown in Fig. 3.30d, has four such switches to connect its four terminals to the voltage force and measure I/O pads (Section 2.5.4).

The wiring within one sub-section of the array (rows 3, 4 and columns 7, 8) is shown in Fig. 3.31. The V_{DDC} and GND distribution is to supply power to the NAND2 and the inverter in the DUT selection circuit in Fig. 3.30b. The voltage levels of the V_{DDC} power supply are not critical. The resistance of the voltage force wires AVF7 and AVF8 is kept to a minimum. The voltage sense wires, ZVM3 and ZVM4, and the RAD and CAD wires may have more resistance as the current flowing in these wires is negligibly small.

The logic gates in each DUT add to the total area and leakage currents in a 2D array. There are several different schemes to minimize the overhead of DUT selection circuit. Three such schemes are shown in Fig. 3.32. In the three-terminal measurement configuration in Fig. 3.32a, one terminal of all the DUTs is connected to the GND wire. This reduces the number of MOSFETs in the DUT by a factor of 2. Correction for the GND wire resistance can be made by shorting the corner DUT locations in the array and using the method described in Section 3.4.3. Single-ended passgates as shown in Fig. 3.32b can be used in place of transmission gates.

The 2D array circuit shown in Fig. 3.30a can be expanded to accommodate several thousand DUTs. The macro architecture is further modified if more DUTs per

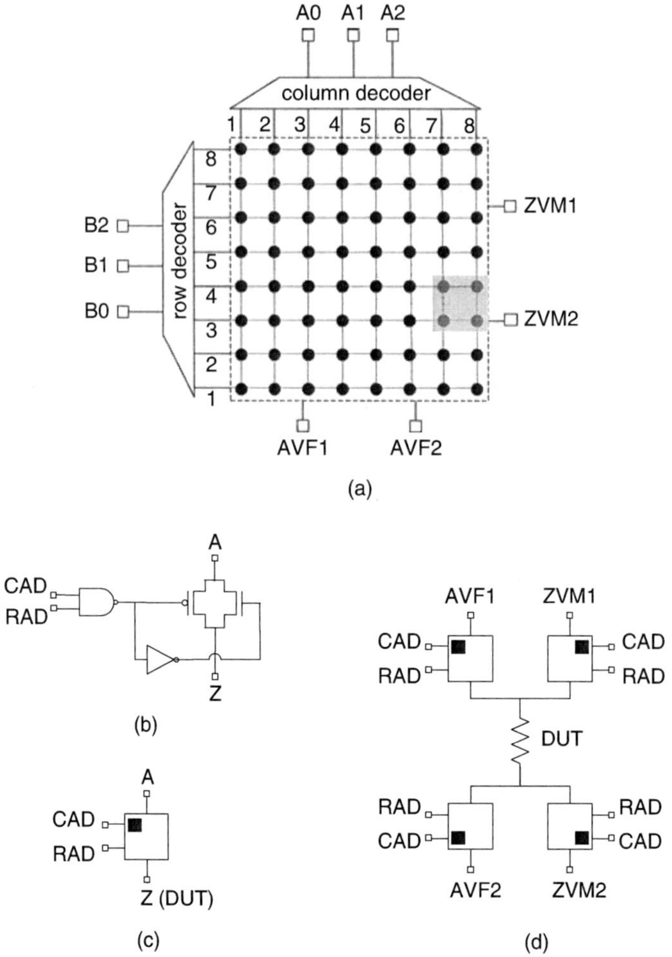

Fig. 3.30 **a** A 2D array with 3-bit column and row decoders to address any one of 64 DUTs. DUT selection circuit with CAD and RAD decode bits: **b** circuit schematic and **c** symbol. **d** Resistor DUT with four switches

macro are desired, as in the case of yield monitors. The design concept shown in Fig. 3.33 is similar to that of a large memory array. A number of 2D sub-arrays are placed in a matrix within the macro. The power supply and signal wires are placed in the space between the sub-arrays. The additional logic required for steering signals is also included. Macros of this complexity are typically designed with sophisticated tools for logic-to-schematic checking and auto wire routing. A test hierarchy to first isolate defective sections, followed by localizing individual defects within sections, is built in the design to reduce total test time.

A number of different 2D array design schemes have been implemented for measuring defect densities, yield [12–15], and mismatch characteristics of resistors [16].

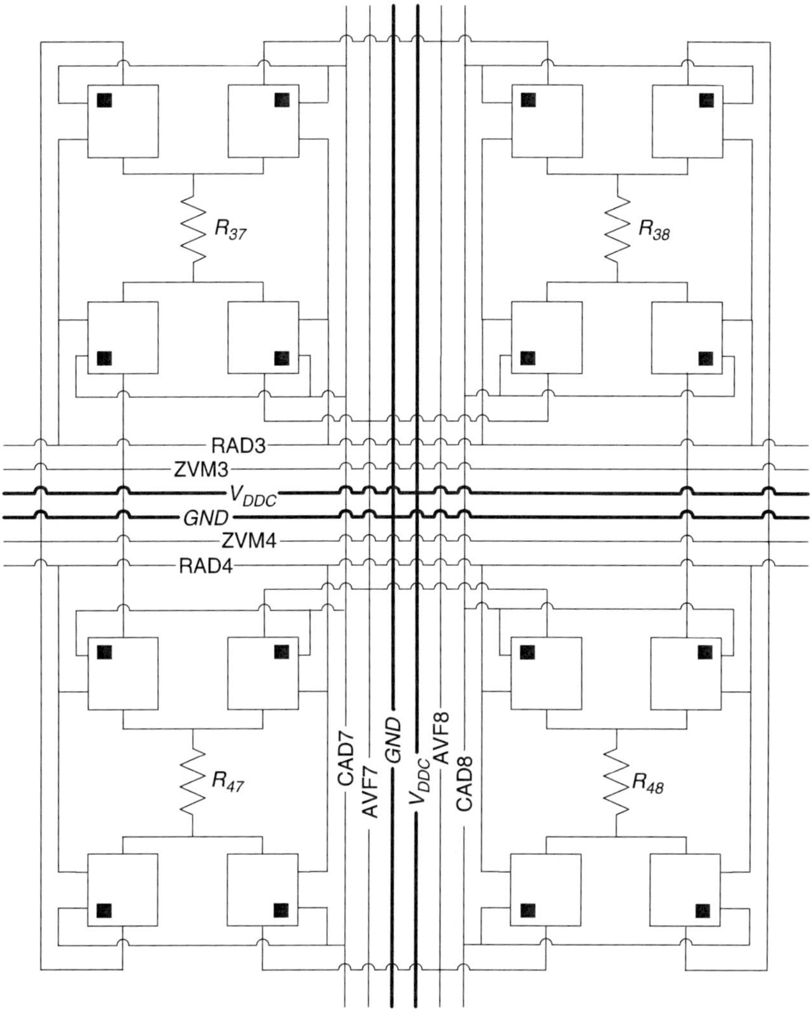

Fig. 3.31 The wiring for a 2D array sub-section highlighted in Fig. 3.30a. The row and column addresses of the DUTs are (3,7), (3,8), (4,7), and (4,8)

Array test structures, incorporating both passive and active devices [17, 18], can be found in references cited therein.

3.5 Test Structures for Metrology Applications

The Greek cross and the bridge structure described in Section 3.2.5 are used extensively for linewidth or critical dimension (CD) measurements for metrology applications [19–21]. The CD measurement accuracy requirements (~0.1%) impose

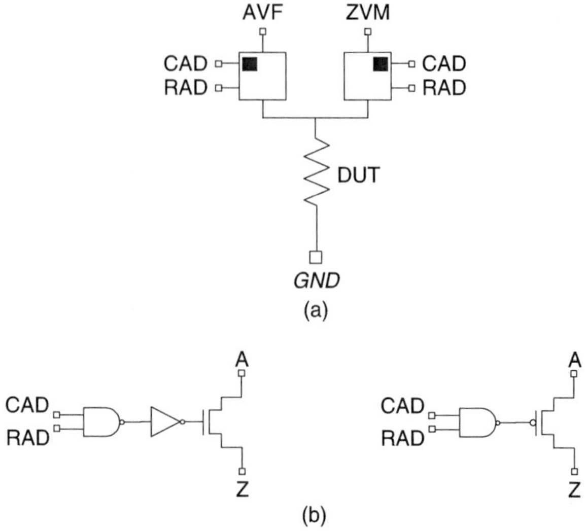

Fig. 3.32 **a** Simplified DUT selection circuits with common GND. **b** Single-ended passgates for connecting GND (n-FET) and V_{DDC} (p-FET)

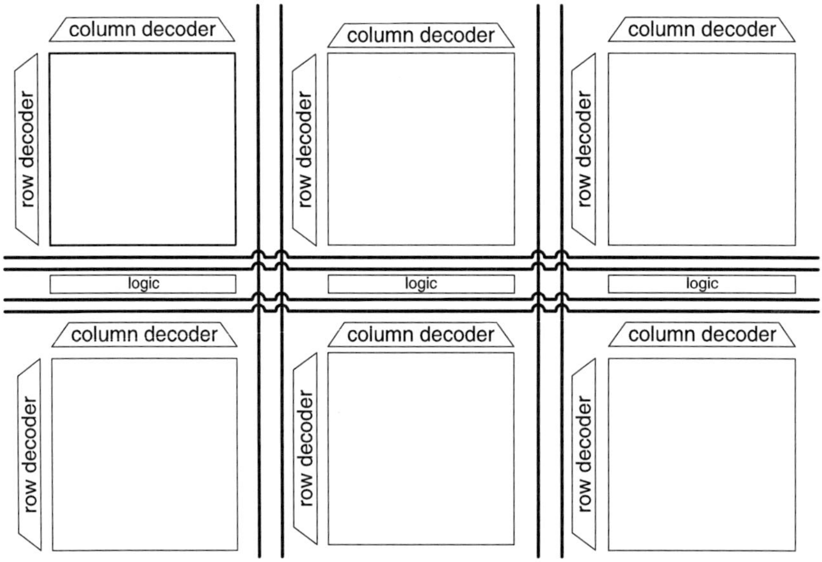

Fig. 3.33 Multiple 2D array sections in a macro

several macro design constraints. The DUTs are isolated with dedicated I/O pads for forcing current and measuring voltage. The area efficiency of these test structures is low and the test time is longer than that for standard resistors and yield monitors. However, the designs are much less complex and typically only one or two conducting layers are used.

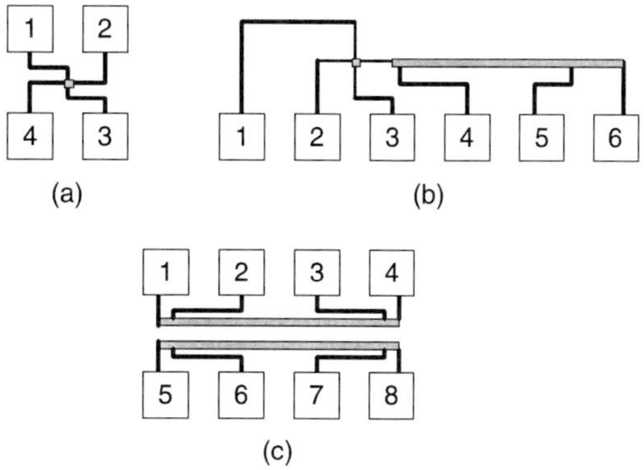

Fig. 3.34 Physical layout of **a** a Greek cross, **b** combined Greek cross and bridge, and **c** a matched resistor pair test structure

The physical layouts of an isolated Greek cross, a combined Greek cross and a bridge, and a bridge test structure for characterizing matched pair resistor DUTs are shown in Fig. 3.34. The absolute linewidth of the bridge is obtained from the combined structure in Fig. 3.34b. The sheet resistance ρ_{sh} of the layer is measured using the Greek cross connected to I/O pads 1, 2, 3, and 4. From this ρ_{sh}, and the measured resistance of the bridge of known length l, its width w is estimated from Eq. (3.1). The electrical linewidth can be calibrated against the linewidth measurements made by optical or atomic force microscopy [20].

Sheet resistance measurements using the Greek cross are carried out with different combinations of current and voltage contact pads as described in Section 3.2.5. In metrology applications, a number of steps are taken to eliminate sources of errors. In a standard resistance measurement, a constant current is forced through the resistor and the voltage across the resistor is measured. In metrology applications, by making a second measurement after reversing the direction of current flow and taking the average of the two readings, any voltage offsets arising from thermoelectric effects are eliminated. Measurements are repeated several times and averaged to reduce errors introduced by measurement noise. Measurements are also made on a pair of matched DUTs shown in Fig. 3.34c. By measuring a number of such pairs, the effect of line edge roughness and other spatial variations arising from processes such as lithography and etching can be established. These types of test structures may also be placed in orthogonal directions on silicon to evaluate line orientation effects.

With six I/O pads for the combined structure shown in Fig. 3.34b, a standard 1×25 macro template can only accommodate four such structures. The Van der Pauw method and measurement accuracy requirements in metrology applications pose a challenge in creating large arrays of these test structures, similar to those for resistor DUTs described in Examples 2, 3, and 4. Generally, an array matrix of isolated discrete test structures is created. Such arrays take up a large area and are more suitable for dedicated test chips. Test time reduction can be achieved by parallel testing of isolated DUTs as long as the ATE can provide the desired measurement accuracy.

References

1. Schroder DK (2006) Semiconductor material and device characterization, 3rd edn. Wiley, Hoboken, NJ
2. Lindley D (2004) Degrees Kelvin. Joseph Henry, Washington, DC
3. Pauw van der LJ (1958) A method of measuring specific resistivity and hall effect of discs of arbitrary shape. Philips Res Rep 13:1–9
4. Versnel W (1979) Analysis of the Greek cross, a Van der Pauw structure with finite contacts. Solid State Electron 22:911–914
5. Application Note 4156-11 Precision measurement of metal line width in sub-quarter micron interconnect systems. Agilent Technologies. Available via http://www.home.agilent.com/agilent/facet.jspx%t=80030.k.1%cc=US%lc=eng%sm=g%pageMode=TM. Accessed 3 Feb 2011
6. Stapper CH, Rosner RJ (1995) Integrated circuit yield management and yield analysis: development and implementation. IEEE Trans Semicond Manuf 8:95–102
7. Kuo W, Kim T (1999) An overview of manufacturing yield and reliability modeling for semiconductor products. Proc IEEE 87:1329–1344
8. Verzi B (2009) Considerations for parallel and array test patterns. In: International conference on microelectronic test structures, Oxnard, 30 Mar 2009
9. Walton AJ, Gammie W, Fallon M, Stevenson JTM, Holwill RJ (1991) An interconnect scheme for reducing the number of contact pads on process control chips. IEEE Trans Semiconduct Manuf 4:233–240
10. Ward D, Walton AJ, Gammie WG, Holwill RJ (1992) The use of a digital multiplexer to reduce process control chip pad count. Proceedings of the 1992 IEEE International conference on microelectronic test structures, 1992, pp. 129–133
11. Ketchen MB, Bhushan M, Costrini G (2009) Addressable arrays implemented with one metal level for MOSFET and resistor variability characterization. Proceedings of the 2009 IEEE International conference on microelectronic test structures, 2009, pp. 13–18
12. Hess C, Inani A, Lin Y, Squicciarini M, Lindley R, Akiya N (2006) Scribe characterization vehicle test chip for ultra fast product wafer yield monitoring. Proceedings of the 2006 IEEE international conference on microelectronic test structures, 2006, pp. 110–115
13. Hess C, Squicciarini M, Yu S, Burrows J, Cheng J, Lindley R et al (2008) High density test structure array for accurate detection and localization of soft fails. Proceedings of the 2008 IEEE International conference on microelectronic test structures, 2008, pp. 131–136
14. Karthikeyan M, Fox S, Cote W, Yeric G, Hall M, Garcia J et al (2006) A 65 nm random and systematic yield ramp infrastructure utilizing a specialized addressable array with integrated analysis software. Proceedings of the 2006 IEEE international conference on microelectronic test structures, 2006, pp. 104–109
15. Cabrini A, Cantarelli D, Cappelletti P, Casiraghi R, Maurelli A, Pasotti M et al (2006) A test structure for contact and via failure analysis in deep-submicrometer CMOS technologies. IEEE Trans Semicond Manuf 19:57–66
16. Tian W, Steinmann P, Beach E, Khan I, Madhani P (2008) Mismatch characterization of a high precision resistor array test structure. Proceedings of the 2008 IEEE international conference on microelectronic test structures, 2008, pp. 11–16
17. Doong KY-Y, Hsieh S, Lin S-C, Shen B, Cheng J-Y, Kwai D-M et al (2001) Addressable failure site test structures (AFS-TS) for CMOS processes: design guidelines, fault simulation, and implementation. IEEE Trans Semicond Manuf 14:338–355
18. Doong KYY, Bordelon J, Chang K-J, Hung LJ, Liao CC, Lin SC et al (2006) Field-configurable test structure array (FC-TSA): enabling design for monitor, model and manufacturability. Proceedings of the 2006 IEEE international conference on microelectronic test structures, 2006, pp. 98–103

19. Enderling S, Brown CLIII, Smith S, Dicks MH, Stevenson JTM, Mitkova M et al (2006) Sheet resistance measurement of non-standard cleanroom materials using suspended Greek cross test structures. IEEE Trans Semicond Manuf 19:2–9
20. Shulver BJR, Bunting AS, Gundlach AM, Haworth LI, Ross AWS, Smith S et al (2008) Extraction of sheet resistance and line width from all-copper ECD test structures fabricated from silicon preforms. IEEE Trans Semicond Manuf 21:495–503
21. Smith S, Tsiamis A, McCallum M, Hourd AC, Stevenson JTM, Walton AJ et al (2009) Comparison of measurement techniques for linewidth metrology on advanced photomasks. IEEE Trans Semicond Manuf 22:72–79

Chapter 4
Capacitors

Contents

Signal propagation delays and signal rise and fall times in CMOS circuits are related to the *RC* time constants of the circuit elements in the path. Determination of capacitance *C* of MOSFETs and parasitic elements is therefore an essential part of building models for circuit simulations. MOSFET capacitance components must be characterized as a function of voltage to capture their behavior during a transient switching cycle. Capacitance measurements are utilized to determine MOSFET gate linewidths and gate oxide thickness. Capacitances of conducting layers with dielectric isolation give a measure of inter-level dielectric properties such as effective dielectric constant and film geometries such as layer thicknesses and linewidths. *C–V* characteristics of diodes are used for profiling carrier densities in silicon devices. Carrier lifetime measurement in silicon utilizes *C–V* and *C–time* characteristics of MOS capacitors.

M. Bhushan, M.B. Ketchen, *Microelectronic Test Structures for CMOS Technology*,
DOI 10.1007/978-1-4419-9377-9_4,

In this chapter, test structures for measuring capacitances of passive elements and for *C–V* characterization of MOSFETs and diodes are described. Properties of capacitors are covered in Section 4.1. In Section 4.2, capacitance measurement techniques are discussed. Capacitor DUT designs are described in Section 4.3, followed by five examples of capacitance macro designs in Section 4.4. A description of inductance and capacitance scaling is given in Section 4.5. Capacitance measurements of logic gates and interconnects using ring oscillators is covered in Chapter 6. High-frequency measurement of coupling capacitance between interconnect wires is described in Example 2 of Chapter 7.

The physics of MOSFET and parasitic capacitances in CMOS circuits is covered in textbooks on semiconductor devices [1, 2]. Capacitance measurement techniques for semiconductor devices are described in [3].

4.1 Properties of Capacitors

The first demonstration of stored charge in a capacitor along with the invention of the Leyden jar took place in 1746. The relationship between capacitance and voltage is attributed to Michael Faraday and the unit of capacitance (Farad, F) is named in his honor. The influence of dielectric properties of insulating materials on signal transmission became a major concern during the installation of trans-Atlantic telegraph cables in the later part of the 19th century. Working with the Atlantic cable company, Lord Kelvin made the important observation that the capacitance and resistance of cables influenced the shape of an electric pulse traveling down a long cable.

In the last 150 years, capacitance measurements have played an important role in characterizing and optimizing the properties of new dielectric materials. Development of planar silicon technology has been greatly facilitated with silicon oxide (SiO_2) as both a high-quality gate dielectric in MOSFETs and a mask against diffusion of donor (n-type) and acceptor (p-type) impurities in silicon. With a low dielectric constant of $\sim$4.2, SiO_2 films also serve as insulting layers in the wiring of CMOS circuits. The push toward scaling wire capacitances further down has led to the development of new dielectric materials to replace SiO_2. This class of materials, known as low-*k* dielectrics, includes fluorine- or carbon-doped SiO_2, porous SiO_2, and spin-on polymeric materials.

The MOS capacitor forms the basis of CMOS circuit operation. A voltage applied to the gate electrode of this capacitor turns on the conduction through the channel of a MOSFET, thereby changing its state from "off" to "on." The switching speed of CMOS circuits is directly related to the characteristics of the MOS capacitor and its associated parasitic capacitance components. Electric charge storage in MOS capacitors is also used to smooth out spikes and dips in power supply voltage as discussed in Section 2.4.6.

4.1.1 Thin-Film Capacitors

A parallel plate capacitor comprising two metal plates, each of length l and width w, is shown in Fig. 4.1. The space between the plates is filled with an insulating material of relative dielectric constant ε. If the plate dimensions are much larger than the separation h between the plates, the capacitance C is given by

$$C = \frac{\varepsilon\varepsilon_0 lw}{h} = \frac{\varepsilon\varepsilon_0 A}{h}, \tag{4.1}$$

where A is the area of the plate and $\varepsilon_0 \left(= 8.854 \times 10^{-12}\ \text{F/m}\right)$ is the vacuum permittivity. From the measured capacitance of parallel plate capacitors of known metal plate geometry, the value of ε or h can be derived from Eq. (4.1).

In a static state, no current flows through an ideal capacitor held at a constant voltage. A transient current flows through such a capacitor only during charging and discharging. While resistance measurements can be carried out with a DC voltage or current source, measurement of capacitance, therefore, requires a time-dependent AC source or measure unit.

The total charge Q stored in a capacitor to raise its voltage V is obtained by integrating the charging current I over time:

$$Q = \int_0^t I\,\mathrm{d}t = CV. \tag{4.2}$$

The energy E stored in a capacitor is

$$E = \frac{1}{2}CV^2. \tag{4.3}$$

Capacitance measurements on a DUT may be carried out by measuring either Q or E, with an AC source operating at a fixed frequency connected across it. These measurement techniques are covered in Section 4.2.

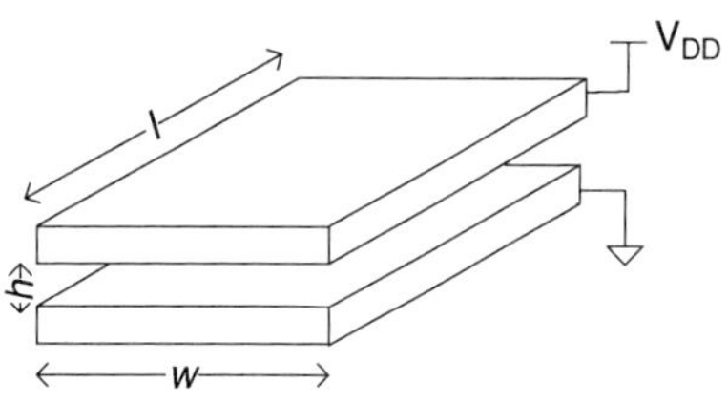

Fig. 4.1 A parallel plate capacitor of plate length l, width w, and plate separation h

The magnitude of the capacitive reactance or the impedance of a pure capacitor $|Z| = Z$ is inversely proportional to the frequency f of the AC source:

$$Z = \frac{dV}{dI} = \frac{1}{\omega C}, \tag{4.4}$$

where $\omega = 2\pi f$ is the angular frequency.

Capacitors in CMOS circuits have parasitic resistances in series and in parallel with the electrodes. The circuit symbol of a pure capacitor and an equivalent circuit schematic of a capacitor with parasitic resistances are shown in Fig. 4.2. The resistance R_s in series with the capacitor is the resistance of the metal electrodes, silicon diffusion areas and wires connecting the DUT to the test equipment. The parallel resistance R_p represents a leakage current path through the dielectric material.

Commercial impedance meters (LCR meters) use an AC source to measure impedance. The effect of parasitic resistances is minimized by selecting an appropriate value of ω such that $R_s \ll 1/\omega C$, while at the same time, $R_p \gg 1/\omega C$. If R_s is the dominant parasitic component, a low frequency of operation is used. On the other hand, if R_p is small because of leakage through the dielectric, a high frequency of operation is necessary. Test equipment and capacitance characteristics are important considerations in designing test structures and for selecting the measurement frequency for different types of capacitor DUTs as described in Section 9.2.3.

4.1.2 Interconnect Wire Capacitors

Interconnect wiring in CMOS circuits is carried out with metal layers M1 and above. In CMOS products, typically, long wires on alternate layers are placed in orthogonal directions for ease of wiring. Cross sections through M1, M2, and M3 layers in two perpendicular planes are shown in Fig. 4.3. Capacitances between adjacent signal wires on the M2 metal level, C_{left} and C_{right}, and to wires above and below, C_{up} and C_{down}, are indicated in Fig. 4.3a. These capacitances are dependent on the layer dimensions and properties of surrounding dielectrics. Capacitance per unit length

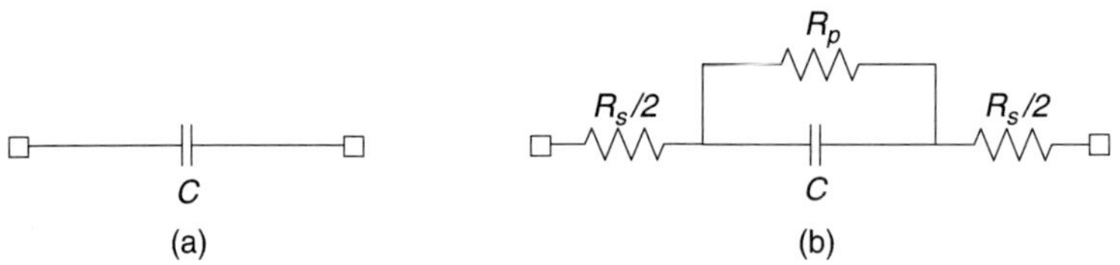

Fig. 4.2 Circuit schematic **a** of ideal capacitor, and **b** of capacitor with parasitic resistances in series and parallel

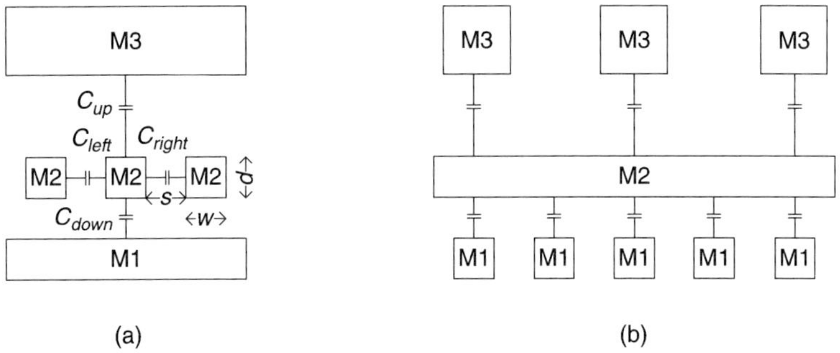

Fig. 4.3 Schematic cross sections of interconnect metal layers M1, M2, and M3: **a** with M2 wire into the plane and **b** with M2 wire parallel to the plane

of minimum width wires is nearly the same for all metal levels, as discussed in Sections 2.4.3 and 4.5.

For dielectric isolation of metal interconnect layers in CMOS technology, the dielectric films are composite of several materials, selected for their mechanical, etching, and chemical mechanical polishing (CMP) characteristics. These materials are layered to minimize capacitance, while providing good process control and reliability. An effective dielectric constant $\varepsilon_{\mathrm{eff}}$ is defined to represent the composite dielectric film. The $\varepsilon_{\mathrm{eff}}$ values of low-k dielectrics range from 3.5 for doped oxide to ~2.7 for organic polymers. Air gaps in dielectrics have been introduced to further reduce $\varepsilon_{\mathrm{eff}}$ to ~2.0.

With scaling of MOSFET dimensions, interconnect wire dimensions are also scaled to achieve high wiring densities. The simplified parallel plate approximation shown in Fig. 4.1 no longer holds for metal wire capacitances in present day CMOS technologies. For narrow wires ($w \approx h$), the fringing electric field lines at the edges add to the line capacitance as indicated in Fig. 4.4. Capacitance modeling of such complex geometries is carried out with 2D or 3D electromagnetic field solvers [4]. Computer programs for capacitance modeling based on electromagnetic finite element analysis are commercially available.

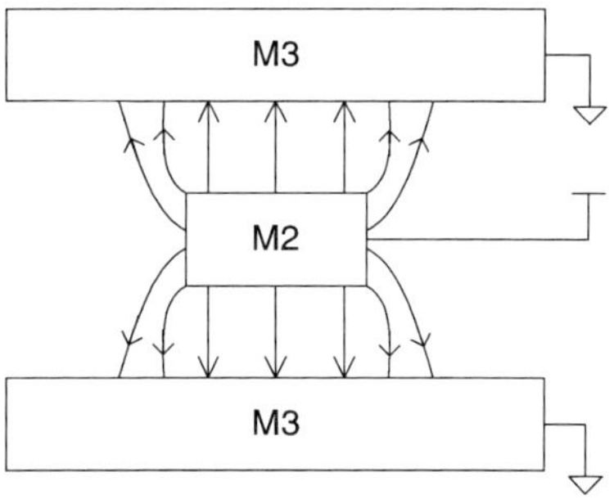

Fig. 4.4 Electric field lines in a narrow wire

As a result of mutual capacitances between wires, a signal propagating on one wire induces voltages and currents in its neighbors:

$$I = C\frac{\mathrm{d}\,(V_1 - V_2)}{\mathrm{d}t}, \tag{4.5}$$

where $V_1(t)$ and $V_2(t)$ are the time-dependent voltage levels in wires 1 and 2. This coupled signal noise or cross talk is more significant in long signal wires, running in parallel, and for voltage levels changing in opposite directions, for example

$$\frac{\mathrm{d}V_1}{\mathrm{d}t} = -\frac{\mathrm{d}V_2}{\mathrm{d}t}. \tag{4.6}$$

Distributed capacitance between two wires in the same plane is shown in Fig. 4.5a along with two cases of time-dependent voltage signal levels in Fig. 4.5b, c. There is no cross talk when both $V_1(t)$ and $V_2(t)$ are rising signals in phase with each other. When $V_1(t)$ is rising and $V_2(t)$ is falling and the signals are 180° out of phase, the net effect is a doubling of the effective capacitance, an example of the so-called Miller effect.

The mutual capacitance between wires on the same level, C_{left} and C_{right}, increases as the spacing s between wires is reduced. Metal process recipe variations may result in wider linewidths on wafers or optical masks. Widening the wires and thereby reducing s has only a small effect on wire RC, as an increase in wire capacitance is at least partially compensated by a decrease in the wire resistance. However, cross talk increases with $1/s$ and may become significant in minimum pitch wires, affecting the noise immunity of the circuits. Test structures are designed to measure and monitor cross talk between wires at all levels, with varying wire geometries and pattern densities, both for tuning fabrication processes such as CMP and for building capacitance models.

4.1.3 MOS Capacitor

The MOS capacitor is formed by a thin oxide film separating the gate electrode from the silicon substrate or body as shown for an n-MOS in Fig. 4.6a. A conducting n-channel is formed by applying a positive voltage at the gate (G) terminal.

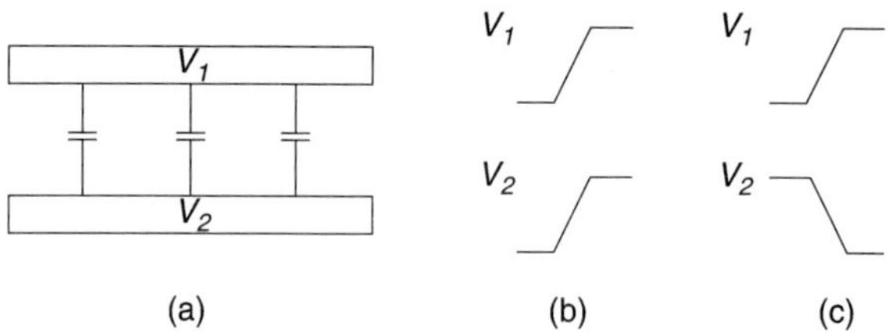

Fig. 4.5 **a** Distributed coupled capacitance between parallel signal wires. Voltage waveforms with signals **b** in phase and **c** out of phase

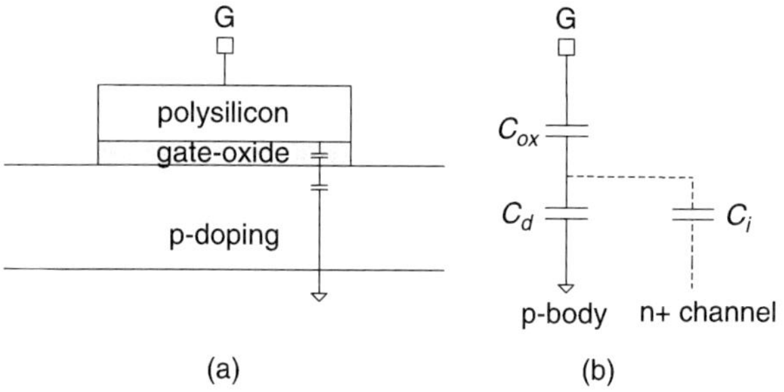

Fig. 4.6 **a** Schematic cross section of an n-MOS capacitor. **b** n-MOS gate capacitance components in depletion and inversion modes

The gate-to-body capacitance C_g can be decomposed into the gate oxide capacitance C_{ox} in series with the silicon layer capacitance in the depletion or in the inversion states, C_d and C_i, respectively, as shown in Fig. 4.6b for an n-MOS capacitor.

The gate capacitance C_g is a function of the voltage applied to the gate electrode and the measurement frequency. In Fig. 4.7, the qualitative behavior of C_g as a function of gate voltage V_g for an n-MOS capacitor is shown. At negative values of V_g, positively charged majority carriers (holes) accumulate at the oxide–silicon interface and the C_g is equal to C_{ox}. At small positive values of V_g, the channel is depleted and the capacitance is equal to that of the depletion layer C_d in series with C_{ox}. As V_g is increased further, an inversion layer begins to form with electron accumulation at the interface. The capacitance increases again, reaching a value of C_{ox} when the n-channel formation is complete.

Note that the C_g behavior shown in Fig. 4.7 for $V_g > 0$ is observed only at frequencies low enough ($\lesssim$100 Hz) for the minority carriers to respond to the changes in V_g. At high frequencies, with the minority carrier lifetime longer than the period of the AC signal, the channel remains in a depleted state even with a positive V_g.

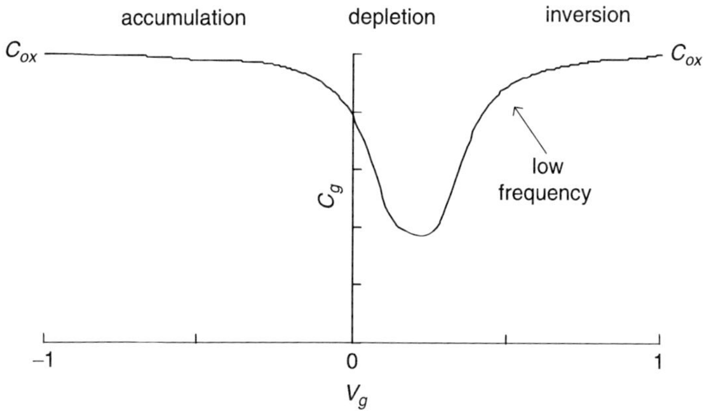

Fig. 4.7 Low-frequency quasi-static C_g vs. V_{gs} plot for an n-MOS capacitor

The capacitance measurements on MOS capacitors are typically made in a quasi-static mode. *C–V* characterization of an n-MOS capacitor can be carried out at high frequencies under intense illumination or when contact with the p-type substrate is made through a heavily doped n^+ region. This n^+ region acts as a source of electrons (minority carriers) to the inversion layer. In a MOSFET structure, heavily doped source and drain regions connect to the conducting channel when in inversion mode.

A brief description of MOSFETs is given in Sections 2.1.1 and 5.1. An in-depth treatment of MOSFET capacitances can be found in textbooks on semiconductor devices [1, 3].

The capacitance components for an n-FET are shown in Fig. 4.8a. In addition to the MOS capacitance, several parasitic capacitances are introduced. Here, C_{gb} is the gate-to-body MOS capacitance, equivalent to the C_g of a MOS capacitor described previously. Parasitic capacitances between the source and the drain to the substrate or the body (B), C_{sb} and C_{db}, are the capacitance contributions of the n^+/p junctions. These capacitances can be further separated into area and perimeter components.

Parasitic capacitances between the gate (G), and the source (S) and the drain (D) are delineated by the gate-to-source and gate-to-drain overlap regions. The overlap capacitance C_{ov} is the sum of gate oxide capacitance in the overlap region C_{do}, an outer fringe capacitance C_{of}, and an inner fringe capacitance C_{if}:

$$C_{ov}\left(V_{gs}=0\right)=C_{do}+C_{of}+C_{if}. \tag{4.7}$$

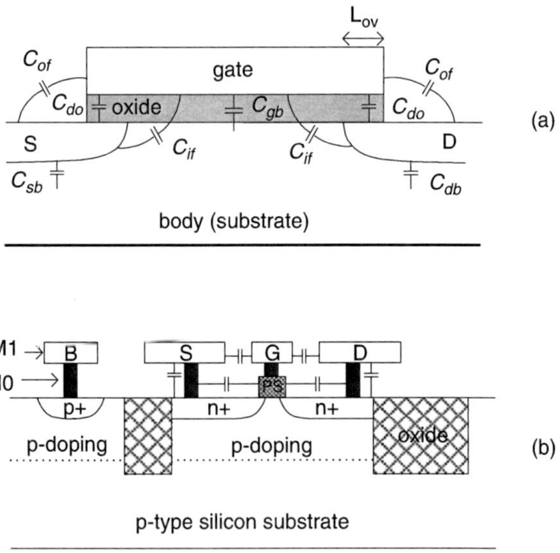

Fig. 4.8 **a** MOSFET capacitances between S, D, G, and B terminals. **b** Schematic cross section of an n-FET, indicating metal parasitic capacitances

The total gate oxide capacitance C_{ox} is the sum of channel capacitance C_{ch} and overlap capacitances in the inversion mode, where

$$C_{ox} = C_{ch} + 2\left(C_{do} + C_{of}\right). \quad (4.8)$$

The gate-to-source C_{gs} and gate-to-drain C_{gd} capacitances, which include the C_{do}, C_{of}, and C_{if}, vary as a function of applied voltage bias.

Metal connections to the gate, source, and drain electrodes contribute additional parasitic capacitances to the MOSFET structure, as depicted in the physical cross section of an n-FET in Fig. 4.8b. The total gate capacitance C_{gT} includes the contributions from the intrinsic MOSFET capacitances and the metal contact-to-gate capacitances as these cannot be easily separated during measurements. Full C–V characterization of C_{gT}, C_{gs}, C_{gd}, C_{sb}, C_{db}, and their sub-components is carried out for building capacitance models for circuit simulations and for engineering MOSFET processes to optimize circuit performance. The capacitance components between the gate and the source and the drain are subject to Miller effect during switching of logic gates (Section 6.3.1) and their contribution to the signal delay can be significant.

MOSFET gate capacitance measurements can be carried out at high frequencies as charge in the channel is exchanged nearly instantaneously with the source and drain regions. Quasi-static MOSFET gate capacitance measurements in the inversion mode are utilized for monitoring gate oxide thickness which can be derived from a large-area, gate oxide capacitor using Eq. (4.1). The dielectric thickness derived by this method includes quantum-mechanical effects influencing the inversion layer depth. The measured gate oxide thickness in the inversion mode t_{inv} is ~0.3 nm larger than physical gate oxide thickness t_{ox}. A calibration of the electrical measurement of t_{inv} is carried out with physical t_{ox} measurements using ellipsometry and transmission electron microscopy. Depletion in the polysilicon gate layer reduces the gate oxide capacitance.

With the scaling of CMOS technology, the measurement of the electrical channel length L_{eff} with dimensions <50 nm has become increasingly challenging. The traditional L_{eff} extraction methods, such as shift-and-ratio technique [1, 3], are now replaced with measurement of metallurgical (physical) gate length L_p from gate capacitance measurements. The value of L_p differs from L_{eff} because of the contributions of the gate-to-source and gate-to-drain overlap regions [5]. The length of the overlap region between the gate and the source and drain L_{ov} can be estimated from the measured values of C_{do} and t_{ox}. The electrical channel length of the MOSFET L_{eff} is then obtained from the metallurgical gate length L_p and L_{ov}:

$$L_{eff} = L_p - 2L_{ov}. \quad (4.9)$$

The values of L_p from electrical measurements are calibrated against the values obtained with more precise physical measurement using optical techniques [3]. Measurement of MOSFET capacitance components and extraction of t_{inv} and L_{eff} are described in more detail in Section 4.4.2 (Example 2).

The capacitance of a reverse-biased metal–semiconductor junction, modeled as a parallel plate capacitor, is also expressed using Eq. (4.1), where h is the depletion layer width of the junction. MOS capacitor is used for carrier density profiling in the MOSFET body and in the polysilicon gate [3]. Frequency dependence of C–V characteristics is used for determining lifetime of charge carriers in the depletion region. Effective dielectric constant $\varepsilon_{\mathrm{eff}}$ of some dielectric materials may be frequency dependent.

4.2 Capacitance Measurements

Capacitance measurements in the pF to μF range may be carried out by directly connecting a capacitor to a commercially available impedance meter. Capacitance measurements in the fF range require on-chip circuitry, increasing the design complexity of the test structures.

4.2.1 AC Impedance Measurement

Off-the-shelf impedance (LCR) meters are ideally suited for capacitance measurements of isolated capacitors. Capacitance measurements for DUT capacitances values $\gtrsim 1$ pF can be made at frequencies below ~50 kHz using DC probes. A special high-frequency test setup is required when the measurement frequency is increased to measure smaller capacitance values.

The principle of operation of an LCR meter is illustrated in Fig. 4.9a. An AC source is used to deliver a time-averaged constant current through a resistor R in series with the capacitor under test, and the voltages V_{r} and V_{z} across the resistor and the capacitor are measured. The capacitance is given by

$$\frac{1}{\omega C} = R\frac{V_{\mathrm{z}}}{V_{\mathrm{r}}} \text{ or } C = \frac{V_{\mathrm{r}}}{\omega R V_{\mathrm{z}}}. \tag{4.10}$$

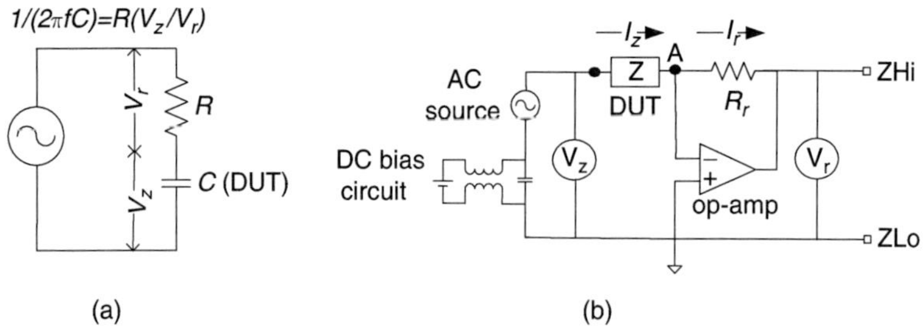

Fig. 4.9 **a** Capacitance measurement circuit using an AC source. **b** Circuit for AC impedance measurement with a DC bias applied to the DUT

The expression in Eq. (4.10) holds for an ideal capacitor. In the presence of a series resistance R_s and a shunt resistance R_p as shown in Fig. 4.2, both the magnitude and the phase of the voltage across the capacitor are measured to take the parasitic resistances into account. For large C and R_p, the series resistance R_s may become significant. In the presence of a leakage path in the dielectric material of the capacitor, because of either process-induced defects or presence of significant tunneling currents in ultra-thin dielectrics, the measurement accuracy is affected as $1/\omega C$ approaches R_p. Higher operating frequencies are selected with smaller capacitances so that the impedance $1/\omega C$ remains $\ll R_p$.

In a three-terminal measurement, the DUT terminals are isolated from the GND. In this scheme, parasitic capacitances may be shunted to GND and eliminated from the measurement. The capacitance of cables and probes connecting the DUT to the LCR meter is determined by making a measurement without any contact with the silicon wafer. This background capacitance is subtracted from the measured DUT capacitance to account for the parasitic capacitance of the test setup. The LCR application manual also contains the recommended frequency and measurement accuracy for a range of capacitance values. A detailed description of LCR meter operation is given in Section 9.2.3.

4.2.2 Charge-Based Capacitance Measurement (CBCM)

Capacitance measurements can be carried out without the use of an external LCR meter using the charge-based capacitance measurement (CBCM) technique [6]. With this technique, a high-frequency source charges and discharges a capacitor under test at a constant rate, and the time average DC current I through the capacitor is measured. The AC source may be an external oscillator or an on-chip clock signal to drive an inverter stage loaded with a capacitor DUT as shown in Fig. 4.10a. The capacitance, from Eq. (4.2), is given by

$$C = \frac{I}{V_{DD}f}, \tag{4.11}$$

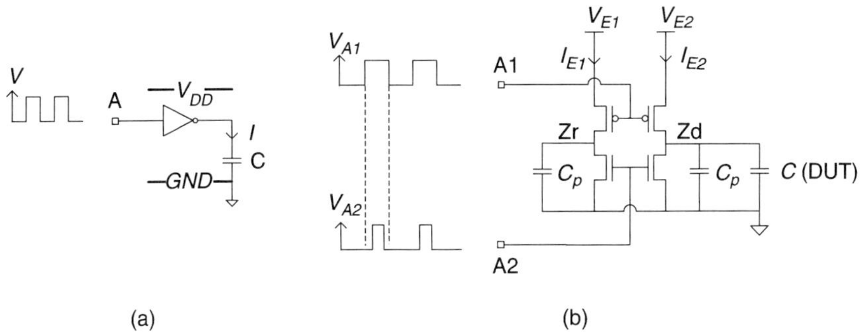

Fig. 4.10 **a** Circuit schematic illustrating the principle of CBCM technique. **b** CBCM circuit comprising a pair of pseudo-inverters with independent non-overlapping clock inputs, V_{A1} and V_{A2}, for the p-FETs and n-FETs, respectively

where f is the frequency of the applied signal. In practice, a differential measurement scheme is implemented, with an on-chip reference circuit to subtract the capacitance contributions of circuit elements other than the capacitor DUT. Because of the need for on-chip CMOS circuitry, this method can only be used for measuring MOSFET capacitances and interconnect wire capacitances in a fully integrated process.

The schematic of a circuit illustrating the CBCM technique is shown in Fig. 4.10b. A pair of matched pseudo-inverters is configured such that the p-FETs are driven by a signal applied at input node A1 and n-FETs by a second independently controlled signal at input node A2. The pseudo-inverters are powered by two independent power supplies, V_{E1} and V_{E2} ($V_{E1} = V_{E2} = V_{DD}$), and the currents I_{E1} and I_{E2} are measured. In this differential measurement scheme, the left pseudo-inverter serves as a reference for subtraction of the parasitic capacitance C_p, which is assumed to be equal for both inverters. The capacitor to be measured, C (DUT), is connected to the output of the pseudo-inverter on the right.

Signal input waveforms at nodes A1 and A2 are also shown in Fig. 4.10b. These are non-overlapping clock edges to ensure that the p-FET and the n-FET are not turned on simultaneously, thus preventing any short-circuit current through the inverters. Initially, with $V_{A1} = V_{A2} =$ "0," the p-FETs are on and the n-FETs are in the off state. This sets the nodes Zr and Zd at "1" and the capacitors C_p and C (DUT) are fully charged. With input signal V_{A1} rising, the p-FETs are switched off and the n-FETs remain in the off state. With input signal V_{A2} rising, the n-FETs are turned on, bringing nodes Zr and Zd to "0" and discharging the capacitors. The measured average DC currents I_{E1} and I_{E2} flowing through the inverters are those for charging the capacitances along with leakage currents of the MOSFETs.

The capacitance C of the DUT is

$$C = \frac{(I_{E2} - I_{E1})}{V_{DD} f} = \frac{\Delta I}{V_{DD} f}, \tag{4.12}$$

where f is the frequency of the clocks and ΔI is the difference in the measured DC current values. A more accurate determination of the capacitance may be made by measuring ΔI at different values of V_{DD}, at a fixed frequency, and extracting C from the slope of ΔI vs. V_{DD} plot. Alternatively, measurements may be made at different values of f at a fixed V_{DD} and the capacitance obtained from the slope of ΔI vs. f plot. In this method, the background leakage current of the MOSFETs is subtracted, thus improving measurement accuracy. Test time is minimized by making measurements at only two frequencies, f_1 and f_2, and the capacitance is calculated as

$$C = \frac{\left(I_{E2\_1} - I_{E1\_1}\right) - \left(I_{E2\_2} - I_{E1\_2}\right)}{V_{DD}\,(f_1 - f_2)}. \tag{4.13}$$

When input signals V_{A1} and V_{A2} are applied externally and contact with the test structure is made using a standard DC probe card, the upper frequency limit is ~1 MHz. For ΔI in the range of ~1 μA, and $V_{DD} = 1.0$ V, the measured minimum capacitance value is ~1 pF. The minimum capacitance can be lowered to 1 fF if ΔI in the range of ~1 nA is measured with an accuracy of $\lesssim 1\%$.

Smaller capacitance values (1 fF) are measured at frequencies in the range of several hundred MHz to ~1 GHz. High-speed probes are required if an external pulse generator is used. Alternatively, a DC probe card may be used if frequency signals are generated within the macro itself. By increasing the measurement frequency to ~1 GHz, gate capacitance of an individual MOSFET C_{gT} can be measured accurately even in the presence of significant gate oxide leakage current I_{gl}. As an example, for a 1.0 μm wide MOSFET, $C_{gT} = 1$ fF and from Eq. (4.12), at $f = 1$ GHz, $\Delta I = 1$ μA. The error in measurement of C_{gT} introduced by an I_{gl} of 1 nA is then ~0.1%.

There are several sources of error in the CBCM technique described above. The two pseudo-inverters, even with identical dimensions and physical layouts, are not perfectly matched because of random variations in MOSFET properties. As a result, there will be a residual parasitic capacitance and leakage current included in the expression in Eq. (4.12). Another source of error is from charge injection when a MOSFET is turned off. This additional charge varies with the load on the inverter and is not fully accounted for by the subtraction method. A number of improvements to the basic CBCM technique have been demonstrated to reduce these sources of error and to improve measurement accuracy in the fF range. A comparison of the different techniques is discussed in [7].

Schematics of circuits utilizing a charge injection error-free CBCM (CIEF-CBCM) scheme [8–10] for MOSFET and wire capacitance measurements are shown in Fig. 4.11a, b. In Fig. 4.11a, the DUT is a MOSFET gate capacitor

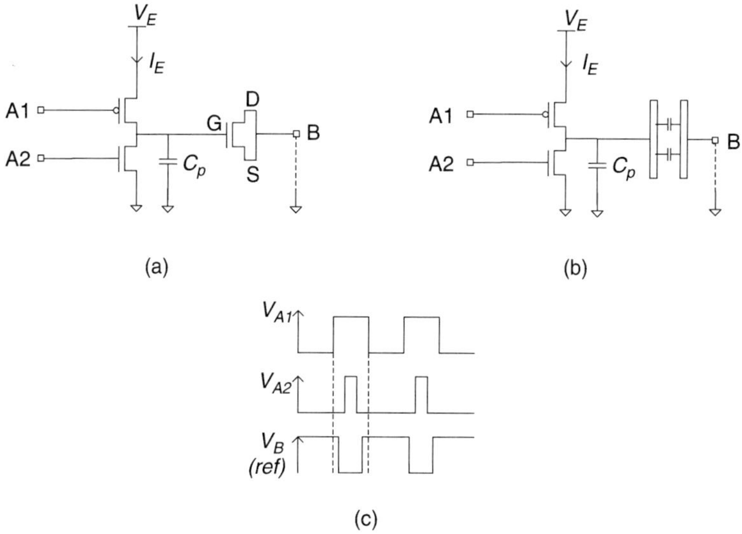

Fig. 4.11 CIEF-CBCM technique to measure **a** MOSFET gate capacitance and **b** interconnect cross talk capacitance. **c** Input signal waveforms

and in Fig. 4.11b, the DUT is formed by capacitively coupled interconnect wires. The capacitor DUT is driven by a pseudo-inverter as in the conventional CBCM technique but the use of a separate reference inverter is eliminated. As with conventional CBCM, the p-FET and the n-FET of the pseudo-inverter are driven with two independent non-overlapping clock signals, V_{A1} and V_{A2}. A third clock signal V_B drives one terminal of the DUT for measuring the reference capacitance. With this approach, the single pseudo-inverter serves as its own ideal reference and many potential sources of error subtracted out. In the case of a fixed capacitive DUT, as in Fig. 4.11b, the charge injection associated with the pseudo-inverter MOSFET channel is also completely eliminated by subtraction.

The clock signal waveforms for V_{A1}, V_{A2}, and V_B are shown in Fig. 4.11c. In operation, initially node B of the DUT is held at GND and the current I_E corresponding to I_{E2} in conventional CBCM is measured for the capacitor DUT. When input signal V_B is applied to node B of the DUT, its terminals B and G are held at the same potential during charging and discharging. In this case, the measured current I_E ($= I_{E1}$), now corresponding to I_{E1} in conventional CBCM, is primarily due to the parasitic capacitance C_p. The DUT capacitance is obtained from Eq. (4.12) in a similar fashion as in the conventional CBCM method. By using waveforms shown in Fig. 4.11c with a range of closely spaced V_B values, full C–V characterization of the MOSFET DUT can be carried out.

A number of different approaches have been advocated to improve the measurement accuracy by accounting for the leakage currents, charge injection, and device mismatch. The circuit for one very attractive variant of CBCM called quadrature-clocked, voltage-dependent capacitance measurement (QVCM) is shown in Fig. 4.12a [7]. The pseudo-inverters are replaced by two n-FET detector elements. Independent voltage signals, 180° out of phase with each other, are applied to the gates of these two n-FETs. The signal applied at node B of the capacitor DUT is ± 90° out of phase with the other two signals. The n-FETs provide a path for the charging and discharging current to the capacitor to flow from and to GND.

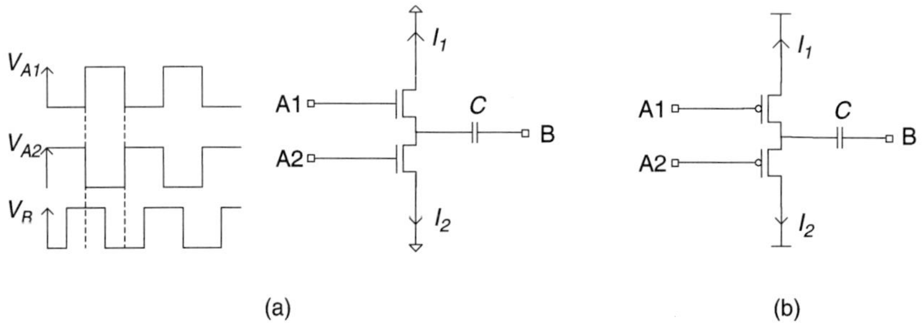

Fig. 4.12 **a** A circuit schematic of QVCM scheme with two n-FETs driven by independent input signals V_{A1} and V_{A2} and a third signal applied to node B of the capacitor. **b** Similar circuit with the two n-FETs replaced with p-FETs

The currents I_1 and I_2 are measured at two different values of V_B, V_{B1} and V_{B2}. The DUT capacitance is then given as

$$C = \frac{(I_{1\_1} - I_{1\_2}) - (I_{2\_1} - I_{2\_2})}{2f\,(V_{B1} - V_{B2})}, \tag{4.14}$$

where I_{1_1} and I_{1_2} are current measurements at V_{B1}, and I_{2_1} and I_{2_2} are current measurements at V_{B2}.

The QVCM method has the advantage that the two n-FETs do not experience a voltage across the source and the drain and hence there is no MOSFET channel leakage contribution to the measured currents. The current induced by charge injection is also cancelled by subtraction. There is a small current contribution from the gate oxide tunneling current which can be minimized by using thick-oxide n-FETs.

In another application of this approach referred to as cross talk-based CBCM (CTCM), coupling capacitance between interconnect wires is measured [8]. This was first implemented with the circuit shown in Fig. 4.12b. The n-FETs in the QVCM circuit are replaced by p-FETs and the flow of charging and discharging currents through the p-FETs is from and to the V_{DD} terminals of the power supplies.

4.2.3 Ring Oscillator-Based Capacitance Measurement

Capacitance components of CMOS circuit elements may be extracted from a ring oscillator circuit by measuring its frequency and its power in the active and quiescent states. The measured capacitance is integrated over the switching cycle of a circuit and gives a true measure of the effective switching capacitance. Capacitance measurement of CMOS circuit elements is carried out by using a differential pair of ring oscillators, differing in design only by the capacitance of the circuit element of interest. An advantage of this method is a self-consistent determination of circuit delays and capacitances at frequencies in the range of product applications. In another approach, full C–V characteristics of MOSFETs are obtained by applying a voltage bias to the gate of a MOSFET load in each stage of a ring oscillator. The ring oscillator-based capacitance measurement technique and associated test structure designs are described in detail in Chapter 6.

4.3 Capacitor DUT Designs

The design of a capacitor DUT is governed by the material properties and geometrical arrangement of its constituent films, the capacitance range, the parasitic resistance components, and the desired measurement accuracy. Representative ranges of capacitances and measurement frequencies for different test techniques are listed in Table 4.1.

Table 4.1 Range of frequencies and capacitances for different measurement techniques

Test method	Frequency range	Capacitance range
LCR meter	1 kHz to 1 MHz	1 pF to 1 F
LCR meter	1 MHz to 3 GHz	10 fF to 1 F
CBCM	1 kHz to 1 GHz	<1 fF to 1 μF
Ring oscillator	~10 MHz to 1 GHz	<1–100 fF per stage

Capacitance measurements with LCR meters at frequencies $\lesssim$1 MHz can be carried out using standard DC probe cards, and no on-chip CMOS circuitry is required. This allows characterization of interconnect wire capacitances and discrete MOSFETs in short-loop test vehicles in early technology development and in manufacturing. However, with a minimum capacitance of ~1 pF, the DUT areas are large and only a small number of unique capacitor designs can be accommodated in a scribe line. The DUT area can be reduced with a higher frequency LCR meter, but special probe card design and bench measurements are required. The DUT area can be reduced by as much as a factor of 1000× or more with CBCM techniques and measurements on such test structures can be made with DC probe cards.

The capacitance per unit length of the metal wires is primarily a function of the ratios of their dimensions and does not change appreciably with scaling. Hence, for DUT sizing purposes, we may assume the capacitance per unit length to be the same for all layers with scaled wire dimensions. The area occupied by a wire capacitor DUT can be estimated as

$$A = (w + s)\, l. \tag{4.15}$$

The DUT area for minimum width wire in a metal layer with 2× dimensions is twice the area of a similar DUT in 1× dimensions, with nearly the same capacitance. The area of a 2.0 pF DUT, excluding connecting wires, for different layer dimensions is shown in Table 4.2. The space of 2,400 μm^2 between two I/O pads in our standard 1×25 padset (Appendix A) is insufficient for placement of most of these DUTs. The macro area is substantially increased when placing DUTs above or below the array of pads.

Metal interconnect DUTs are generally stacked to reduce the DUT area. This is illustrated in Fig. 4.13 for metals M1–M4. In Fig. 4.13a, wires are connected

Table 4.2 Approximate area of a 2.0 pF capacitor in different conducting layers

Layer	W (μm)	s (μm)	C (fF/μm)	~Area (μm^2)
MOSFET C_{gT}	0.05	0.15	1.00	400
Thin metal (M1)	0.10	0.10	0.20	2,000
Thin metal (M1)	0.10	0.20	0.15	4,000
Thick metal (M3)	0.20	0.20	0.20	4,000

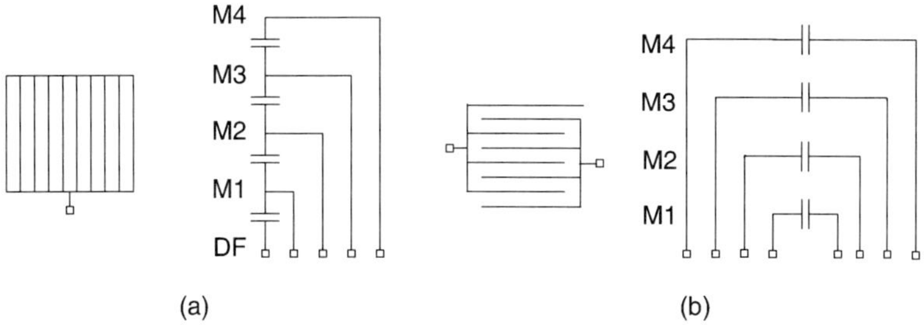

Fig. 4.13 Top view line drawing and stacked capacitor schematic of **a** wire mesh to measure $C_{\mathrm{up}} + C_{\mathrm{down}}$ and **b** wire comb to measure C_{w}

in parallel at each layer and the capacitance between any two layers is measured. This type of capacitor DUT is useful for determining dielectric properties such as thickness. In Fig. 4.13b, a wire comb structure is used to measure the $C_{\mathrm{w}}\left(= C_{\mathrm{up}} + C_{\mathrm{down}} + C_{\mathrm{left}} + C_{\mathrm{right}}\right)$. If the metal density is 50% or more, the metals above and below serve as nearly perfect ground planes. The wire width and spacing is varied in accordance with technology GRs. The wire segment lengths are limited to maintain $R_{\mathrm{s}} \ll 1/\omega C$ in the measurement frequency range. An estimation of dissipation factor and capacitance measurement frequency ranges is included in Section 9.2.3.

Metal wires may be of different width, spacing, and pitch. DUTs for process monitoring of lithography, etching, and CMP generally have uniform dimensions. The wire dimensions and spacing may be varied within a DUT for yield monitoring as shown in Fig. 4.14a. It is also useful to measure both resistance and capacitance of DUTs designed with minimum width wires and different spacing values to study process sensitivity to wire dimensions. With constant pitch, the wire resistance decreases with increase in w, while the capacitance increases with the corresponding reduction in s. The design of a serpentine wire with an interdigitated comb structure shown in Fig. 4.14b provides a measurement of both R_{w} and C_{w} for the same wire.

A DUT design for measuring C_{gT} of an n-FET using an LCR meter is illustrated in Fig. 4.15a. The width of the PS fingers and the number of H0 via contacts are adjusted to minimize the series resistance and to ensure that $R_{\mathrm{s}} \ll 1/\omega \mathrm{C}$ in the range of measurement frequencies. This type of layout can be extended to get a capacitance of ~1 pF. Independent contacts with S and D terminals can be provided for measurement of other MOSFET capacitance components.

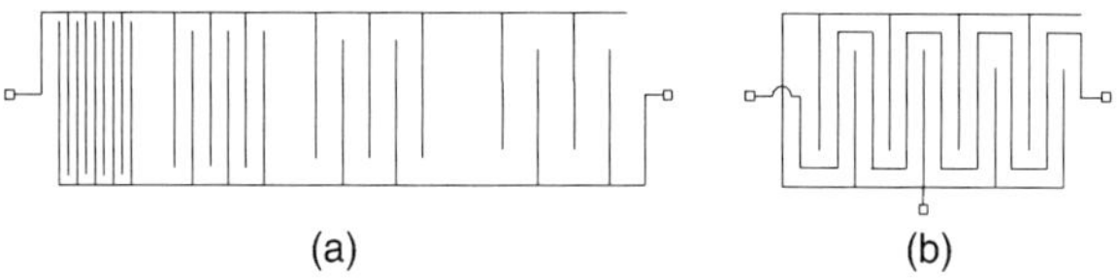

Fig. 4.14 Top view line drawing of metal wire capacitor DUT **a** with variable spacing and **b** serpentine to measure R_{w} and C_{w}

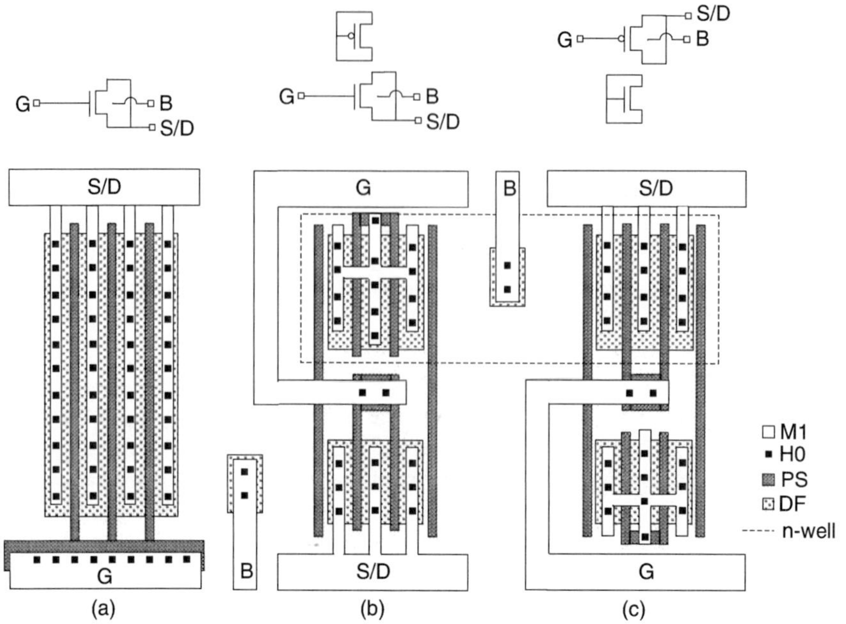

Fig. 4.15 Physical layouts of MOSFET gate capacitor DUTs: **a** for an n-FET, **b** and **c** for an n-FET and a p-FET derived from an inverter logic gate

The physical layout of the capacitor DUT as shown in Fig. 4.15a is not truly representative of CMOS circuit layouts. Sensitivity of device characteristics to the details of PS and DF layer dimensions, the surrounding environment, and the pattern densities becomes important as CMOS technologies are scaled to smaller dimensions. In the 65 nm technology node and beyond, compressive or tensile stress layers are used for mobility enhancement and the MOSFET characteristics may depend on the proximity of these layers to the DF layers for both n-FET and p-FET. It is highly desirable to closely match the physical layouts of the DUTs to circuit layouts on a product and to measure the *I–V* and *C–V* characteristics of MOSFETs on similar DUT designs. A closer approximation to the layout styles in CMOS circuits is obtained with layouts as shown in Fig. 4.15b, c for an n-FET and a p-FET, respectively. These layouts are derived from that of a standard inverter and discussed in greater detail in Section 5.3. Such designs are more suitable for smaller capacitor DUTs measured using CBCM techniques.

4.4 Capacitor Macro Designs

Capacitor macros are designed for process development and monitoring of metal wire properties, MOSFET gate length, and gate oxide thickness measurements, and for modeling *C–V* characteristics of MOSFET capacitance components. Macros designed for placement on short-loop test vehicles for metal layers comprise discrete capacitors directly connected to I/O pads. MOSFET capacitance components may also be measured using discrete capacitor DUTs. Measurements on such discrete capacitors are carried out with an LCR meter and the DUT capacitance values are typically >1 pF. Compact 1D array macros incorporate CBCM (or QVCM) circuits and the DUT capacitances are substantially reduced. Large 2D capacitor arrays are used for monitoring sensitivity to process and random variations. Measurements on the CBCM macros are made using standard parametric test equipment in a laboratory or parametric ATE in a silicon manufacturing line.

The first two examples in this section describe discrete capacitor macros for conducting layers and MOSFETs. In Example 3, implementation of CBCM and QVCM techniques for measuring small capacitor structures with a single level of metal is described. In Example 4, on-chip clock generation circuitry is included with QVCM, eliminating the need of external pulse generators. Some ideas on designing 2D array capacitor macros are discussed in Example 5. The integration schemes in 1D and 2D arrays, DUT selection circuits such as the use of decoders and scan chains, and on-chip clock generation using ring oscillators are covered in more commonly used resistor, MOSFET, and ring oscillator macros in the examples given in Chapters 3, 5, and 6.

4.4.1 Example 1: Discrete Passive Capacitor Macros

Macro configurations for measuring discrete passive capacitors with an LCR meter are shown in Fig. 4.16. In short-loop test vehicles or in early stages of technology development, such designs are useful for metal wire capacitance measurements. The measurement frequency for in-line measurements is recommended to be in the 10–100 kHz range and the DUT capacitance is designed to be $\geq$1 pF. The

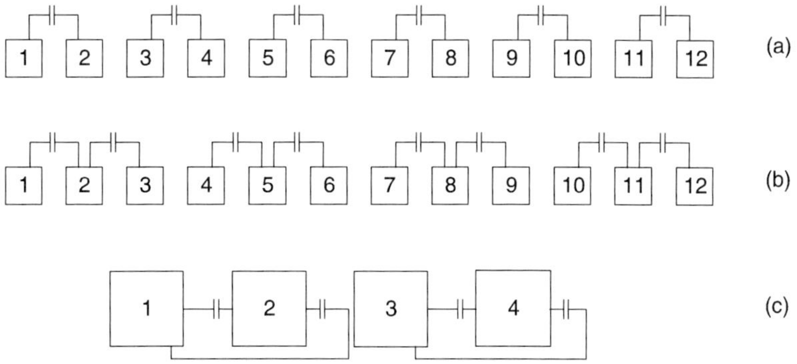

Fig. 4.16 I/O pad assignments for capacitor macros with **a** isolated pads and **b** shared pads. **c** Placement of capacitors in the space between I/O pads

approximate DUT areas for 2.0 pF capacitors formed with different conducting layers are given in Table 4.2. Metal capacitors typically occupy areas larger than the 2,400 μm^2 space between two I/O pads in the standard 1×25 padset. These capacitor DUTs are typically placed, at least in part, above the I/O pads, thereby increasing the height of the macro.

The resistance of the wires can be neglected by appropriate DUT design and selection of test frequency. Consider, for example, an M1 capacitor designed with $w = s = 0.10\,\mu m$ having a total capacitance of 2 pF and a wire length of 10,000 μm. With R_w of 2.0 $\Omega/\mu m$, the wire resistance for a serpentine is then 20 kΩ. If measurements are made at 100 kHz, $1/\omega C_w = 800$ kΩ, the dissipation factor D_f is 0.03, and the measurement error due to parasitic resistance is <0.01% (Section 9.2.3).

Capacitor DUTs may be isolated or share an I/O pad as shown in Fig. 4.16a, b. The shared I/O pad is connected to the LO terminal of the LCR meter. The macro height can be reduced by placing capacitors connected in parallel between the I/O pads as illustrated in Fig. 4.16c. Metal wire capacitor DUTs shown in Fig. 4.17 can be configured to measure only C_{up} and C_{down} capacitances or to measure the total capacitance per unit length C_w. In the comb structure in Fig. 4.17b, with one additional I/O pad, the resistance of one side of the comb section can also be measured. Linewidth narrowing is indicated by an increase in wire resistance R_w and a decrease in C_w.

The capacitance of cables, probes, I/O pads as well as that of wire connections to the DUT must be subtracted from the measured capacitance where feasible. Capacitance of the cables and probes can be measured directly by lifting the probes off the silicon wafer surface. The I/O pad capacitance in bulk silicon technology can be significant (~1 pF). A subtraction technique to account for this parasitic capacitance must be included in the macro design.

Three different examples of macro designs for the measurement of the parasitic capacitances on silicon are shown in Fig. 4.18a–c. In the first scheme shown in (a), one pair of I/O pads in a macro is dedicated for measurement of parasitic pad capacitance. It is assumed that this capacitance is the same for all other pairs of I/O pads. In a second approach shown in Fig. 4.18b, one macro in a reticle field with only I/O pads may be dedicated for measuring capacitance of all of the I/O

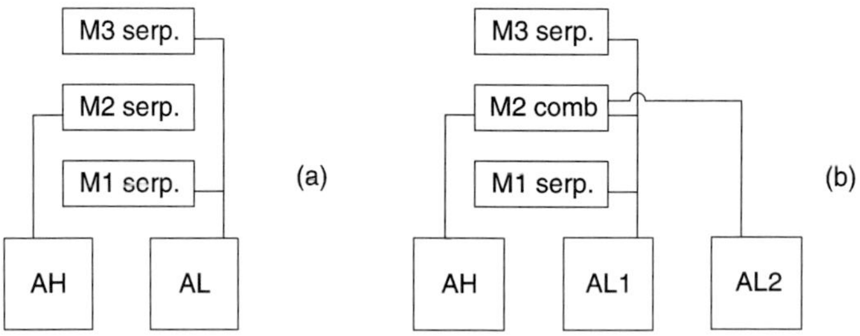

Fig. 4.17 Interconnect wire capacitors for measurement of **a** C_{up} and C_{down} and **b** C_w and wire resistance R_w

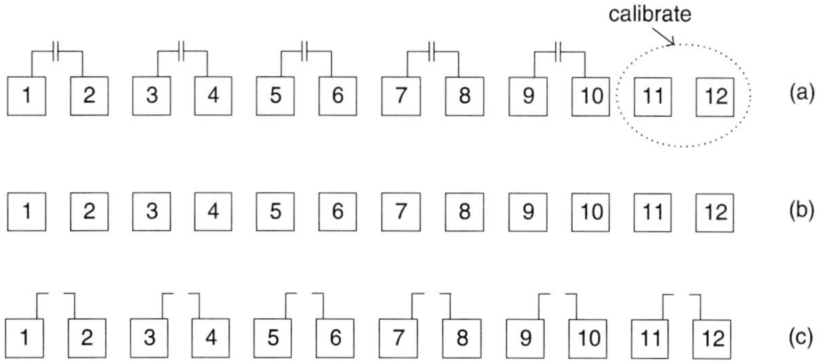

Fig. 4.18 Macro designs for parasitic capacitance calibration: **a** one pair of I/O pads dedicated for calibration in each macro, **b** only I/O pads, and **c** I/O pads and wires for DUT connections

pads. The measured values of parasitic capacitance in this macro are applied to all macros on a reticle field or a wafer. In Fig. 4.18c, the macro design includes the wire connections to the capacitor DUTs and the parasitic capacitance of the wires is also subtracted from the measured capacitance of the DUTs in a corresponding macro. This approach gives a higher accuracy but requires more macros and thereby more silicon area if multiple capacitor macros with different DUT designs are placed in a reticle field.

4.4.2 Example 2: MOSFET Capacitor Macros

Measurement of MOSFET capacitances is typically carried out at the M1 metal level for early technology feedback. Measurements are made on different DUTs for extraction of each of the key MOSFET parameters such as t_{ox}, C_{gT}, C_{ov}, and L_p. These parameters are used for process control and for modeling MOSFET *I–V* characteristics. It is therefore important to ensure that the physical layouts and the surrounding layer pattern densities of these capacitor DUTs are similar to each other and to the layouts used in CMOS circuits as described in Section 4.3.

Gate oxide thickness and gate-tunneling currents are measured on a MOS capacitor (comprising a number of large-area parallel plates each with dimensions $\gtrsim 10\ \mu m$). Contact with gate and diffusion areas are made with M1 metal through H0 vias, appropriately designed to maintain $R_s \ll 1/\omega C$ and a low dissipation factor at the measurement frequency. Measurements of C_{gT} and C_{ov} are made on MOSFET DUTs with PS length comparable to that in representative MOSFET structures, or better still, on DUT layouts derived from CMOS logic gates.

A commonly used technique for measuring C_{gT} and C_{ov} is referred to as the split *C–V* method [3, 11]. This method has also been adapted to measure carrier mobility. Cross sections of an n-FET in accumulation and inversion modes are shown in Fig. 4.19a, b, respectively. In the accumulation mode, a high density of holes (e^+) is present near the oxide–silicon interface. The capacitance between the gate (G)

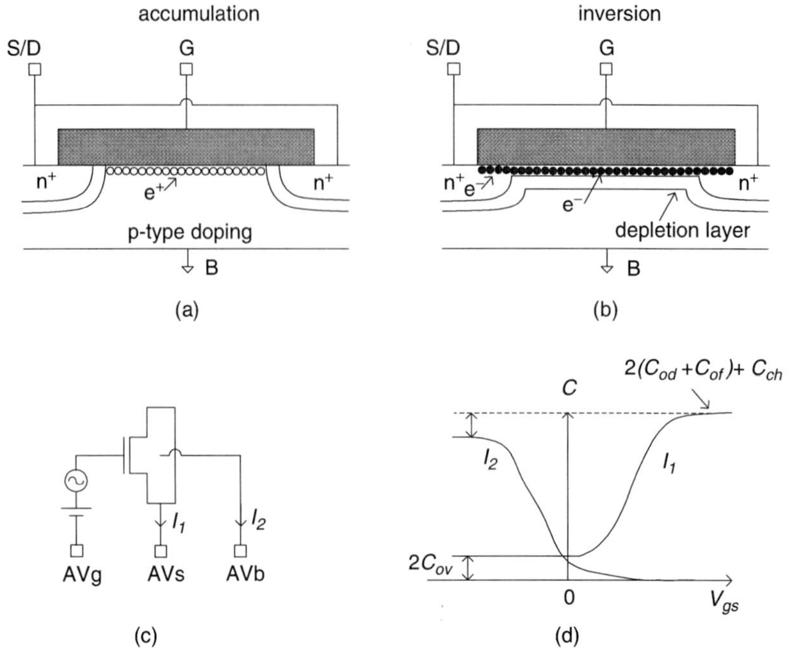

Fig. 4.19 An n-FET in **a** accumulation and **b** inversion. **c** A measurement setup for split *C–V* for an n-FET. **d** *C–V* plots obtained from measuring I_1 and I_2 in **c**

and the source/drain (S/D) terminals is the overlap capacitance ($= 2C_{ov}$) and the capacitance between the G and the body (B) is the channel capacitance C_{ch}. In the inversion mode, with the presence of the n^+ channel, the capacitance between the G and S/D terminals is the gate capacitance C_{gT}. Once the inversion layer formation is completed, the body is shielded from the gate and the G-to-B capacitance is negligible.

The test setup for an n-FET is shown in Fig. 4.19c. The currents and hence the capacitances from G to S/D and G to B are measured independently. The split *C–V* curves are shown in Fig. 4.19d. The sum of the two curves is similar to the MOS *C–V* curve shown in Fig. 4.7.

I/O pad assignments for *C–V* and *I–V* characterization of discrete n-FETs are shown in Fig. 4.20. Individual DUTs are designed for C_{ov}, C_{gT} (for L_p extraction), C_{sb}, C_{db}, and *I–V* characterization. The capacitance for the n^+ diffusion-to-body is separated into an area and a perimeter component. These components can be

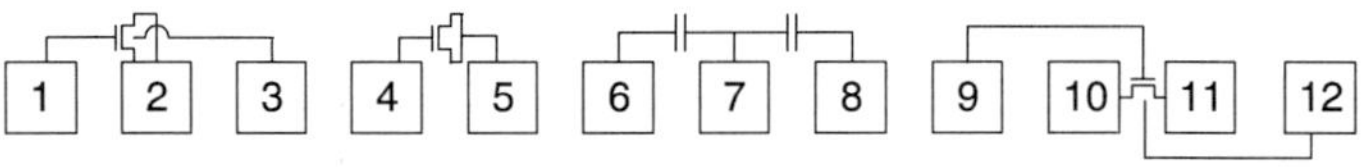

Fig. 4.20 I/O pad assignments in a section of a macro for MOSFET *C–V* and *I–V* characterization

extracted with two capacitors of equal area but with different perimeter values. An n-FET for *I–V* characterization is included in the same macro. A standard 1×25 padset macro can accommodate one set of n-FET and p-FET DUTs of similar layouts.

4.4.3 Example 3: CBCM (QVCM) Macros Testable at M1

The CIEF-CBCM and QVCM techniques are well suited for MOSFET capacitance measurements. As capacitance values of the order of a few fF can be measured accurately, the capacitor DUT area is reduced by a factor of ∼ 1000 compared with the DUTs in the discrete capacitor macros in Example 2.

A basic scheme for implementation of the CIEF-CBCM technique for MOSFET gate and overlap capacitances at the M1 metal level is shown in Fig. 4.21a, b. In Fig. 4.21a, a pseudo-inverter drives the gate of an n-FET for gate capacitance measurement. In Fig. 4.21b, the n-FET gate terminal is connected to I/O pad B instead. In this configuration, the overlap capacitance including the outer fringe capacitance can be extracted [9]. An external pulse generator drives the clock inputs A1, A2, and B. DC current delivered by the power supply connected to I/O pad VE is measured with input B at GND. The parasitic capacitance determination is made by repeating the measurement with the clock signal applied to I/O pad B as described in Section 4.2.2. Five I/O pads are required for measuring one capacitor and five such DUTs can be accommodated in a standard 1×25 padset macro.

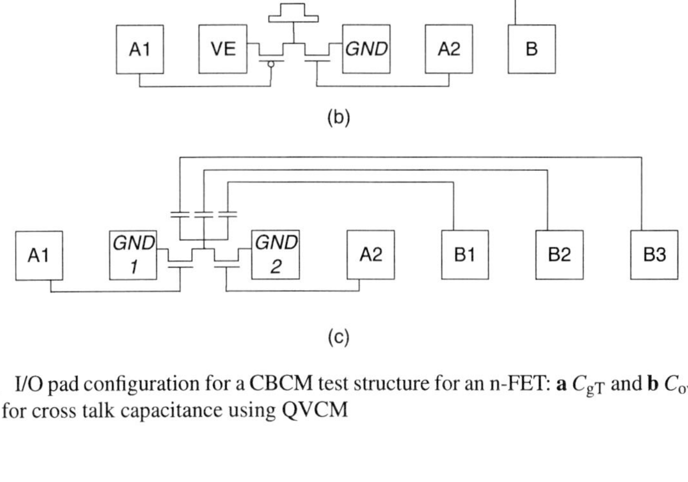

Fig. 4.21 I/O pad configuration for a CBCM test structure for an n-FET: **a** C_{gT} and **b** C_{ov}. **c** Shared I/O pads for cross talk capacitance using QVCM

The circuits shown in Fig. 4.21a, b can be used for implementing the QVCM (CTCM) technique by replacing the p-FET in the pseudo-inverter with an n-FET and connecting VE and GND to independent SMUs. A pad sharing scheme for measuring cross talk capacitance using QVCM is shown in Fig. 4.21c. Each pair of n-FETs connects to three capacitors, which could be different components of an interconnect wire (C_{up}, C_{down}, and $C_{left/right}$) [8]. The second terminal of each capacitor DUT is connected to an independent I/O pad which is used to null the capacitance of the DUT not under test.

With QVCM, macro efficiency is further improved with the use of common clock signals for all n-FET pairs as shown in Fig. 4.22. Two pads per macro are dedicated to the clock signals for the n-FETs. The number of capacitors in a 1×25 padset macro is maximized with six n-FET pairs loaded with 11 capacitors, bringing the number of capacitors per macro to 66. This number can be increased to 121 if one terminal of all the n-FETs is held at a common GND. In this case, the current I_2 is measured using inverted clock signal at the second node after I_1 measurement is completed. Although the area efficiency of the macro is improved with this scheme, the use of separate GND pads is preferred as it preserves the wiring symmetry. The number of DUTs per macro with these different configurations is listed in Table 4.3.

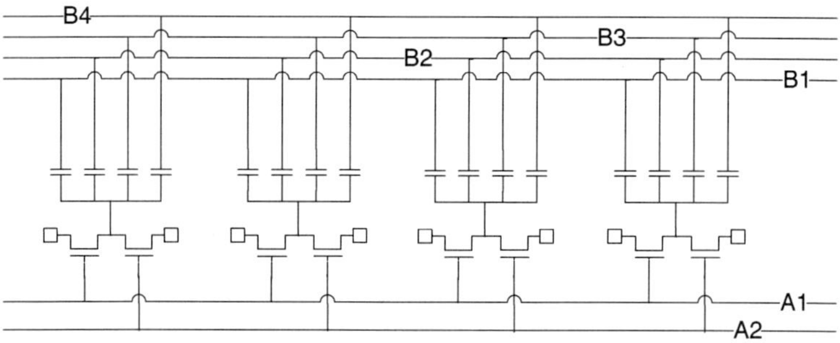

Fig. 4.22 QVCM macro with shared input clock signals for implementation at the M1 metal level

Table 4.3 Total number of capacitor DUTs per 1×25 padset macro with different QVCM integration schemes

I/O pad configuration	Number of DUTs per FET pair	Total no. of DUTs
Isolated	1	5
Multi-capacitor units	9	36
Shared A1 and A2	11	66
Shared A1, A2, and one GND	11	121

4.4.4 Example 4: QVCM Macros with On-chip Clock

The frequency of operation in the CBCM designs described in Example 3 is limited by the probe card and the maximum operating frequency of the pulse-generating equipment. A circuit to generate all three clocks for the QVCM (CTCM) method is shown in Fig. 4.23. With this on-chip circuit, measurements in the GHz frequency range can be made with a standard DC probe card, thereby providing a capability for measuring sub-fF capacitors in the manufacturing line along with other parametric measurements.

The ring oscillator in Fig. 4.23a is enabled by raising the input voltage at the EBL node from "0" to a "1." The output of the ring oscillator is tapped at two nodes, roughly dividing the stages in two equivalent sections. The true and complementary outputs of the ring oscillator, A1 and A2, are used to drive the gates of the two n-FETs in the QVCM circuit. The second output of the ring oscillator B, ~90° out of phase with the previous two signals, drives the capacitor DUT. With multiple capacitors connected through a common n-FET pair, a decoder is used to direct the B signal to a selected DUT as indicated in Fig. 4.23b.

The floorplan of a section of a macro with an on-chip clock generation scheme is shown in Fig. 4.24. A two-input decoder is used to accommodate four capacitors per n-FET pair. If a three-input decoder is used instead, with eight capacitor DUTs

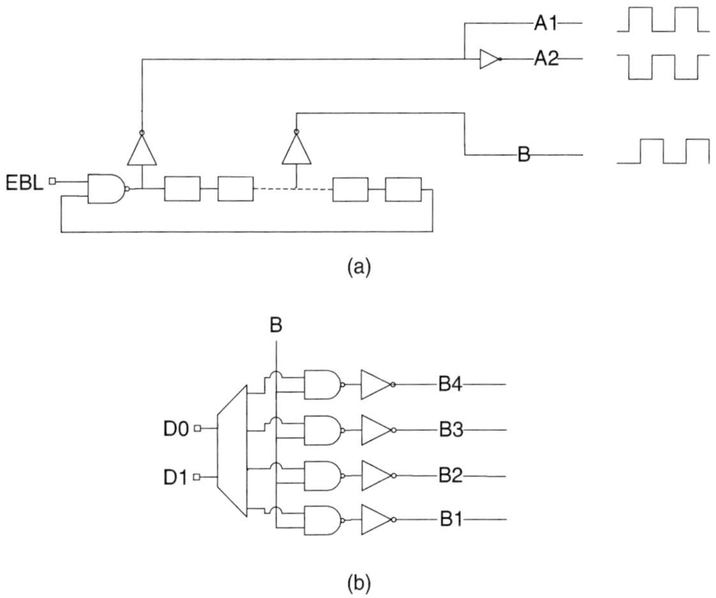

Fig. 4.23 **a** Schematic of a circuit to generate 180° and 90° out-of-phase clocks for QVCM scheme. **b** A decoder circuit to direct the B signal to the desired DUT

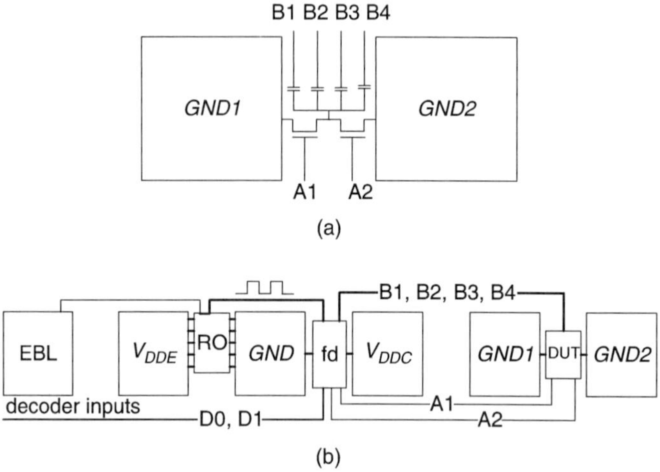

Fig. 4.24 **a** Placement of QVCM DUT between I/O pads. **b** Floorplan of a section of a QVCM macro with an on-chip clock generator using an RO and a frequency divider (fd) block

per n-FET pair, the total number of DUTs in a 1×25 padset macro is 72 compared with 66 in Example 3.

4.4.5 Example 5: 2D Capacitor Array Macros

2D capacitor arrays are utilized for rapid characterization of a large number of capacitor DUTs [12]. Such macro designs are useful as CMP process monitors and MOSFET gate length extraction for a large variety of physical layouts. The design of a 2D array for capacitance measurements described here follows the 2D arrays described for resistors and MOSFETs in Chapters 3 and 5, respectively. The resistor DUTs require a single voltage force source. The MOSFET DUTs require at least two voltage sources to bias the gate and drain terminals. The capacitor DUT in a CBCM configuration, on the other hand, requires three or more voltage sources. The exact value of the voltage levels applied to the gates of the pseudo-inverters in the CBCM circuit is not critical for wire capacitance measurements.

The circuit schematic and the symbol of a CIEF-CBCM DUT along with control switches used in a 2D CBCM array are shown in Fig. 4.25a–c. A 2×2 section of the CBCM array is shown in Fig. 4.25d. The switches shown in Fig. 4.25b, c steer the clock and DC signals to a DUT when its CAD input is a "1." The voltage levels to the unselected DUTs (CAD = "0") are set at negative clamp values, applied at nodes AVCp and AVCn, to minimize leakage currents and to ensure a high degree of isolation. Additional current-steering circuitry is included in the 2D array circuit to eliminate residual background currents (see MOSFET 2D array example in Section 5.4.4). This design can be easily adapted for CIEF-CBCM or other CBCM configurations.

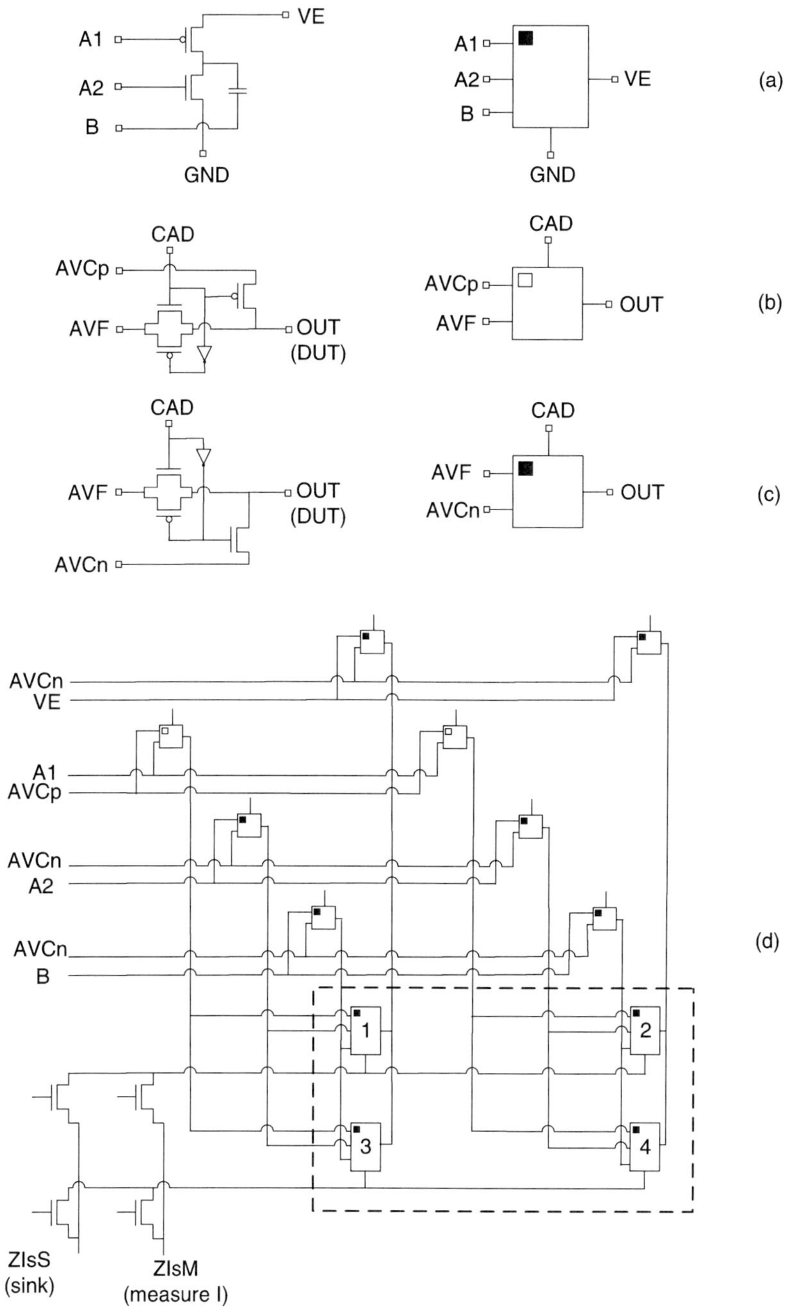

Fig. 4.25 **a, b** Circuit schematic and symbol of a CBCM DUT. **c** Switch configurations and symbols. **d** 2×2 section of a 2D CIEF-CBCM array

4.5 Capacitance and Inductance: A Closer Look

In the case of wires, there is an interesting and important connection between capacitance and inductance. We will briefly address this, both to provide more insight into the behavior of capacitance and also to give a perspective into where inductance fits into the circuit and test structure design picture. As mentioned in Section 2.4.3, inductance does not usually play a significant role with circuit interconnects on a local length scale. However, it can be a very important consideration when it comes to power distribution and also for long low resistance lines carrying signals on-chip or through a package or I/O system. We will first discuss some basic properties of inductance and then describe how capacitance and inductance are intimately connected to each other in an elegant way.

Maxwell's equations give a complete description of the magnetic and electric fields around any set of conductors in response to the presence and flow of charge [13, 14]. That said the actual calculation of these fields can be a very complicated problem, often requiring extensive numerical simulation. On the other hand, from Maxwell's equations there are two expressions that can be derived which are extremely helpful in determining electric and magnetic fields in the presence of stationary and moving charges in a number of classic symmetric situations. Related capacitance and inductance values can then also be determined in a straightforward manner. One of these expressions is Gauss's law which states that the surface integral of $\boldsymbol{E} \cdot \mathrm{d}\boldsymbol{a}$ over a surface enclosing a charge q is $q/\varepsilon\varepsilon_o$, or $\int \boldsymbol{E} \cdot \mathrm{d}\boldsymbol{a} = q/\varepsilon\varepsilon_0$, where $\boldsymbol{E}$ is the vector electric field and $\mathrm{d}\boldsymbol{a}$ is a vector area element of the enclosing surface. The other is a corresponding expression for magnetic fields known as Ampere's law stating that the line integral of $\boldsymbol{B} \cdot \mathrm{d}\boldsymbol{l}$ around a path that encloses a wire carrying a current I is $\mu_o I$, or $\int \boldsymbol{B}.\mathrm{d}l = \mu_o I$, where $\boldsymbol{B}$ is the vector magnetic field, $\mathrm{d}\boldsymbol{l}$ is a vector length element along the encircling path, and μ_o $\left(= 4\pi \times 10^{-7}\mathrm{H/m}\right)$ is the permeability of free space.

The self-inductance L of a closed loop can be expressed as $L = \varphi/I$, where φ is the magnetic flux threading the loop in response to current I flowing through it. The magnetic flux φ is in turn given by the expression $\varphi = \int \boldsymbol{B} \cdot \mathrm{d}\boldsymbol{a}$, where $\boldsymbol{B}$ is again the vector magnetic field, $\mathrm{d}\boldsymbol{a}$ is a vector area element, and the integral is taken over any complete surface that bounds the loop. For illustration purposes it is convenient and relevant to consider small sections of long symmetric transmission line structures that reduce to two-dimensional problems from which the inductance per unit length can be determined.

The first structure to be considered is a segment of length l of a symmetric stripline as shown in Fig. 4.26. A blow-up of the cross section in the vicinity of segment l is shown in Fig. 4.27. This transmission line structure consists of two long parallel metal plates, each of width w, separated by distance h and embedded within a uniform dielectric of relative dielectric constant ε. Let us assume that $w/h >> l$. With currents flowing as described in the caption of Fig. 4.27, the magnetic field $\boldsymbol{B}$ is directed from right to left in the gap as indicated. Furthermore, $\boldsymbol{B}$ is essentially constant in the gap and is nearly zero everywhere outside the gap region. Applying Ampere's law around the path indicated in the figure gives $|\boldsymbol{B}| = B = \mu_o I/w$. If we

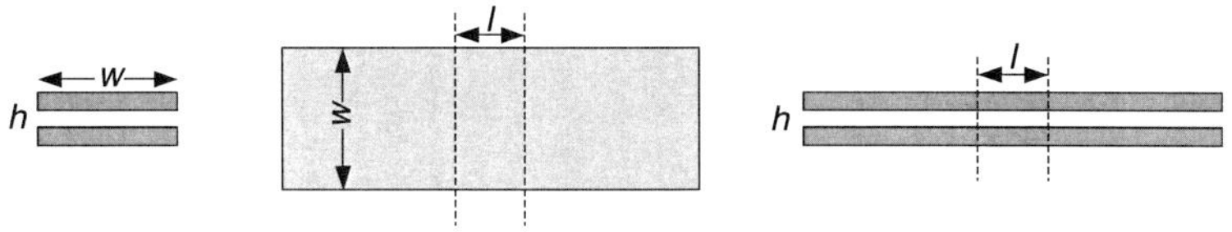

Fig. 4.26 Parallel plate transmission line (stripline) structure. **a** Cross-sectional view, **b** top view, and **c** side view. The magnetic field in a central segment of length l is considered

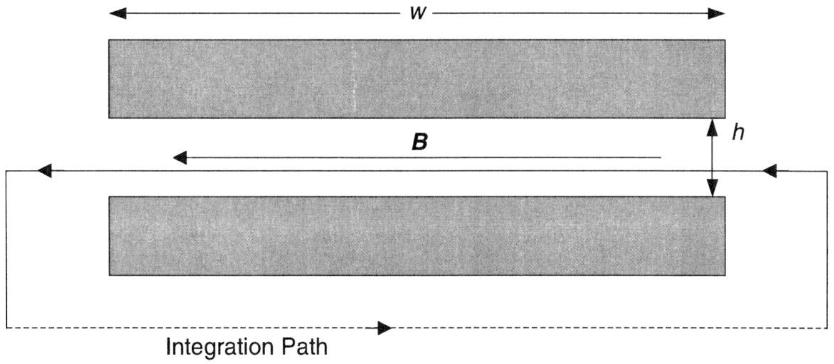

Fig. 4.27 Blow-up of the cross section of the parallel plate transmission line. Current I is flowing into the top electrode and equal ground return current I is flowing out of the bottom electrode. It is assumed that the current is flowing only on the inside surfaces of the plates (characteristic of very high frequency) and is uniform across the width w. The magnetic field $\boldsymbol{B}$ in the gap is indicated along with the integration path used in conjunction with Ampere's law

then calculate the magnetic flux passing between the plates of the section of length l of the transmission line, we find $\varphi = Bhl = \mu_0 Ihl/w$. Dividing by I, this gives $L = \mu_0 hl/w$, or an inductance per unit length of

$$L_w = \frac{\mu_0 h}{w}. \tag{4.16}$$

Referring now back to Eq. (4.1), note that the capacitance per unit length for this same structure can be expressed as

$$C_w = \frac{\varepsilon\varepsilon_0 w}{h}. \tag{4.17}$$

Equation (4.1) not only was simply asserted but also can be derived using Gauss's law, much as Ampere's law was used to obtain L_w. Thus C_w increases linearly with w/h, while L_w increases linearly with h/w. Furthermore the propagation speed v for a transient signal going down this line is given by

$$v = \frac{1}{\sqrt{L_w C_w}} = \frac{1}{\sqrt{\varepsilon\varepsilon_0\mu_0}} = \frac{c}{\sqrt{\varepsilon}}, \tag{4.18}$$

where c is the speed of light in vacuum. It can be shown that this fundamental relationship between L, C, and c holds for any constant cross-sectional transmission line embedded in a uniform dielectric, provided the currents are constrained to flow on the surface of the conductors. The value of ε for silicon is ~11.7, while that of SiO_2 is ~3.9, and low-k dielectric materials may have $\varepsilon < 3$. In addition, current and magnetic fields penetrate beyond the surface, tending to raise L_W. Practical resultant values of v in silicon circuits are in the range of $c/2$ to $c/5$. However, as previously mentioned in Section 2.4.3, it is the RC time constant and not v that determines signal delay in CMOS circuits.

The linear dependence of L and C on dimensions, on the other hand, is a special result that only holds for stripline geometries. We will next examine a coplanar geometry that has a very different and more typical dimensional dependence. Consider the transmission line geometry shown in Fig. 4.28. This transmission line consists of two parallel metal wires, each of diameter $2r_o$ with a center-to-center spacing of R_o and embedded in a uniform dielectric of relative dielectric constant ε. This is obviously a much more open structure than is the stripline, and electromagnetic fields will occupy an extended region between and around the conductors as indicated in Fig. 4.29. Ampere's law can again be used to calculate the magnetic field and the inductance per unit length.

Referring to Fig. 4.29 consider first the magnetic field in the vicinity of a long straight wire of radius $2r_o$ carrying current I out of the page. The magnetic field will be in the circumferential direction as indicated. Using any circular field line as a path at a distance r from the center of the wire, Ampere's law tells us that $2\pi rB = \mu_o I$, or $B = \mu_o I/2\pi r$. In the case of two parallel current-carrying wires in the form of a coplanar wire transmission line as shown in Fig. 4.28a, b, the magnetic field is well approximated, by superposition, as the sum of the magnetic fields from two isolated wires. The magnetic flux between the wires for a section of length l of this transmission line is then calculated as $\varphi = 2\left[\mu_o Il/2\pi\right] \int r^{-1}\, dr$, over the range from r_o to R_o, the pre-factor of 2 accounting for the equal contributions from both wires. This gives $\varphi = \left[\mu_o Il/\pi\right] \ln\left[R_o/r_o\right]$, which when divided by Il leads to

$$L_W = \frac{\pi}{\mu_o} \ln\left[\frac{R_0}{r_0}\right]. \tag{4.19}$$

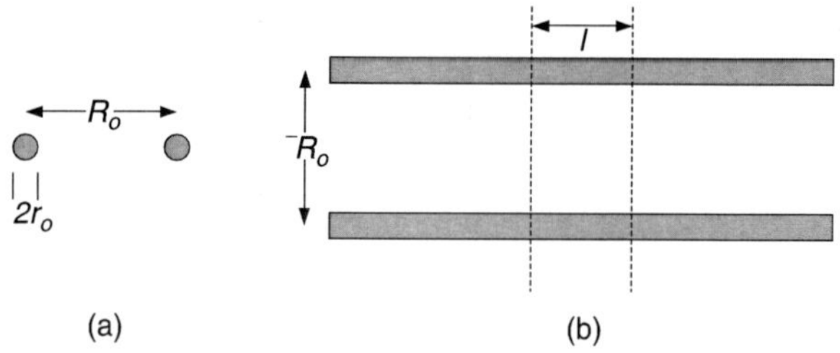

Fig. 4.28 Coplanar wire transmission line. **a** Cross-sectional view showing the two wires, each of radius r_o, with center-to-center separation of R_o, and **b** top view indicating a short section of length l far from the ends

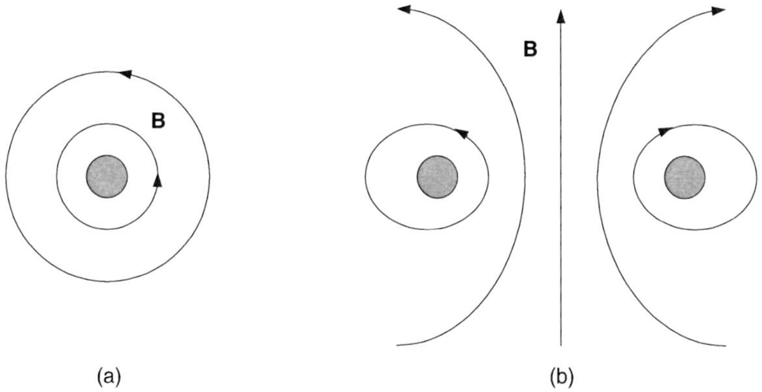

Fig. 4.29 Magnetic fields in the vicinity of long straight current carrying wires. **a** Single isolated wire of diameter $2r_o$ with current I out of the page and **b** two similar wires in the form of a coplanar wire transmission line, corresponding to Fig. 4.28, with current I out of the page (*left*) and return current I into the page (*right*). It is assumed that the current is flowing only on the surface of the conductors

An analogous calculation for capacitance using Gauss's law leads to the expression for capacitance per unit length as

$$C_W = \frac{\pi \varepsilon \varepsilon}{\ln\left[\frac{R_0}{r_0}\right]}. \tag{4.20}$$

As in the case of the stripline configuration, $v = 1\sqrt{L_W C_W} = 1/\sqrt{\varepsilon \varepsilon_o \mu_o} = c/\sqrt{\varepsilon}$, where c is the speed of light in vacuum. However, unlike in the stripline case, both C_W and L_W exhibit logarithmic rather than linear dependence on transverse dimensions. As the separation between the wires is increased, L_W slowly increases, while at the same time, C_W slowly decreases. This weak dependence on transverse dimensions is in fact characteristic of the situation for most on-chip wiring which is why assuming that, for example, $C_W = 0.2$ fF/μm is a good approximation over a fairly wide variety of actual wiring situations. In the case of inductance, it should be noted that, expressed in convenient units, $\mu_o = 1.26$ pH/μm. It follows that a quick rough estimate of the inductance of open structures (such as the coplanar wire transmission line) can typically be made as L(pH) = size of the structure in μm.

Sophisticated numerical packages are widely available for accurate calculations of self-inductance and capacitance, as well as mutual inductances and capacitances which can play a significant role in, for example, cross talk between different signal paths. While such packages are necessary for a precise design, one should keep in mind the general behavior and magnitudes discussed here which are often sufficient for small-scale designs.

References

1. Taur Y, Ning TH (2009) Fundamentals of modern VLSI devices, 2nd edn. Cambridge University Press, New York, NY
2. Weste NHE, Eshraghian K, Smith MJS (2000) Principles of CMOS VLSI design, 2nd edn. Addison Wesley, Reading, MA
3. Schroder DK (2006) Semiconductor material and device characterization, 3rd edn. Wiley, Hoboken, NJ
4. Wong S-C, Liu PS, Ru J-W, Lin S-T (1998) Interconnection capacitance models for VLSI circuits. Solid State Electron 42:969–977
5. Fleury D, Cros A, Romanjek K, Roy D, Perrier F, Dumont B et al (2008) Automatic extraction methodology for accurate measurements of effective channel length on 65-nm MOSFET technology and below. IEEE Trans Semicond Manuf 21:504–512
6. Chen JC, McGaughy BW, Sylvester D, Hu C (1996) An on-chip, attofarad interconnect charge-based capacitance measurement (CBCM) technique. IEEE international electron device meeting (IEDM) technical digest, 1996:69–72
7. Polansky S, Solomon P, Ketchen M, Shiling E, Economikos L, Bhushan M (2009) Back-end-of-line quadrature-clocked voltage-dependent capacitance measurements. 35th international symposium for testing and failure analysis (ISTFA), San Hose, 2009
8. Vendrame L, Bortesi L, Cattane F, Bogliolo A (2006) Crosstalk-based capacitance measurements: theory and applications. IEEE Trans Semicond Manuf 19:67–77
9. Chang Y-W, Chang H-W, Lu T-C, King Y-C, Ting W, Ku Y-HJ et al (2006) Charge-based capacitance measurement for bias-dependent capacitance. IEEE Electron Device Lett 27: 390–392
10. Sylvester D, Chen J, Hu C (1998) Investigation of interconnect capacitance using charge-based capacitance measurement (CBCM) technique and three-dimensional simulations. IEEE J Solid-State Circuits 33:449–453
11. Koomen J (1973) Investigation of MOST channel conductance in weak inversion. Solid State Electron 16:801–810
12. Hayes JD, Agarwal K, Nassif S (2009) Technique for the rapid characterization of parametric distributions. IEEE Trans Semicond Manuf 22:66–71
13. Jackson JD (1975) Classical electrodynamics, 2nd edn. Wiley, New York, NY
14. Purcell EM (1985) Electricity and magnetism, Berkeley physics course, vol 2, 2nd edn. McGraw-Hill, New York, NY

Chapter 5
MOSFETs

Contents

The switching speed of a MOSFET is a function of its current drive strength, its internal capacitances, and the *RC* load it drives. In Chapter 4, test structures for characterization of MOSFET capacitances are described. To first order these capacitance components, within a technology node, scale with the device dimensions. The current drive of a MOSFET, on the other hand, can be modulated over a much wider range by engineering the doping in the channel, using carrier mobility enhancement techniques, modifying the properties and dimensions of constituent layers, and varying its physical layout. Statistical fluctuations in the number of dopant atoms in the channel and local linewidth variations lead to variability in parameters of nominally identical MOSFETs. Hence, the focus on MOSFET macro design and test in this chapter primarily concerns DC I–V and variability characterization.

The physics of semiconductor devices including MOSFETs is treated in detail in standard textbooks [1–2]. In this chapter, basic MOSFET properties and parameter definitions are given in Section 5.1. Parametric test equipment and measurement techniques for MOSFET characterization are covered in Section 5.2. MOSFET DUT designs and physical layouts are discussed in Section 5.3. Five examples of macro designs for measuring the characteristics of individual MOSFETs and matched pairs and their statistical distributions are presented in Section 5.4.

M. Bhushan, M.B. Ketchen, *Microelectronic Test Structures for CMOS Technology*,
DOI 10.1007/978-1-4419-9377-9_5, © Springer Science+Business Media, LLC 2011

5.1 MOSFET Properties

The principle of the MOSFET device was first proposed by Julius Edgar Lilienfeld in 1925. For over 35 years, practical application of this concept was held up by difficulties in processing thin gate oxides. The first major breakthrough in CMOS technology development is attributed to Frank Wanlass, who successfully fabricated an n-FET and a p-FET on the same substrate. CMOS device fabrication continues to challenge scientists and engineers as the technology is scaled to ever smaller dimensions and new materials, processes, and device configurations are developed. In a CMOS technology node, many different types of MOSFETs are offered for circuit applications: high performance and low power digital logic, analog, and memory. These MOSFETs may differ in their *I–V* characteristics and capacitances as well as in channel lengths and widths. For mixed signal applications and I/O circuits, MOSFETs with thicker gate oxides are made available to meet reliability criteria at higher operating voltages.

In technology development and manufacturing, a major part of MOSFET characterization is carried out at a test stop immediately following M1 metal delineation to allow early feedback in the process cycle. Test structures typically comprise single MOSFETs wired for *I–V* and *C–V* measurements. With a large number of MOSFET offerings, the area occupied by such MOSFET structures testable at the M1 metal level and the time to test them become significant. Corresponding macro designs configured for parallel testing at the M1 metal level with multiplexing schemes to reduce the I/O pad count are preferred. However, yield considerations very early in the technology development cycle limit design complexity.

Here, we briefly describe basic MOSFET characteristics and key parameter measurements for tracking their behavior. Intrinsic variability in MOSFET parameters arising from random dopant fluctuations and local linewidth variations increases as MOSFETs are scaled to smaller dimensions. The nature of statistical distributions of MOSFET parameters and data analysis techniques are covered in Chapter 10. Extraction of MOSFET properties from circuit delay measurements is described in Chapter 6.

5.1.1 MOSFET DC I–V Characteristics

The physical cross sections and electrical representations of an n-FET and a p-FET, and their *I–V* characteristics are shown in Fig. 5.1. Each MOSFET has four terminals: source (S), gate (G), drain (D), and substrate or body (B) as shown in Fig. 5.1a, b. The B terminal may be set at a fixed potential unless independent body bias is desired and the n-well and p-well are isolated from the substrate. The voltage at the G terminal V_{gs} is measured with respect to the potential at the S terminal. The drain-to-source current I_{ds} is plotted against the drain-to-source voltage V_{ds} for a set of fixed gate-to-source voltages, V_{gs1}, V_{gs2}, and V_{gs3}, in Fig. 5.1c, and against V_{gs} for a fixed V_{ds} value in Fig. 5.1d. The voltages and currents in Fig. 5.1c, d are normalized to the maximum V_{ds} and V_{gs} ($=V_{DD}$) values and the corresponding maximum current, I_{on}, respectively.

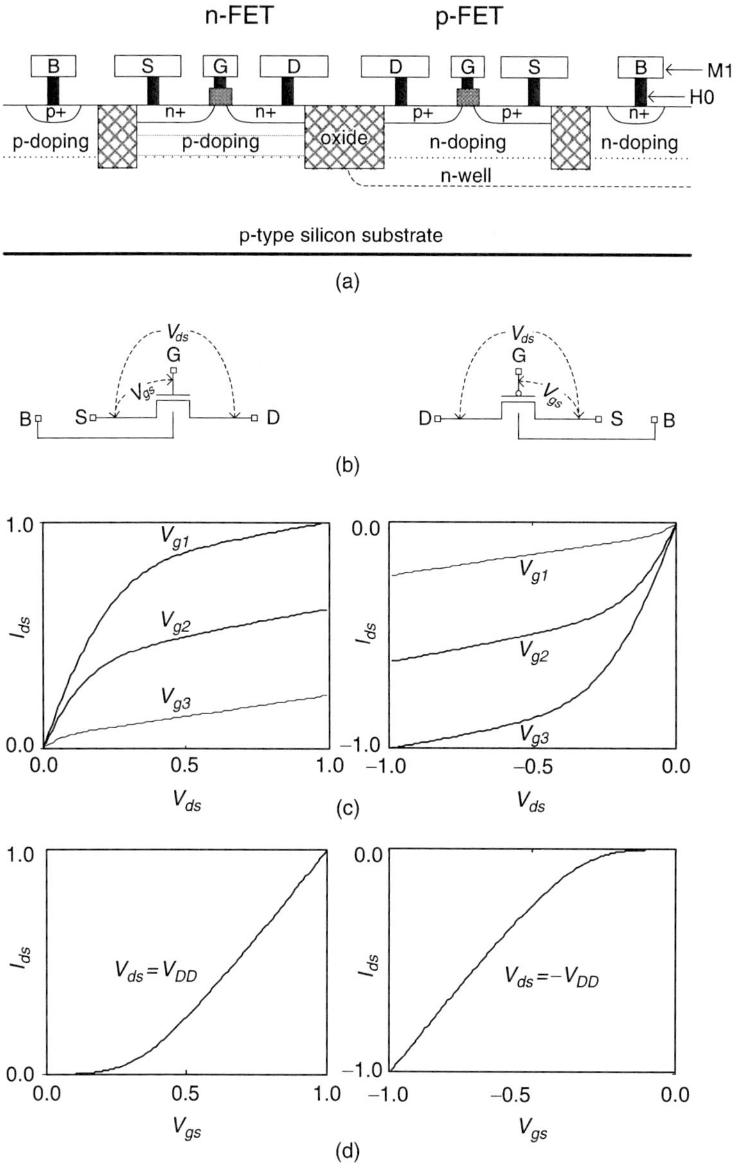

Fig. 5.1 An n-FET and a p-FET: **a** schematic cross sections, **b** symbols, **c** $I_{ds}-V_{ds}$ characteristics for different V_{gs} values, V_{gs1}, V_{gs2}, and V_{gs3}, and **d** I_{ds}–V_{gs} characteristics

The source supplies negatively charged electrons for current conduction through the channel of an n-FET and positively charged holes for current conduction through the channel of a p-FET. The gate-to source voltage V_{gs} and drain-to-source voltage V_{ds} as well as the drain-to-source current I_{ds} are of opposite signs for the n-FET and p-FET, as illustrated in Fig. 5.1c, d. For I–V characterization, the S terminal is typically held at GND for the n-FET and at V_{DD} for the p-FET. Since the external power supply voltages are measured with respect to GND, in test structures for n-FET and p-FET characterization, the polarity of voltage levels in the peripheral circuits needs to be carefully considered.

DC characterization of MOSFETs is an essential part of modeling CMOS circuit behavior. MOSFET characteristics are obtained by setting V_{gs} or V_{ds} at a fixed value and measuring I_{ds} while sweeping V_{ds} or V_{gs}, respectively, to obtain *I–V* curves such as those shown in Fig. 5.1c, d. The non-linear *I–V* behavior is separated into different operating regions as follows:

- subthreshold or "off" region,
- linear region,
- saturation or "on" region.

The threshold voltage of a MOSFET V_t is defined as the value of V_{gs}, for a particular fixed V_{ds}, at which the channel is turned "on" and I_{ds} meets a threshold criteria. In the subthreshold region, with $V_{gs} \le V_t$, I_{ds} varies exponentially with V_{gs}, and it is convenient to plot $\log_{10} I_{ds}$ as a function of V_{gs}. The slope of the $\log_{10} I_{ds}$ vs. V_{gs} plot, the subthreshold slope (SS) is defined as

$$\mathrm{SS} = \frac{\mathrm{d}V_{gs}}{\mathrm{d}\left(\log_{10} I_{ds}\right)}. \tag{5.1}$$

At 25°C, SS is in the range of 70–100 mV/decade. It varies with temperature and doping concentration in the channel.

In the linear region, I_{ds} of an ideal MOSFET is expressed as

$$I_{ds} = \beta\left[\left(V_{gs}-V_t\right)V_{ds} - \frac{V_{ds}^2}{2}\right],\ 0 < V_{ds} < \left(V_{gs}-V_t\right) \tag{5.2}$$

while in the saturation region

$$I_{ds} = \beta\frac{\left(V_{gs}-V_t\right)^n}{2},\ 0 < \left(V_{gs} - \mathrm{V}_t\right) < V_{ds}, \tag{5.3}$$

with $n = 2$ in long-channel MOSFETs. The expression in Eq. (5.3) is modified in short-channel MOSFETs due to velocity saturation. In an ideal MOSFET, β is a function of MOSFET channel length L_p, and width W, effective mobility μ_{eff}, gate oxide dielectric constant ε, and thickness t_{ox}:

$$\beta = \frac{\varepsilon\varepsilon_0\mu_{eff}}{t_{ox}}\frac{W}{Lp}. \tag{5.4}$$

The relationships expressed in the above equations are modified by several effects that come into play when MOSFET geometries are scaled to sub-μm dimensions.

Subthreshold characteristics are affected by gate-induced drain leakage (GIDL) and gate oxide tunneling current I_{gl}. Drain-induced barrier lowering (DIBL), velocity saturation, and other short- and long-channel effects modulate V_{t} and I_{ds}. The introduction of mobility enhancement by engineering strain in the MOSFET channel has resulted in the dependence of effective mobility on MOSFET physical layout and on the relative placement and dimensions of strain enhancement layers.

The effect of MOSFET geometry and physical layout coupled with different offerings for low power, high performance, maximum operating voltage, and an option of independently controlled substrate (well) bias for n-FETs and p-FETs has led to the need for characterizing many types of MOSFETs and capturing these effects in SPICE simulation models. On a routine basis, full I–V sweeps with a few hundred current measurements may be prohibitive from a test time perspective. Hence, the DC characterization of MOSFET is simplified by focusing on a few key locations in the I_{ds}–V_{ds} and I_{ds}–V_{gs} landscape.

The key parameters in the subthreshold region are I_{off}, V_{tlin}, and V_{tsat}, indicated by solid circles on the I_{ds}–V_{gs} plots for an n-FET in Fig. 5.2. With $V_{\mathrm{gs}} = 0.0$ V the n-FET is in the off-state, and its I_{ds} ($= I_{\mathrm{off}}$) gives a measure of the leakage power in CMOS circuits. The threshold voltage V_{t} is a measure of V_{gs} at which the MOSFET channel begins to conduct. It is a very useful parameter for device design and process optimization as well as for estimation of circuit leakage and signal propagation delays.

As the onset of channel conduction is not a step function, a number of different definitions of V_{t} can be found in the literature [1–4]. In routine MOSFET characterization, two of these definitions are commonly used. For the onset of conduction in the linear region, a full I–V sweep is made with V_{ds} of ~ 0.05 V. A voltage value V_{gse} is obtained by linear extrapolation of the I_{ds}–V_{gs} curve to $I_{\mathrm{ds}} = 0$, and V_{tlin} is given by

$$V_{\mathrm{tlin}} = V_{\mathrm{gse}} - \frac{V_{\mathrm{ds}}}{2}.$$

In the presence of parasitic series resistance and mobility degradation at high I_{ds}, the I_{ds}–V_{gs} curve may deviate from linearity. In this case, extrapolation is carried out from the slope in the I_{ds} vs. V_{gs} plot at the point of maximum transconductance g_{m}, as illustrated in Fig. 5.2a, where $g_{\mathrm{m}} \left(= \mathrm{d}I_{\mathrm{ds}}/\mathrm{d}V_{\mathrm{gs}}\right)$.

For the onset of the saturation region, V_{tsat} is obtained by extrapolating the $\sqrt{I_{\mathrm{ds}}}$–V_{gs} plot to $I_{\mathrm{ds}} = 0$ at $V_{\mathrm{ds}} = V_{\mathrm{DD}}$ to give $V_{\mathrm{tsat}} = V_{\mathrm{gse}}$.

A second method for determining V_{t}, the constant I_{ds} method, is indicated in Fig. 5.2b. Here V_{t} is defined as the V_{gs} to achieve a specified I_{ds} in the region of constant subthreshold slope. A standard definition for MOSFETs of different geometries is

$$I_{\mathrm{ds}} = I_{\mathrm{dsvt}} \frac{W}{L_{\mathrm{p}}}, \tag{5.5}$$

where I_{dsvt} is constant for a MOSFET type.

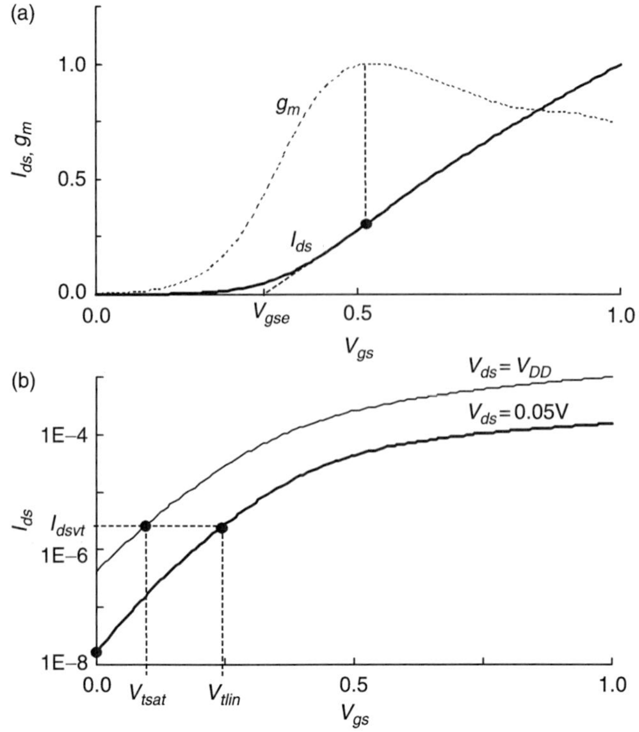

Fig. 5.2 An n-FET I_{ds} vs. V_{gs} plots indicating methods to determine V_{tlin} and V_{tsat}: **a** by linear extrapolation and **b** by constant I_{ds} current method

There may be small differences between the values of the threshold voltages in both the linear and saturation regions obtained by these two different methods. Because of somewhat arbitrary definitions of V_t, it is important to take the V_t measurement algorithm into account when comparing MOSFET parameter data from different sources.

In the linear and saturation regions, the locations of key measurement points on the I_{ds}–V_{ds} curves are indicated in Fig. 5.3. These are distributed in different areas of interest in the I_{ds}–V_{ds} space. The value of I_{dlin} $\left(V_{ds} = 0.05\text{ V}, V_{gs} = V_{DD}\right)$ is indicative of the parasitic resistance in series with the MOSFET channel resistance. The other current values, I_{hi}, I_{lo}, and I_{mid}, are indicative of MOSFET drive strengths during switching of CMOS circuits. Static CMOS gate delays are generally correlated with I_{on}. A strong correlation of logic gate delays is obtained with I_{eff} [5], where

$$I_{eff} = \frac{(I_{hi} + I_{lo})}{2}. \tag{5.6}$$

In thin gate oxides ($t_{ox} < 3$ nm), the gate-tunneling current I_{gl} becomes significant, increasing by a decade with a ~0.25 nm reduction in t_{ox}. It increases rapidly

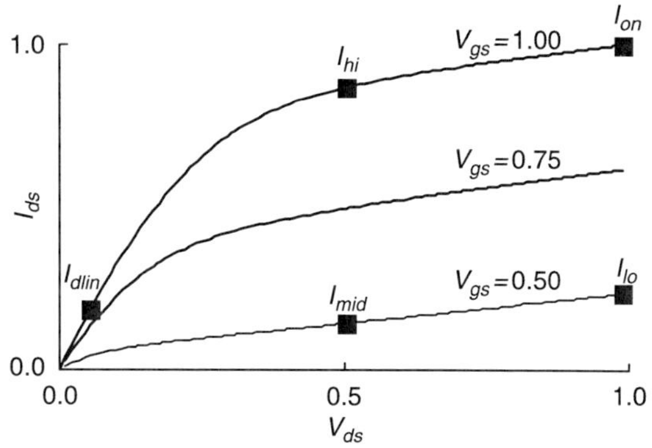

Fig. 5.3 I_{ds} measurement locations on an n-FET I_{ds}–V_{ds} plots; I_{mid} and I_{lo} are measured at $V_{gs} = 0.5$ and I_{dlin}, I_{hi}, and I_{on} at $V_{gs} = 1.0$

with increase in gate voltage but has weak temperature dependence. A maximum value of I_{gl} for an n-FET is obtained with S and D terminals at GND and the gate at V_{DD}. Its contribution to the total leakage current in the off-state of an n-FET $(V_{gs} = 0,\ V_{ds} = V_{DD})$ is through the drain overlap region. In very low-leakage MOSFETs or at very low temperatures, the gate-tunneling current in thin-oxide devices may dominate over the current through the channel in the off-state.

The measurement locations for the key MOSFET parameters in the subthreshold and saturation regions are summarized in Table 5.1. For convenience, the I_{ds} values are normalized to the width of the MOSFET ($= I_{ds}/W$) and have units of A/μm. A correction may be applied to the values of W and L of the MOSFETs, to capture any difference in the MOSFET dimensions as specified in the design data, to those obtained on a silicon wafer. These bias corrections are empirically determined from I_{ds} measurements on MOSFETs of varying W and L_p.

The effect of body bias on MOSFET characteristics is captured by measuring V_t as a function of V_{bs}. This relationship for an n-FET is shown in Fig. 5.4. There is an increase in V_t and a corresponding decrease in I_{off} as the body is reverse biased ($V_{bs} < 0$). This property may be exploited in reducing leakage power under

Table 5.1 Source, drain, and gate voltages for measuring n-FET parameters with the p-well tied to GND

Parameter	Source voltage	Drain voltage	Gate voltage
I_{off}	GND	V_{DD}	0
I_{dlin}	GND	0.05–0.10 V	V_{DD}
I_{mid}	GND	$V_{DD}/2$	$V_{DD}/2$
I_{lo}	GND	V_{DD}	$V_{DD}/2$
I_{hi}	GND	$V_{DD}/2$	V_{DD}
I_{on}	GND	V_{DD}	V_{DD}
I_{gl}	GND	GND	V_{DD}
V_{tlin}	GND	0.05–0.10 V	–
V_{tsat}	GND	V_{DD}	–

Fig. 5.4 An n-FET V_t as a function of body bias V_{bs}

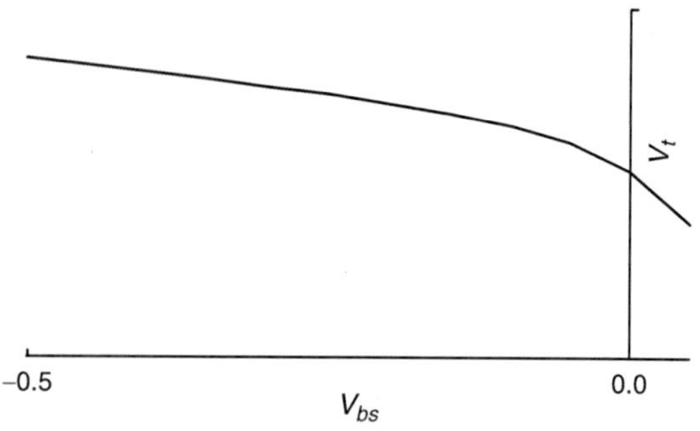

high-voltage stress conditions during reliability tests. The body-effect coefficient γ is a function of doping in the channel and t_{ox} and is defined as

$$\gamma = 1 + \frac{dV_t}{dV_{bs}}, \text{ at } V_{bs} = 0. \tag{5.7}$$

The V_{tsat} of MOSFETs is lowered when the channel length becomes comparable to the depletion width at high drain bias, primarily due to DIBL. This short-channel effect (SCE) is also referred to as V_t roll-off. Values of V_{tsat} and V_{tlin} are plotted as a function of L_p to track the roll-off signature. MOSFETs are engineered in an attempt to maintain a constant DIBL $(= V_{tlin}–V_{tsat})$ in the range of allowed channel lengths. In analog circuit applications, two parameters of interest are the output conductance g_{ds} $(= dI_{ds}/dV_{ds})$ and transconductance g_m $\left(= dI_{ds}/dV_{gs}\right)$. The definitions of parameters calculated from measured data for an n-FET are listed in Table 5.2.

As MOSFET characteristics vary with temperature, measurements are made over a wide range of temperatures (typically –40 to 140°C) for modeling CMOS circuit behavior. MOSFET channel leakage increases and thereby V_t is lowered as the temperature is increased. However, because of mobility degradation at higher temperatures, I_{ds} in the saturation region is lowered. This trend is dependent on the MOSFET design and may be reversed over some temperature range.

Table 5.2 Definitions of calculated electrical parameters of an n-FET

Parameter	Definition	Comments
I_{eff}	$(I_{hi} + I_{lo})/2$	Correlate to circuit delay
V_{tlin}	V_{gs} at $I_{ds} = 0$, $V_{ds} = 0.05 - 0.10$ V	Extrapolate I_{ds}–V_{gs} plot
V_{tsat}	V_{gs} at $I_{ds} = 0$, $V_{ds} = V_{DD}$	Extrapolate $\sqrt{I_{ds}}$–V_{gs} plot
SS	$dV_{gs}/d\left(\log_{10} I_{ds}\right)$	$0 < V_{gs} < V_t$
γ	$1 + dV_t/dV_{bs}$	$V_{bs} = 0$
DIBL	$(V_{tlin}–V_{sat})$	Varies with L_p (SCE)
g_m	dI_{ds}/dV_{gs}	Saturation region
g_{ds}	dI_{ds}/dV_{ds}	Saturation region

In typical circuit applications, MOSFETs are symmetric with respect to S and D terminals and these terminals can be interchanged. This feature provides flexibility in physical layouts and in I/O pad sharing in test structures. A higher drain current can be obtained by engineering the channel doping on the S side and in these asymmetric MOSFETs, the S and D terminals are uniquely defined, imposing constraints on I/O pad sharing as discussed further in Section 5.3.

When a MOSFET is in its on-state, its current drive may slowly degrade over time. There are two sources of this degradation. The hot-electron (Hot-e) effect is more pronounced at low temperatures under conditions of high gate and drain bias. In CMOS circuits, this condition is present only during switching. The positive or negative bias temperature instability (PBTI in n-FETs or NBTI in p-FETs) occurs at high V_{gs} and low V_{ds} and an increase in V_t is observed over time. In an inverter configuration, the n-FET undergoes PBTI stress when the input is a "1" and the p-FET undergoes NBTI stress with the input at "0." Characterization of these effects is carried out under accelerated stress conditions over many hours, typically at $1.5\times V_{DD}$ and temperatures in the range of 120–140°C. For such measurements, macros are generally packaged and placed in a temperature-controlled oven.

5.1.2 Systematic and Random Variations

The MOSFET I_{ds} values described by Eqs. (5.2), (5.3), and (5.4) scale linearly with its width W and inversely with its channel length L_p. Deviations from this ideal behavior occur as device dimensions are reduced to the sub-μm regime. The V_t at high V_{ds} values ($\sim V_{tsat}$) decreases as L_p is reduced and short-channel effects (SCEs) mentioned in the previous section, such as DIBL and velocity saturation, become significant. Linewidth variation in L_p across a wafer, influenced by local pattern density and other processing steps, results in variation in current drives of MOSFETs of identical design dimensions. Another source of variations in L_p is the linewidth variation within a MOSFET, especially near the edge of the DF island. In very narrow MOSFETs, the width may vary because of variations in the dimensions of the DF island itself.

Systematic variations in nominally identical MOSFETs arising from the variations in the optical mask used for photoresist exposure occur within a reticle field but have the same relative differences on different reticle fields. Local systematic variations are related to linewidth variations arising from process bias and local environment such as pattern density of PS and DF layers. These types of systematic variations may depend on the location of the reticle field on a wafer. As an example, the reticle fields located near the edges of a wafer may experience a larger variation because of a stronger sensitivity to lithography and etch bias.

Variations in the gate linewidths (PS level) in MOSFETs of the same total width are illustrated in Fig. 5.5. In Fig. 5.5a–c, three MOSFETs of the same total width but different number of PS fingers connected in parallel are shown. The inset in Fig. 5.5 shows channel length broadening from process bias at the DF edge. The

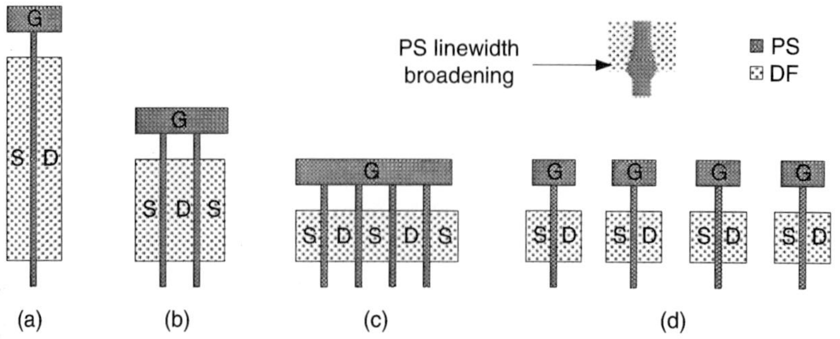

Fig. 5.5 Physical layouts of MOSFETs with same total width: **a** single PS finger, **b** two parallel PS fingers, **c** four parallel PS fingers, and **d** four isolated PS fingers. *Inset* shows channel widening at the DF edge

MOSFET with four PS fingers is more affected by a weaker section at the edge as it has eight such sections than the MOSFET with a single finger with only two such sections. Systematic bias in DF width would also cause the four PS finger MOSFET characteristics to be different from the single PS finger MOSFET. In Fig. 5.5d, each of the four fingers of the MOSFET in Fig. 5.5c can be measured independently and all fingers can also be measured in parallel.

With a large number of identical MOSFETs arranged as in Fig. 5.5d, the parameter mean and standard deviation of the population is obtained. Macro designs for variability characterization are more complex as data are collected on a large number of MOSFETs and trade-offs in measurement accuracy and test time may need to be considered. The basic concepts in statistics and the nature of MOSFET parameter distributions are covered in Chapter 10. Because of the non-linear nature of I–V characteristics in the subthreshold region, a geometric mean for V_t and I_{off} is used instead of an arithmetic mean as discussed in the next section.

In addition, random variations in properties of MOSFETs that are identical by design arise from statistical variations in the number of dopant atoms in the channel. As both L_p and W are reduced with CMOS scaling, the total number of dopant atoms is reduced and the statistical variations from random dopant fluctuation (RDF) become significant. The device-to-device variability is larger in the subthreshold region than in the saturation region where more charge carriers are present in the channel. A commonly used parameter to track random variability is the standard deviation of V_t, $\sigma(V_t)$, for nominally identical MOSFETs. The $\sigma(V_t)$ value is related to MOSFET properties as given in Eq. (5.8) below:

$$\sigma\,(V_t) = 3.2 \times 10^{-10} \frac{t_{eq} N_d}{\sqrt{W L_p}} \approx \frac{4\,mV}{\sqrt{W L_p}}, \tag{5.8}$$

where N_d is the dopant concentration and t_{eq} is the equivalent SiO_2 thickness of the gate dielectric, which may be SiO_2 or a high-k (HK) dielectric material.

For our standard MOSFET dimensions of $L_p = 0.04$ μm and $W = 1.0$ μm in Appendix A, $\sigma(V_t)$ from Eq. (5.8) is 0.020 V. This corresponds to an I_{off} variation of ~50% ($SS = 100$ mV/decade) and an I_{eff} variation of ~2%. In a narrow MOSFET ($W = 0.2$ μm) $\sigma(V_t)$ increases to 0.045 V and the variation in I_{off} is ~150% of the mean value. The effect is even larger for smaller MOSFET dimensions (~0.1 μm) in SRAM cells. Other sources of random variations include line edge roughness, non-uniformity in t_{eq}, and H0 via resistances. Their impact on variability is more pronounced in devices with short channels, narrow widths, and single H0 vias in S and D contacts with the M1 metal.

5.2 *I–V* Measurements

The standard procedure for measuring MOSFET DC parameters is to use the VFIM method, similar to measuring a resistance. However, because of the strong non-linear nature of the *I–V* characteristics, a number of measurements are required for full characterization. The sensitivities of I_{ds} on V_{ds} and V_{gs} vary widely in different regions of operation, setting different constraints on the accuracy of I_{ds} measurements. In addition, random variations in MOSFET characteristics may necessitate measuring a large number of DUTs of each type.

We first consider the measurement accuracy requirements in different operating regions of the MOSFET. In the subthreshold region, I_{ds} has a high sensitivity to a change in V_{gs}. Assuming SS = 100 mV/decade in Eq. (5.1), a 1 mV change in V_{gs} produces a 2.3% change in I_{ds} and a 10 mV change in V_{gs} changes I_{ds} by 25%. Correspondingly, to measure I_{off} with an accuracy of ±1%, any error in V_{gs} should be $\leqslant \pm 0.5$ mV. The lower limit on I_{ds} measurement error is set by the tester force voltage and current measure accuracies in the appropriate ranges.

In the linear region, I_{dlin} $(V_{gs} = V_{DD})$ varies linearly with V_{ds}. In a 1 μm wide MOSFET, with an I_{dlin} of 0.25 mA/μm ($V_{ds} = 0.05$ V), the resistance in the linear region is 200 Ω and a parasitic contact resistance of 2 Ω produces a 1% error. In a 10 μm wide MOSFET, with the same parasitic contact resistance, the error increases to 10%. In the saturation region, I_{ds} increases with V_{gs} but has a smaller sensitivity to V_{ds}. However, with higher currents flowing through the MOSFET, errors introduced by the voltage drops in parasitic series resistances of wires, vias, and probes may become significant.

The measurement of MOSFET *I–V* parameters is further improved with four-terminal measurements across the S and D terminals. In discrete element macros, four-terminal measurements are not area efficient because of I/O pad overhead. Mushroom pads described in Section 2.4.1 may be used for eliminating contact and probe resistance by landing two probes on the S and D pads. The I/O pad area, in this case, is increased at higher metal levels.

In 1D and 2D MOSFET array macros, four-terminal measurements can be used to correct for the *IR* drops across parasitic series resistances and area efficiency is achieved by pad sharing of force and sense terminals for the DUTs. For accurate I_{off} measurements, a correction is applied for the background leakage of MOSFETs in parallel with the active DUT. To improve area and test efficiency, the designs of such complex macros may be preferentially optimized for measurements in either the saturation region or the subthreshold region.

As discussed in Section 5.1.2, the random variations in parameters of nominally identical MOSFETs arise from RDF, line edge roughness, and variations in t_{ox} and H0 resistance. There are a number of different ways to collect MOSFET statistics. In one commonly used method, a matched pair of MOSFETs placed in close proximity is measured on a large number of chips. The parameter values of all the pairs are used to determine the standard deviation (sigma) and the mean is calculated as the average value of all the measurements. This method is both area and test time efficient (only two devices of each flavor per chip) but has the drawback of including systematic variations among chips and wafers in the mean value. In order to get more accurate values of local mean and sigma, a 1D array (Section 5.4.3) is a good choice as it gives a moderate sample size (~30) at the M1 test level.

If only mean parameter values are desired, measurements are made on MOSFETs wired in parallel as shown in Fig. 5.5c. The upper limit on the total width of the parallel combination is set by the *IR* drop in interconnect wires and accuracy requirements for I_{ds}. The I_{ds} in the linear and saturation regions is the arithmetic average of the individual MOSFETs. However, for a constant SS, the measured V_{t} of all MOSFETs in parallel using the "constant I_{ds}" method is smaller than the arithmetic average of the V_{t} of the individual MOSFETs.

Consider n MOSFETs with $W = L_{\mathrm{p}} = 1$. The V_{t} and I_{off} of the jth MOSFET are denoted by $V_{\mathrm{t}\text{-}j}$ and $I_{\mathrm{off}\text{-}j}$, respectively. Using Eq. (5.1) for the $(V_{\mathrm{gs}},\ I_{\mathrm{ds}})$ locations of $(0,\ I_{\mathrm{off}})$ and $(V_{\mathrm{t}},\ I_{\mathrm{dsvt}})$, with $\mathrm{d}V_{\mathrm{gs}} = V_{\mathrm{t}}$

$$V_{\mathrm{t}\text{-}j} = \mathrm{SS}\left[\log_{10}\left(\frac{I_{\mathrm{dsvt}}}{I_{\mathrm{off}\text{-}j}}\right)\right] \tag{5.9}$$

and the average threshold voltage $V_{\mathrm{t\_av}}$ of n individual measurements on MOSFETs at a fixed I_{dsvt} is

$$V_{\mathrm{t\_av}} = \frac{1}{n}\left(V_{\mathrm{t}-1} + V_{\mathrm{t}-2} + \cdots + V_{\mathrm{t}-n}\right)$$

or

$$V_{\mathrm{t\_av}} = \mathrm{SS}\left[\log_{10}\left(\frac{I_{\mathrm{dsvt}}}{\left(I_{\mathrm{off}\text{-}1}I_{\mathrm{off}\text{-}2}\cdots I_{\mathrm{off}\text{-}n}\right)^{\frac{1}{n}}}\right)\right]. \tag{5.10}$$

If these same n MOSFETs are connected in parallel, and a single V_{t} measurement, $V_{\mathrm{t\_all}}$ of the parallel combination is made:

$$V_{\mathrm{t\_all}} = \mathrm{SS}\left[\log_{10}\left(\frac{nI_{\mathrm{dsvt}}}{I_{\mathrm{off}\text{-}1} + I_{\mathrm{off}\text{-}2} + \cdots + I_{\mathrm{off}\text{-}n}}\right)\right]. \tag{5.11}$$

In Eq. (5.10), the expression $\left\{(I_{\text{off-1}}\ I_{\text{off-2}} \ldots I_{\text{off-}n})^{1/n}\right\}$ is the geometric mean of the off-currents and in Eq. (5.11), $\{(I_{\text{off-1}} + I_{\text{off-2}} \ldots + I_{\text{off-}n})\ /n\}$ is the arithmetic mean of the off-currents of all the MOSFETs in parallel. As (geometric mean) $\leq$ (arithmetic mean), the average of individually measured V_t of all MOSFETs, V_{t_av}, $\geq$ measured V_t of all MOSFETs in parallel, V_{t_all}.

In 1D and 2D MOSFET arrays, DUTs are connected in parallel and the unselected DUTs are biased to reduce the background leakage current. This techniques works well with a limited number of MOSFETs connected in parallel when making measurements in the subthreshold region (I_{off} and V_t). In the saturation region, the measured currents are several orders of magnitude higher than the background leakage currents, and a larger number of DUTs may be connected in parallel.

Pulsed *I–V* measurements are made to eliminate the effects of transient trapped charges in HK gate oxides and self-heating in MOSFETs built in PD-SOI technology. This technique is described in Chapter 8.

5.3 MOSFET DUT Designs

In a test structure for a discrete MOSFET with a single PS finger, or multiple PS fingers connected in parallel, the MOSFET may be placed in the space between two I/O pads connected to its S and D terminals as shown in Fig. 5.6. Two additional I/O pads are required for the G and B terminals. In this arrangement, with an independent control of the voltage at the B terminal, I_{ds}–V_{ds} and I_{ds}–V_{gs} characteristics may be obtained for different V_{bs} values. The S and D terminals may be interchanged, and measurements carried out for either symmetric or asymmetric MOSFETs. The

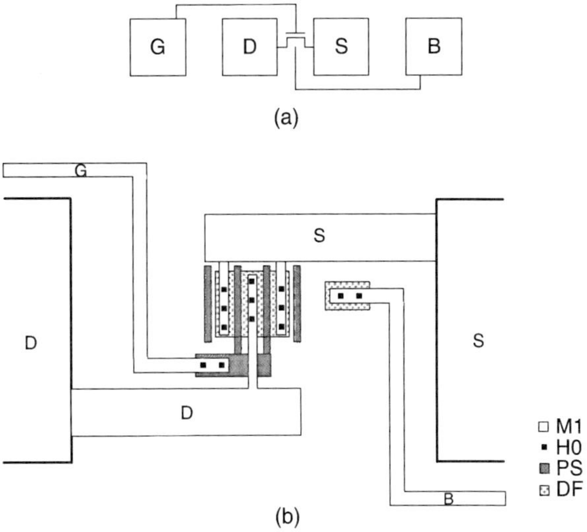

Fig. 5.6 An n-FET test structure: **a** schematic showing I/O pad connections for G, D, S, and B terminals, and **b** physical layout including metal wiring

I/O pad count per DUT is reduced to three if the B terminal is tied to S. In this configuration, *I–V* characterization is carried out only for symmetric MOSFETs at a $V_{bs} = 0$.

The parasitic series resistance of the S and D connections to the I/O pads is minimized with the use of redundant H0 vias and wide metal wires. However, in a two-terminal measurement, probe-to-pad contact resistances and probe series resistances are of concern. Consider an n-FET with $W_n = 10\ \mu m$ situated between two I/O pads. From Table A.1 in Appendix A, I_{on} is 10 mA at 1.0 V. A parasitic resistance of 2 Ω in series with both S and D terminals reduces V_{gs} by 20 mV and V_{ds} by 40 mV. Assuming I_{ds} in the saturation region to be varying linearly with V_{gs}, and nearly independent of V_{ds}, this results in at least a 2% error in I_{on} measurement. The error in measured I_{on} is reduced to ~0.2% for $W_n = 1\ \mu m$. A rule of thumb is to select the MOSFET width in a DUT to limit its I_{on} to ~1 mA. Alternatively, a four-terminal measurement may be made by either landing two probes each on the S and D pads as described in Section 2.4.1 or by assigning two additional pads to the S and D terminals. The measurement error in the subthreshold region for $W_n = 10\ \mu m$ is negligible (for $1\ \mu A < I_{ds} \lesssim 50\ \mu A$, *IR* drop in $V_{gs} < 0.2$ mV). Hence, wide MOSFETs may be used in macros limited to I_{off} and V_t characterization.

The currents flowing in the wires connecting the G and B terminals to their respective I/O pads are very small and these wires can be more resistive. With $W_n = 10\ \mu m$, and an I_{gl} of 1 nA/μm, the maximum current in the G wire is 10 nA. A series resistance of 5 kΩ in this wire creates a voltage drop of 0.05 mV and the impact on the measurement of I_{on} is <0.1%.

The properties of a MOSFET may depend on the physical layout and surrounding environment. For performance tuning and technology benchmarking, it is preferable to use layouts representative of circuit designs on the product. A MOSFET layout may be derived from a logic gate layout in a standard cell design library of a product. A template of a MOSFET DUT is then created by standardizing the spatial separation of the metal wires connected to the S and D terminals, preferably matching the power grid pitch on a product. As an example, the inverter circuit schematic and layout in Fig. 2.7 is modified to isolate the n-FET and the p-FET, as shown in Fig. 5.7a, b. The PS fingers of the n-FET and p-FET are physically separated and the S, D, and G terminals of the MOSFET not under test are electrically shorted. This layout preserves the proximity of the two MOSFET types found in logic gates and resulting influence of stress enhancement layers as well as any systematic linewidth variations in PS, DF, and H0 layers.

The surrounding PS and DF pattern density in a DUT layout can also be configured to match circuit blocks in a product or that of a performance benchmarking ring oscillator. Dummy PS fingers may be included on either side of the MOSFET. A filler cell is created by shorting the S, D, and G terminals of the n-FET and the p-FET in an inverter layout as shown in Fig. 5.7c. In Fig. 5.7d, the circuit schematic of a matrix of filler cells surrounding the MOSFET under test is shown. In this arrangement, the DUT is placed in a local environment similar to the environment in a product. It also corresponds to the physical layout of an inverter ring oscillator used for characterizing the AC behavior of MOSFETs and provides a means for

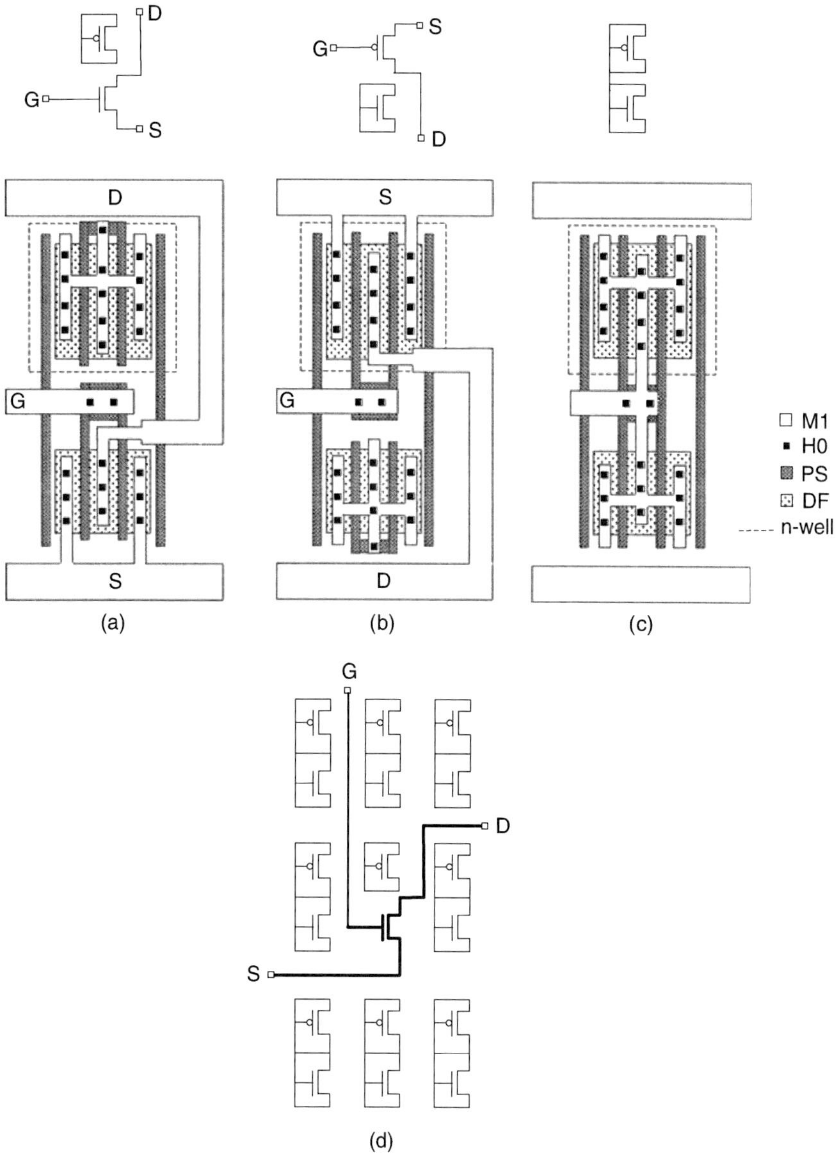

Fig. 5.7 Logic gate representative schematic and layout of **a** an n-FET, **b** a p-FET, and **c** a MOSFET filler cell. **d** Schematic of an n-FET surrounded by filler cells

correlating the DC and AC characteristics of DUTs situated in similar local environments. The DUT may be placed in the center or at the edge of a large array of filler cells and the properties of a set of such DUTs compared to study the impact of local environment. In the macro design discussion to follow, only the schematics of the MOSFET under test, excluding the filler cells, are shown for convenience.

The maximum PS finger width in a logic gate is specified to keep the RC time delay of the PS finger much smaller than its signal propagation delay. As an example, for a 1 μm PS finger width ($L_p = 0.04$ μm and $\rho_{sh} = 10\ \Omega/\square$), the series resistance of the PS finger contacted at one end is 250 Ω. Its contribution to the RC time constant for $C_g = 1.0$ fF/μm is ~0.12 ps, which is ~3% of the logic gate delay (FO = 1) of 4 ps. Hence, in typical logic gate designs used in CMOS products, the PS finger width is restricted to minimize this parasitic delay. Large MOSFET widths are accommodated by placing a number of PS fingers in parallel.

The DUTs for characterization of MOSFETs in an SRAM cell are designed to match the physical layout of the parent SRAM cell. Each SRAM cell has six or eight MOSFETs each (6 or 8T) and a number of different SRAM cell layouts of each type may be offered in a CMOS technology node. These SRAM cells differ in area and packing density. The MOSFET characteristics and the relative p-FET and n-FET widths, W_p and W_n, are tuned for optimizing noise margins. To achieve high memory densities, SRAM device widths are narrower than those used in logic gates. Custom optical proximity correction (OPC) algorithms are used to meet the lithography challenge with such small feature sizes. It is therefore important that physical layouts for characterization of these MOSFETs closely follow those of SRAM cells in a large array.

The circuit schematic of a 6T SRAM cell is shown in Fig. 5.8a. In Fig. 5.8b–d, one n-FET and one p-FET in the latch and one n-passtransistor are wired for I–V characterization, respectively. The terminals of the unused MOSFETs are shorted together. An SRAM filler cell is created by shorting all the terminals of the MOSFETs, in a similar fashion as a logic gate filler cell shown in Fig. 5.7c. The

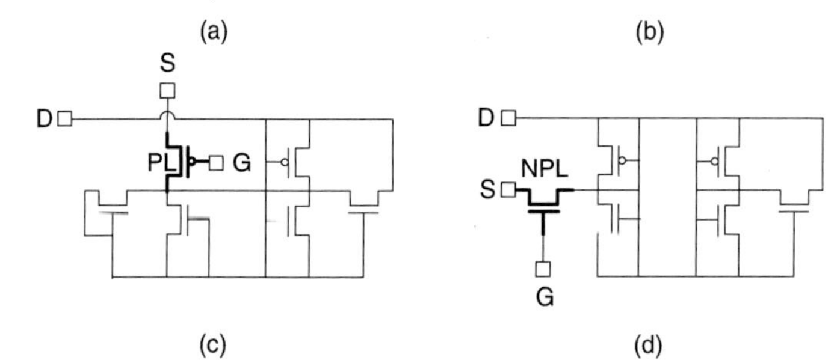

Fig. 5.8 Circuit schematic of **a** a 6T SRAM cell. DUT wiring for **b** an n-FET in the latch, **c** a p-FET in the latch, and **d** an n-FET passtransistor in the SRAM cell

filler cells surround the MOSFET under test to reproduce, as closely as possible, the local environment of an SRAM memory block. With this scheme, product design verification tools may be used without filtering errors originating from floating-gate terminals.

5.4 MOSFET Macro Designs

Test structure designs for MOSFET characterization range from isolated discrete devices to complex 2D arrays for statistical variability studies. The number of *I–V* measurements on each MOSFET is typically >5. The measurement range for current on a single MOSFET may span six or more decades. Parametric testers are preferred for accurate measurements, especially in the low current range (nA to fA). Measurement of V_t requires either a binary search algorithm or a partial *I–V* sweep with >20 measurements. Characterization of random variability is carried out on matched pairs of MOSFETs or on a group of 10 or more nominally identical MOSFETs. The total number of measurements being large, test time in manufacturing becomes an important factor in overall test structure efficiency.

Characterization of macros for MOSFET model build requires full flexibility in voltage bias controls and high accuracy ($< \pm 1\%$) in all current ranges. Typically, *I–V* sweeps (50–200 measurements) covering the full range of bias voltages are made on each MOSFET. As these measurements are carried out on limited hardware, test time for model build is generally not a major concern.

In this section, five different types of macro designs for measuring DC characteristics of MOSFETs are described. The macros with discrete MOSFET DUTs and 1D addressable arrays in Examples 1, 2, and 3 are testable at the M1 metal level. Complex 2D arrays in Examples 4 and 5 require several metal layers. Probe cards are often customized for such very large array designs. In 1D and 2D arrays, the number of DUTs sharing I/O pads is limited by the leakage currents of unselected DUTs and peripheral circuits, and by the *IR* drops in wires when making measurements in the saturation region. It is sometimes convenient to have different 2D array design strategies for measurements of leakage currents and V_t than for *I–V* measurements in the saturation region. The macro designs described here are for n-FETs. Design modifications and test requirements for p-FET DUTs are included wherever applicable.

5.4.1 Example 1: Discrete MOSFET Macros

Macros for discrete MOSFETs are basic designs, useful for detailed characterization at the M1 metal level. The MOSFETs are placed either in the space between two I/O pads or adjacent to the pads. A number of different schemes for I/O pad assignments and pad sharing to improve area efficiency are illustrated in Fig. 5.9 with 12 I/O pads. Two different n-FET configurations are shown, with either the B terminal tied

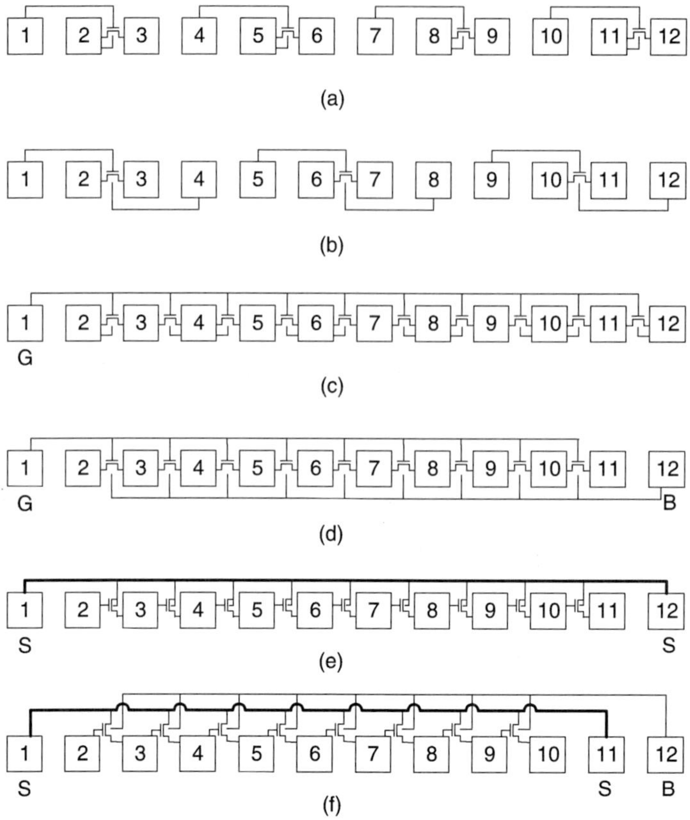

Fig. 5.9 I/O pad assignments for n-FETs: **a** isolated with B tied to S, **b** isolated with independent B terminals, **c** with common G terminal, and B tied to S, **d** common G and common B terminals, **e** common S terminal with B tied to S, and **f** common S and common B terminals

to S or having independent B and S pads. When the B terminal is connected to an I/O pad, the n-FET has four terminals, and its V_{bs} can be independently varied. These configurations along with the total number of n-FETs in a 1×25 padset macro (Appendix A) are listed in Table 5.3.

Table 5.3 Total number of DUTs (n-FETs or p FETs) in a 1×25 padset macro for the configurations shown in Fig. 5.9

Macro	Total DUTs	B terminal	Common terminals
a	8	Tied to S	None
b	6	Isolated	None
c	23	Tied to S	G
d	22	Shared	G, B
e	22	Tied to S	S
f	21	Shared	S, B

In Fig. 5.9a, b, unique I/O pads are assigned to each of the three and four terminals of the n-FETs, respectively. This arrangement has the highest degree of flexibility with independent bias voltage controls for each n-FET terminal. Measurements can be made serially, or in parallel to reduce test time. Four-terminal measurements can be made with mushroom pads as described in Section 2.4.1.

Pad sharing for three- and four-terminal n-FETs is introduced in Fig. 5.9c, d with the use of a common G pad. The current flowing in the wire connecting the G terminals of the DUT to pad 1 (common G pad) is the sum of the gate leakage currents of all the n-FETs. As this current is in the pA to nA range, the wire resistance can be >10 kΩ, while maintaining $\lesssim$1 mV voltage shift in V_{gs}. One limitation of this design is that the same gate voltage gets applied to all the n-FETs. Hence, for MOSFETs with different oxide thicknesses and application voltages, sharing a common G terminal may not be appropriate.

In the arrangement shown in Fig. 5.9e, f, S terminals share two common I/O pads and the G terminal of each n-FET has an independent I/O pad. In this configuration, a different value of V_{gs} may be applied to each n-FET. The S wire travelling across the macro should have a resistance low enough to limit the parasitic voltage drop to <10 mV for I_{on} measurements. In an M1 testable macro design, two (or more) I/O pads may be used to reduce the S wire resistance. With shared B and S terminals, the same V_{bs} value appears across all the n-FETs.

In symmetric MOSFETs, S and D terminals are interchangeable and either common G or common S configurations may be used. A common G configuration with a common B connection, as shown in Fig. 5.9d, is more suitable for characterization of any asymmetric effects in nominally symmetric MOSFETs, as S and D terminals can be exchanged in test.

Macros designs shown in Fig. 5.9 are applicable to p-FETs as well. In case of isolated DUTs, a macro may contain both n-FETs and p-FETs. With shared I/Os and a parallel test approach, it is preferable to have either all n-FETs or all p-FETs in a single macro, although any combination of the number of n-FETs and p-FETs may be accommodated in a macro by isolating each MOSFET group.

The I/O pad assignments to enable parallel measurements are shown in Fig. 5.10. In Fig. 5.10a, all MOSFETs are of the same type (n-FETs or p-FETs) and share I/O pads for S and G terminals. The maximum number of MOSFET tests in parallel is limited by the parasitic series resistance of the S wire. Typically 2–8 DUTs may

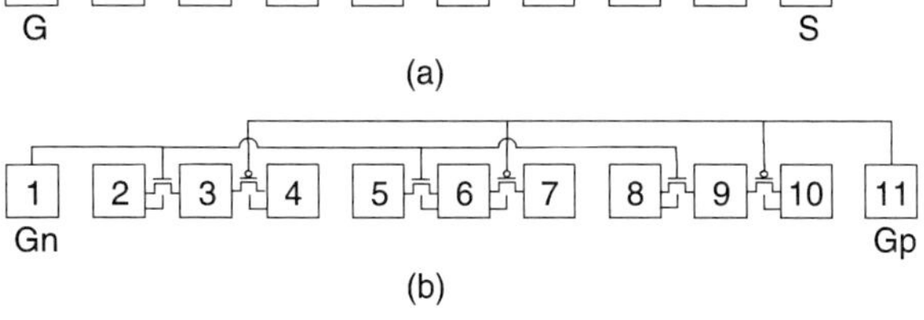

Fig. 5.10 I/O pad assignments for parallel measurements of **a** all n-FETs having common G and S terminals, and **b** one n-FET or one p-FET in each pair

be placed in one group sharing S terminals. In Fig. 5.10b, an n-FET and a p-FET share a common S or D terminal. The G terminals of all the n-FETs and p-FETs are connected to I/O pads, Gn and Gp, respectively. Parallel measurements are carried out in two passes, once for measuring all n-FETs with the S, D, and G terminals of p-FETs tied together and next for all p-FETs with the S, D, and G terminals of the n-FETs tied together.

5.4.2 Example 2: Multiple DUT Unit (md-unit) MOSFET Macros

In the second example, a pad sharing scheme to increase the number of DUTs in a minimum size macro is described. A number of DUTs (either all n-FETs or all p-FETs) are placed between two I/O pads and their S and D terminals are connected in parallel to form an md-unit (multiple DUT unit). The G terminal of each MOSFET in the md-unit is connected to an independent I/O pad which is shared with a corresponding G terminal in all other md-units in the macro. An md-unit with 10 n-FETs is shown in Fig. 5.11, and a section of a macro is shown in Fig. 5.12. This design may be implemented at the M1 metal level.

Measurements are made on one n-FET DUT in each of the md-units simultaneously, while all the other DUTs are biased in the off-state. Hence within an md-unit, there is a background current equal to the sum of the I_{off} values of all the DUTs

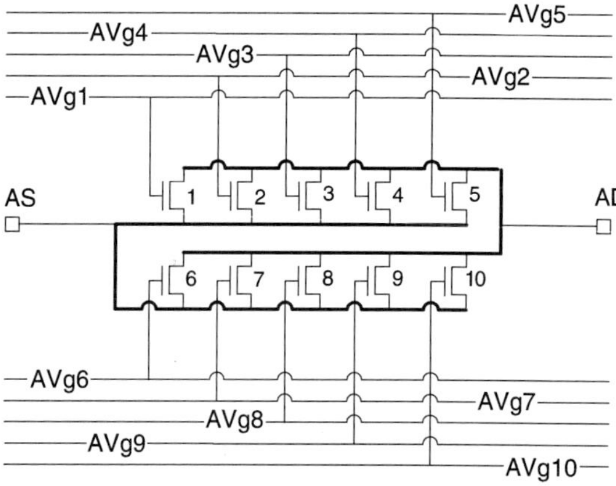

Fig. 5.11 Circuit schematic of an md-unit with S and D terminals of 10 n-FETs connected in parallel and G terminals connected to independent I/O pads

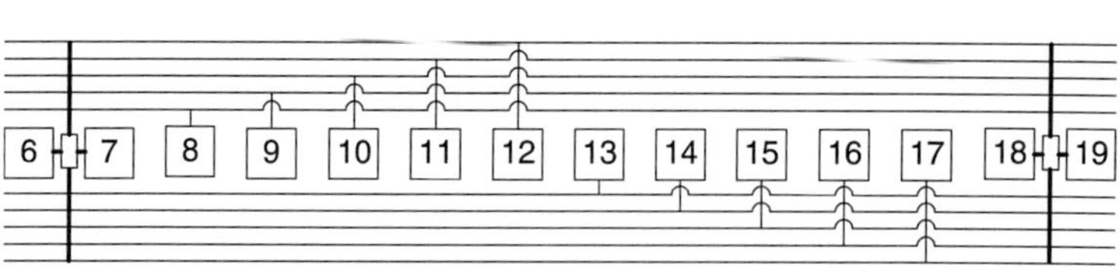

Fig. 5.12 A section of a macro with two md-units. Additional md-units, each requiring two I/O pads, are added to the left and right sides

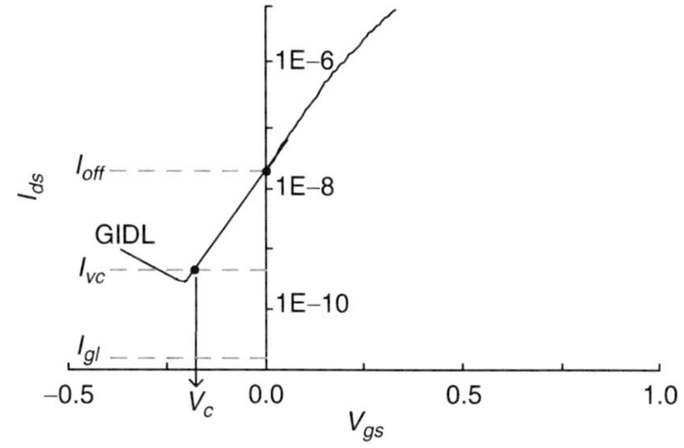

Fig. 5.13 $I_{ds} - I_{gs}$ plot for an n-FET extended to negative V_{gs} region. I_{vc} is the current at a clamp voltage, V_C (= −0.2 V). The GIDL contribution to I_{ds} is negligible for $V_{gs} > V_c$ and I_{gl} is assumed to be << I_{off}

in the off-state. This background current is negligible for I_{ds} measurements in the saturation region but compromises the measurement of I_{off} of an individual DUT.

The current contribution of the devices in the off-state is reduced by applying a negative gate bias V_C. Here we assume that the I_{gl} and GIDL contributions to I_{ds} are negligible. This is illustrated in Fig. 5.13 with an example I_{ds}–V_{gs} plot for an n-FET, extended into the negative V_{gs} region. With increasingly negative values of V_{gs}, I_{ds} first decreases following the subthreshold slope and then starts to increase as the GIDL contribution begins to dominate. The change in slope of the I_{ds}–V_{gs} plot in the negative V_{gs} region is dependent on the relative contributions of S–D diffusion current and GIDL. If the gates of the n-FETs are set at a negative clamp voltage, with $V_{gs} = V_C = -0.2$ V for the case shown in Fig. 5.13, the I_{ds} values are lowered by a factor of ~100 below I_{off} (SS = 100 mV/decade). With 10 nominally identical n-FETs in parallel, the background current would then be 9% of the actual I_{off} of the DUT. The measured I_{off} of the DUT can be corrected by subtracting this amount. In practice, all DUTs in an md-unit are not identical, either by design or because of random variations, and the error in measured I_{off} will be different for each DUT.

The error in I_{ds} in the subthreshold region $(V_{gs} \geq 0)$ is substantially reduced by making two additional measurements, together with the use of a correction algorithm [6]. Let us consider the case with N DUTs in an md-unit. The G terminals of all DUTs in an md-unit are first biased at V_C and next at a desired test V_G (such as $V_{gs} = 0$), and the respective currents $I_{\text{all-VC}}$ and $I_{\text{all-VG}}$ are measured. We define a ratio, χ, as follows:

$$\chi = \frac{I_{1\text{-VC}} + I_{2\text{-VC}} + \cdots + I_{N\text{-VC}}}{I_{1\text{-VG}} + I_{2\text{-VG}} + \cdots + I_{N\text{-VG}}} = \frac{I_{\text{all-VC}}}{I_{\text{all-VG}}}. \tag{5.12}$$

Here χ gives a measure of the average SS of all DUTs in an md-unit. Assuming the SS to be the same for all DUTs, the I_{ds} for DUT of index K at a clamp voltage of V_C is related to its I_{ds} at a bias V_G by the expression

$$I_{K\text{-VC}} = \chi I_{K\text{-VG}}. \tag{5.13}$$

The measured I_{ds} of the Kth DUT, at a bias of V_G ($I_{K\text{-VG-meas}}$), is the sum of its true I_{ds} at a bias of V_G ($I_{K\text{-VG}}$) and the I_{ds} values of the remaining $(N-1)$ DUTs at a bias of V_C:

$$I_{K\text{-VG - meas}} = I_{K\text{-VG}} + (I_{\text{all-VC}} - I_{K\text{-VC}})$$

and rearranged using Eq. (5.12) to give the corrected value:

$$I_{K\text{-VG}} = \frac{1}{(1-\chi)}\,(I_{K\text{-VG - meas}} - I_{\text{all-VC}})\,. \tag{5.14}$$

With this correction algorithm for I_{ds} in the subthreshold region, accurate I_{ds}–V_{gs} measurements can be made over the entire $V_{gs} > 0$ region. This correction scheme continues to work for DUTs of different types within an md-unit, provided all I_{ds}–V_{gs} characteristics have approximately the same SS values.

In the md-unit macro designs, the number of DUTs in a macro is equal to the number of DUTs per md-unit times the number of md-units. The DUTs may be tested serially or in parallel. In case of parallel tests, voltage biases are applied to the S and D terminals of all the md-units and the desired G bias to one element in each unit, while all other DUTs are biased at V_C. Hence, the number of DUTs tested in parallel is equal to the number of md-units in a macro and the number of sets of such parallel measurements is equal to the number of independent G pads. The number of DUTs in a macro can be further increased by sharing the S terminal between two adjacent md-units. In this approach, the number of S and D pads per md-unit is decreased from 2 to 1.5.

The total number of DUTs in each md-unit can be optimized for area and test time efficiency. This is graphically illustrated in Fig. 5.14 by plotting the number of

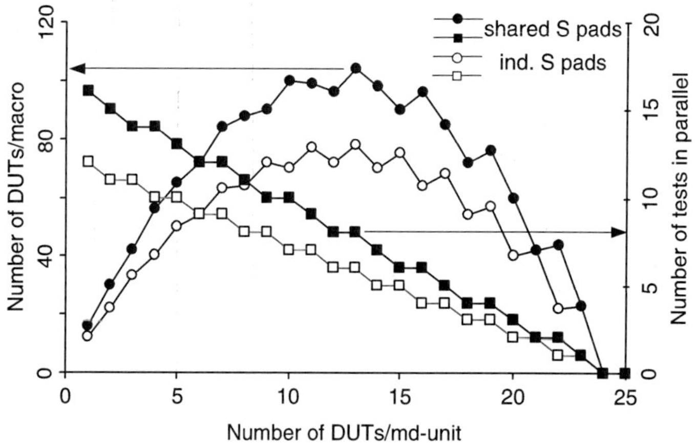

Fig. 5.14 Number of DUTs/macro and number of tests in parallel as a function of number of DUTs in each md-unit

Table 5.4 Number of DUTs (n-FETs or p-FETs) in a standard 1×25 padset macro with discrete DUTs, and with md-units

DUT type	DUT unit	Comments	Number of DUTs
Single DUT	Isolated	B tied to S	8
Single DUT	Common G	B tied to S	23
md-unit	Isolated md-unit	13 DUTs/md-unit	78
md-unit	Shared S	13 DUTs/md-unit	104

DUTs per macro and the number of DUTs tested in parallel as a function of number of DUTs in an md-unit for a standard 1×25 padset macro. Maximum area efficiency is achieved with 13 DUTs per md-unit and a total of 8 md-units which can be tested in parallel.

The number of DUTs for the type of macro designs in Examples 1 and 2 is listed in Table 5.4. The md-unit is clearly more area efficient, but a background correction must be applied in the vicinity of I_{off} for subthreshold characterization.

5.4.3 Example 3: 1D Addressable MOSFET Array Macros

In this third example, the design of a 1D addressable MOSFET array is described [6]. The area efficiency of the macro is improved over that in Example 2 by adding additional circuitry. A decoder is used to select a DUT from a linear array of MOSFET DUT elements, with S and D terminals connected in parallel, as in the previous example. As many as 30 DUTs may be accommodated between two pads in a physical layout implemented at the M1 metal level. This design is similar to the 1D array macro for resistors covered in Example 3 of Chapter 3.

An array unit comprises an N-bit decoder, voltage steering circuitry, and 30 n-FET or p-FET DUTs. There are eight array units in a macro. The number of DUTs is (2^N –2), the remaining two of the decoder outputs are used to turn all the DUTs off or on, a useful capability for applying I_{off} correction, as discussed in Example 2. All array units within a macro share decoder input signals and can be tested in parallel. The decoder implementation allows the number of DUTs in a standard 1×25 padset macro to be increased by a factor of ~2.5× over the md-unit design in Example 2, while maintaining testability at the M1 metal level.

The basic idea of a 1D array unit is depicted in Fig. 5.15 for a 4-bit decoder, 14 n-FET DUT design. The voltage bias pads for V_{gs} and clamp voltage V_{c} are AVg and AVc, respectively. The decoder output signal, CAD, serves as the input signal to a voltage steering circuit, shown in Fig. 5.15a. A symbol for this voltage steering circuit is shown in Fig. 5.15b. The output of the voltage steering circuit is connected to the gate of an n-FET DUT in the array. Its output voltage is equal to V_{gs} if CAD = "1" and V_{C} if CAD = "0." Therefore, the DUT selected by the decoder for I_{ds} measurement gets a gate bias equal to V_{gs} and all other DUTs have their gates clamped at a gate voltage of V_{C}. The voltage steering circuit uses three

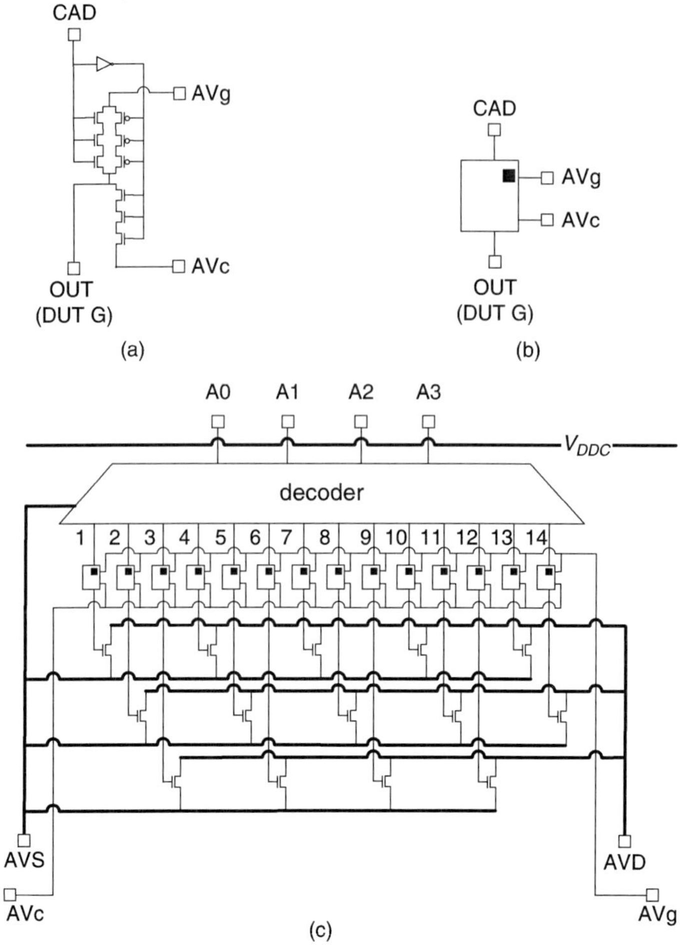

Fig. 5.15 Voltage steering circuit: **a** schematic and **b** symbol. **c** Configuration of a 1D n-FET array unit, with a 4-bit decoder, between I/O pads AVS and AVD

n-FET passtransistors in series to reduce the leakage current and thereby to keep the *IR* drop in the V_G voltage bus to 1 mV. Alternatively, thick-oxide or lower leakage n-FET offerings in the technology of interest may be used.

In Fig. 5.15c, the placement of an array unit between two I/O pads is illustrated for a 4-bit decoder and 14 DUTs. The DUTs are physically arranged in three rows and five columns for this electrical 1D linear array, although any suitable arrangement can be made based on the area of the DUTs and space between the pads. The S and D terminals of all n-FETs are connected to the closest I/O pads. Interdigitated M1 metal combs, emanating from the I/O pads, are used to minimize the *IR* drop in these wires drawing current of the order of ∼1 mA. The I/O pads for DC input signals AVg and AVc are further away, as these wires carry very small currents (∼ nA range or less), and can tolerate higher resistances. An upper limit on the resistance of the AVg and AVc wires is set to limit the worst case *IR* drop to <1 mV, which corresponds to a change of <2% in I_{off}.

The decoder is powered by V_{DDC} and its GND terminal is shared with the source I/O pad (AVS) of the DUTs. The decoder draws a significant current (~mA) during switching, which does not interfere with DUT measurements made after the decoder has switched. Its leakage current, flowing through the AVS pad, also does not impact I_{ds} measurements made on the drain terminal of the DUTs. The V_{DDC} power supply is shared by all the decoders within a macro and any transient *IR* drop within the power supply line is not critical, as long as decoder functionality is maintained.

The circuit schematic of a 4-bit decoder for an n-FET array is shown in Fig. 5.16a to illustrate the decoder implementation. The structure of the schematic, which is replicated in the layout, has been compressed in the vertical direction so that all of the active devices fit in just four rows. This decoder comprises inverter and NAND logic gates, with vertical wires on the PS level for underpassing M1 wires travelling in the horizontal direction. Arranged in this fashion, a 5-bit decoder with ~100 logic gates can be wired at the M1 metal level. The decoder input bits "0000" and "1111" are used for selecting none ("0") or all ("1") of the DUTs for the background current subtraction algorithm. Hence the 4-bit decoder can select 14 unique DUTs. Similarly, a 5-bit decoder is wired to select 30 unique DUTs. For 45 nm technology and beyond, a 5-bit decoder can be accommodated in the 40 μm space between the I/O pads.

The physical layout of a section of the decoder (shaded area in Fig. 5.16a) is shown in Fig. 5.16b. This section has four NAND2 gates with single PS finger n-FETs and p-FETs. The input and output signal wires of the NAND2 gates are drawn primarily on the PS level with the data flow predominately along the direction of the PS rather than orthogonal to it, as is usually the case. The PS signal wires travel under the horizontal M1 V_{DDC} and GND busses. These busses are split with signal wires on minimum width M1 sections situated in the gaps. Additional short sections of PS are used for connections between M1 wires within the gaps as needed. The resistance of the PS wiring segments in the design is of the order of a few hundred to a thousand Ohms or more. This increases the switching delays of the logic gates, which is of little consequence in this application. The type of layout used here is especially useful for compact, custom, delay-insensitive designs where the direction of travel of PS lines is restricted by technology GRs.

Area efficiency of the macro is optimized by sharing the AVS pad between two adjacent array units. A section of the macro layout with this common AVS feature is shown in Fig. 5.17. Four pairs of array units can be accommodated in a standard 1×25 padset macro. The I/O pad assignments for this configuration are shown in Fig. 5.18. Decoder inputs A0–A4 serve all array units and there are two pads for the common V_{DDC} power supply, one at each end of the macro. I/O pad 15 is assigned for measuring the drain voltage of array unit 5 to validate the design. If desired, this pad and one of the V_{DDC} pads may be used for making contact with the n-well and the p-well instead.

A macro template is created using a hierarchical design methodology. Once the template design and the test are validated, many copies of the macro may be created

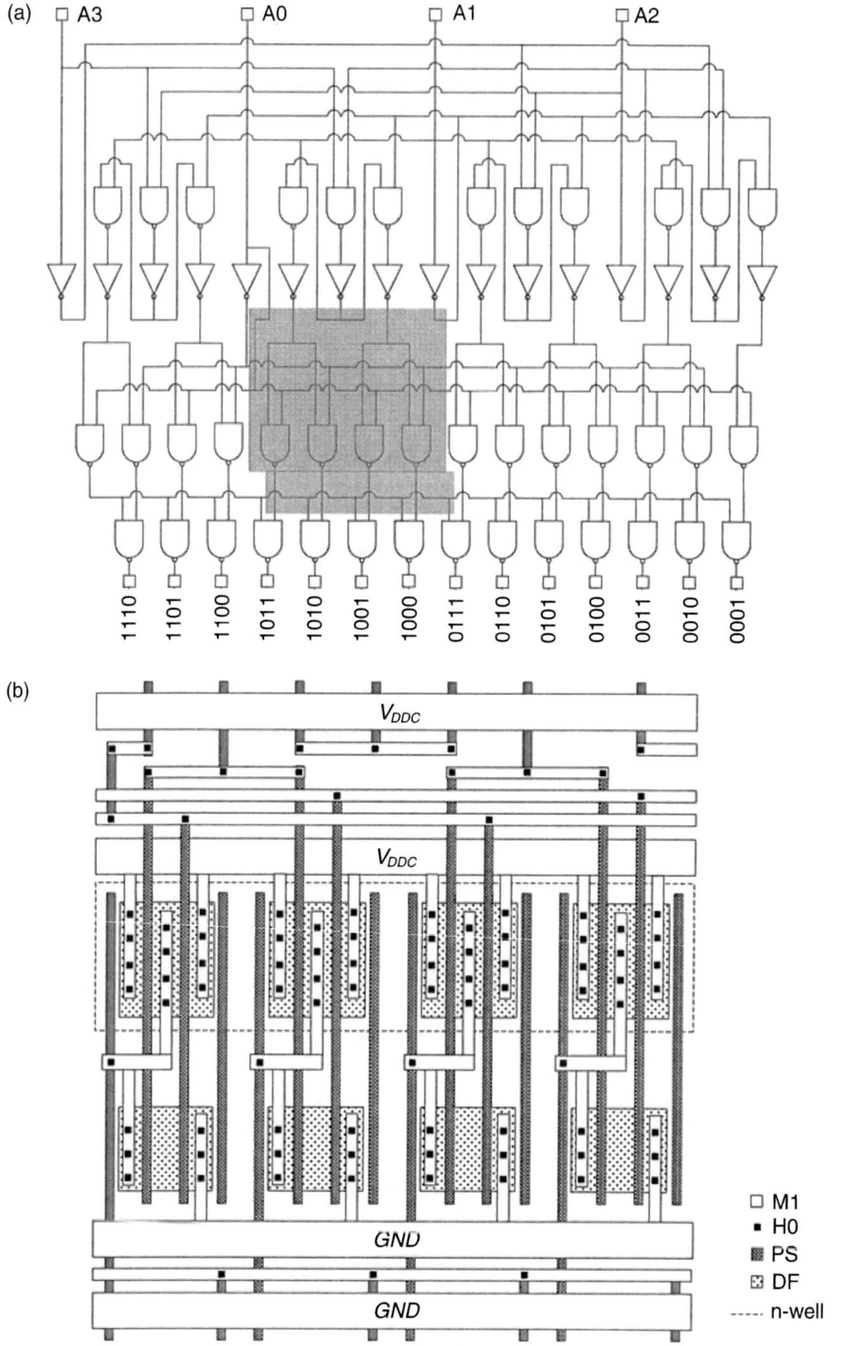

Fig. 5.16 **a** Circuit schematic of a 4-bit decoder with 14 outputs. **b** Physical layout of the *shaded region* of the decoder circuit in **a**

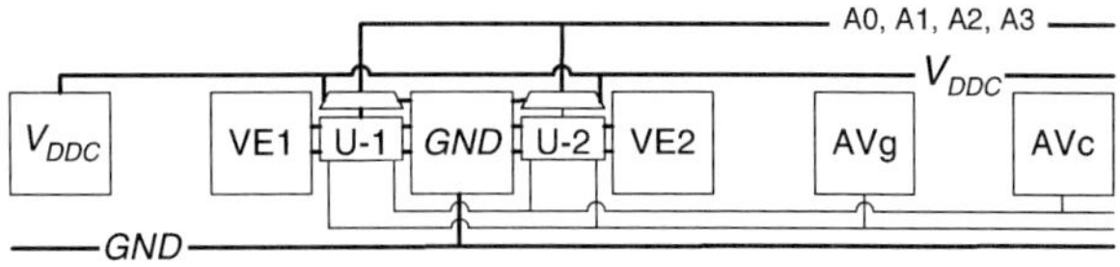

Fig. 5.17 Physical layout schematic of a section of a macro with two 1D array units sharing a GND I/O pad

1	2	3	4	5	6	7	8	9	10	11	12	13	14	15	16	17	18	19	20	21	22	23	24	25
V	A	G	A	A	A	A	G	A	A	A	A	A	A	Z	A	G	A	A	A	A	G	A	V	W
D	V	N	V	V	V	V	N	V	0	1	2	3	4	V	V	N	V	V	V	V	N	V	D	C
D	E	D	E	g	c	E	D	E						5	E	D	E	c	g	E	D	E	D	
C	1		2	L	L	3		4						m	5		6	R	R	7		8	C	

Fig. 5.18 I/O pad assignments for a 1D MOSFET array macro with eight array units using a 5-bit decoder and 30 MOSFETs each

and populated with different DUTs, simply by substituting the DUT blocks in the array units. If each array unit carries nominally identical DUTs, the mean and the standard deviation of their parameters provide a measure of random and systematic variability.

Measurements on n-FETs follow the standard convention of V_{DD} and GND. For a p-FET array macro, the "V_{DDC}" of the decoder and the S terminals of p-FETs are held at GND potential and the "GND" and D terminals of the p-FETs are held at $-V_{DDC}$. The decoder inputs follow the standard convention, now with the high level at "0" and low level at "–1."

A number of checks can be made to validate the macro design. The IDDQ of the decoder is measured and compared with the model to detect any shorted paths causing the decoder to malfunction. The array units in the macro template, when populated with DUTs of different dimensions, are used for validating decoder functionality, and output wiring and the test code. With nominally identical DUTs in an array unit, any systematic differences in measured parameters across an array unit may be detected.

As described here, an array unit macro with a 5-bit decoder can accommodate 240 DUTs in a 1×25 padset macro. This is an $\sim2.5\times$ improvement in area efficiency over the md-unit design in Example 2. Both the array units and md-units can be tested in parallel with digital ATE. With a 1×50 padset, the number of DUTs in the array unit design is 600 (20 array units) compared with 437 for a macro with 19 md-units. However, in very early technology development, because of the complexity in the decoder design and layout, an array unit-based design may have a lower yield than an md-unit-based macro design.

5.4.4 Example 4: 2D MOSFET Array Macros

An alternative to placing a large number of 1D array macros, for characterization of random and systematic variations in MOSFETs in a technology node, is to use 2D array test structures. The area efficiency in 2D arrays can be considerably improved over 1D arrays. Generally, four or more metal levels are required to wire high-density complex logic circuitry for addressing the DUTs and multiplexing the output voltages and currents. Such designs are more suitable for implementation in a mature technology with high circuit yield. Measurement of large 2D arrays is carried out using digital ATE or custom-designed test boards engineered to improve overall test efficiency.

With an increase in the number of DUTs in a 2D array, leakage current contribution from the unselected DUTs and voltage drops in interconnecting wires become significant. These must be accounted for in accurate characterization of MOSFETs over their full range of operation. Low-leakage isolation switches and control circuitry to operate the switches are added to correct for background leakage current and *IR* drops in interconnects. Array design can be simplified if characterization is targeted only in the subthreshold region (I_{off} and V_t) or in the saturation region.

With the availability of multiple metal layers, a number of design options become feasible. Column and row decoders may be used to select a DUT in the array as discussed in Section 2.5.4. This arrangement for an 8×8 array is shown in Fig. 5.19a. In large arrays with tens of thousands of DUTs, it is more efficient to use scan chains for DUT selection and steering input and output voltages. There are a number of published reports on 2D MOSFET array test structures. A few representative articles [7–10] are cited at the end of this chapter.

Here, we give a brief description of a macro design for full *I–V* characterization. This design has been implemented with 10 levels of metal in the 65 nm technology node with 96,000 DUTs in a 1000×96 matrix [7]. The area efficiency (number of DUTs per unit area) of the macro for narrow width MOSFETs is 200× greater than that of the 1D array in Example 3. Measurement accuracy is improved over the full range of *I–V* characterization by forcing and measuring gate and drain voltages at both ends of the selected column, limiting the parasitic *IR* drop to $\lesssim$1 mV. Current is measured on the source side of the selected MOSFET. The source voltage is measured at both ends of the selected row to calibrate the *IR* drop in the row interconnect wires. Measurements are made in serial fashion by selecting one DUT at a time. Level-sensitive scan chains are arranged surrounding the DUT matrix to address the columns and rows in this fashion as shown in Fig. 5.19b. The rectangular form factor of the macro is suitable for placement in a scribe line.

A schematic of a 3×3 sub-section of the 2D array design is shown in Fig. 5.20 [7]. For clarity, column and row control circuitry and switches are shown only at one end. The DUTs (n-FETs) in a column share the G and the D terminals, and the S terminals are shared with DUTs in the same row. A control switch to steer force or clamp voltage to the DUT terminal and the output voltage or current (sense or measure) to the measurement port is shown in Fig. 5.20a and its symbol in Fig. 5.20b. Thick-oxide, low-leakage MOSFETs are used in the transmission gates in the switch to reduce background leakage. The inverter in the voltage steering circuit can be eliminated if both the true and complementary signals to the switches are supplied by the control circuitry.

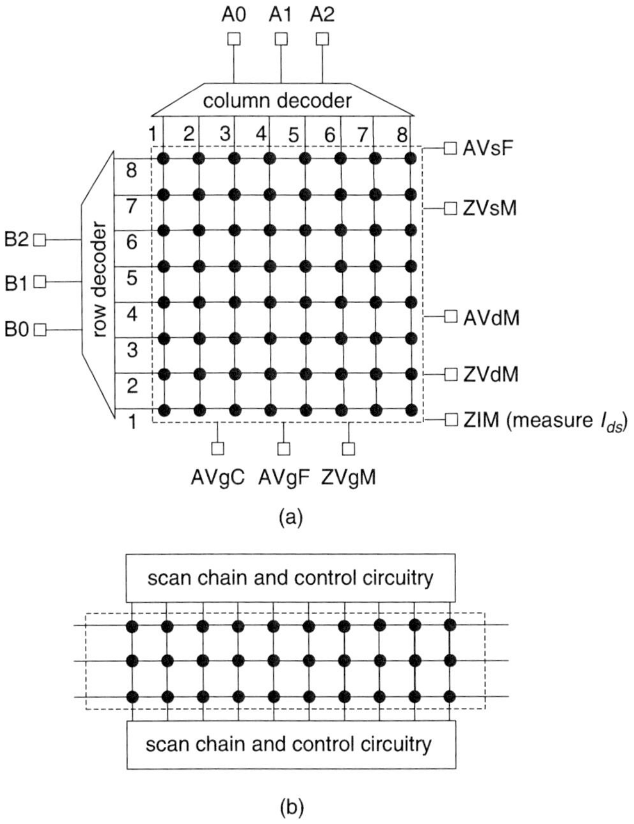

Fig. 5.19 Schematics of 2D MOSFET arrays: **a** with row and column decoder addressing scheme, and **b** with column addressing and control circuitry using scan chains

The column address bit, CAD (= "1"), steers a force voltage applied at the AVF node of the selected column. The CAD bit is set to a "0" in the unselected columns, steering a clamp voltage applied to the AVC node to the DUTs. The clamp voltage V_c is set at a negative V_{gs} value to reduce the background leakage current contribution of the unselected DUTs, as described in Example 3. A second column address bit, CSN, steers the applied voltage to the external measurement equipment. The number of DUTs in a column is chosen to limit the error in V_{gs} and V_{ds} measurements to $\lesssim$1 mV. The design is validated by measuring the gate and drain voltages at both ends of a column to obtain an upper bound on the *IR* drop in a column.

The DUT current is measured on the source side and all DUTs in a row share the S terminal connection. The RAD switches for selecting the row and steering the voltages and currents to the S terminal are constructed from n-passgates. Leakage currents of the unselected rows are steered to the sink node, ZIsS. In Fig. 5.20c, the source voltage is measured on the right side of the selected row, while the current

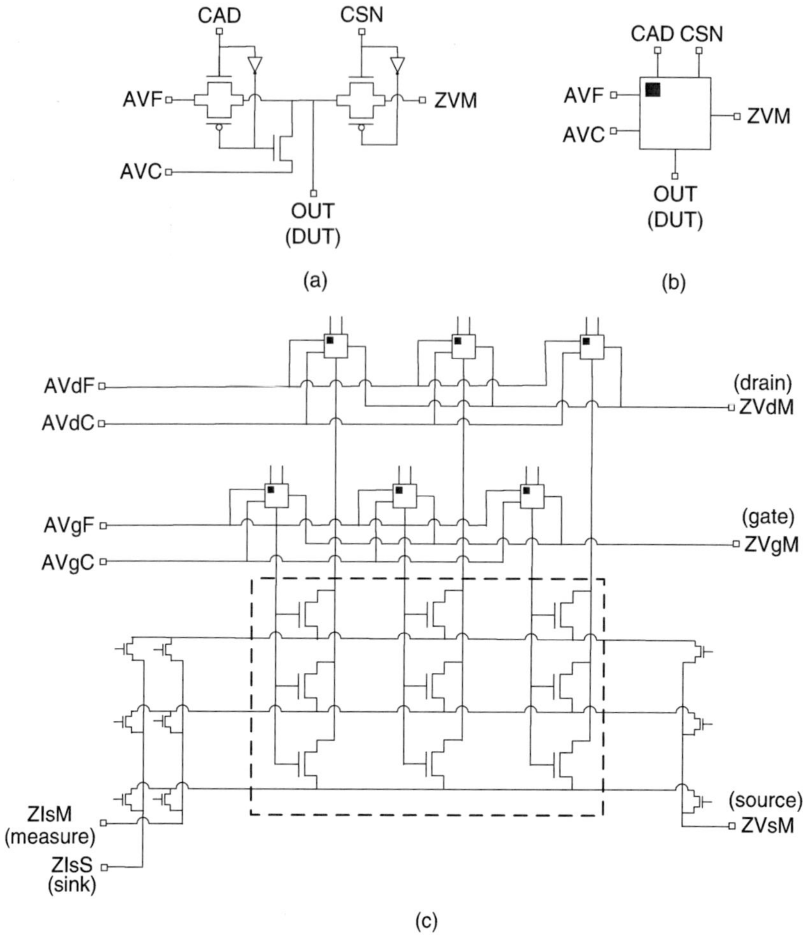

Fig. 5.20 A switch to steer gate and drain voltages in a column: **a** circuit schematic and **b** symbol. **c** Circuit schematic of a 3×3 2D n-FET array

measurement is carried out on the left side of the row. This provides a three-terminal measurement, eliminating error due to *IR* drop in the row interconnect wires. As the bias voltages on the terminals of different MOSFETs being compared may not be identical, postprocessing of the data is required to extract standard MOSFET parameters. Macro design validation, data acquisition, and analysis of all the MOSFETs in a large 2D array can be time consuming.

With shared S, D, and G inputs for all the MOSFETs, this 2D array macro can be accommodated in the standard 1×25 padset for placement in the scribe line. Typically such large arrays are populated with a variety of MOSFET types and dimensions. It is often convenient to create a few DUT templates of different dimensions and assign a specific DUT template to an entire row or a column.

5.4.5 Example 5: 2D Array Macros for Rapid V_t Measurements

A significant reduction in the test time for measurements of statistical variation in V_t values of a large number of MOSFETs can be obtained with the addition of some special features to a standard 2D array [11]. Instead of performing an *I–V* sweep or a binary search, the V_t of each MOSFET is determined by directly measuring V_{gs} at a fixed I_{ds}, controlled by an on-chip operational amplifier (op-amp). The decoder inputs are set by a periodically driven on-chip counter to sequentially select the DUTs. An output voltage within a known constant voltage of V_{gs} (= V_t) for all MOSFETs is read by a digital voltmeter with built-in trigger and statistical functions. The time to measure the V_t value of a MOSFET in the array is $T_c/2$, where T_c is the internal clock period.

Schematics of circuits that use an op-amp to measure V_t are shown in Fig. 5.21a, b for an n-FET and a p-FET, respectively. The voltage at the drain of the MOSFET is set to a value V_f. The V_t at $V_{ds} = V_f$ is defined as the V_{gs} value at a fixed I_{ds}, as given by Eq. (5.5). This current is forced through a precision resistor R_L connected in series with the MOSFET. The op-amp acts to maintain the voltage at its lower input, the same as that at its upper input which has an externally applied value of V_{set}. The op-amp feedback loop thereby adjusts V_{out} such that the current I_{ds} through the MOSFET and the resistor R_L is set to

$$I_{ds} = I_{dsvt}\frac{W}{L_p} = \frac{V_{set}}{R_L}.$$

In the case of the n-FET, V_t is then given by (V_{out}–V_{set}) and is measured by connecting the V_{out} pad directly to a digital voltmeter. An analogous arrangement holds for the p-FET.

Figure 5.22 shows a 3×3 section of an n-FET array, configured for measuring V_t with the op-amp scheme. The current forced through R_L is steered through the selected n-FET. For a small array, up to ~30×30, the complexity needs to be no

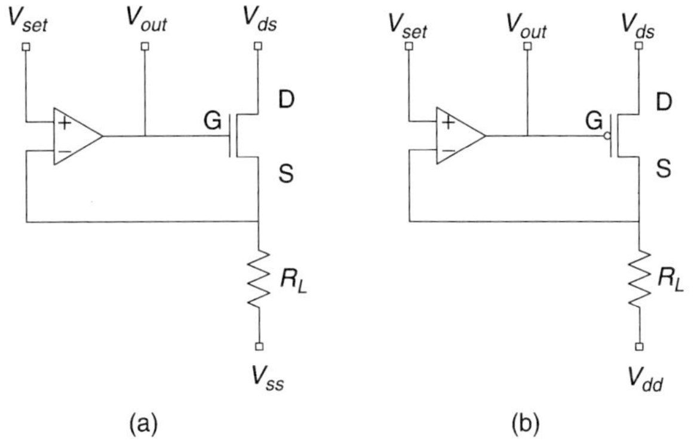

Fig. 5.21 Circuit schematic for V_t measurement at a constant current for **a** an n-FET and **b** a p-FET. Reproduced from [11], with permission, © 2008 IEEE

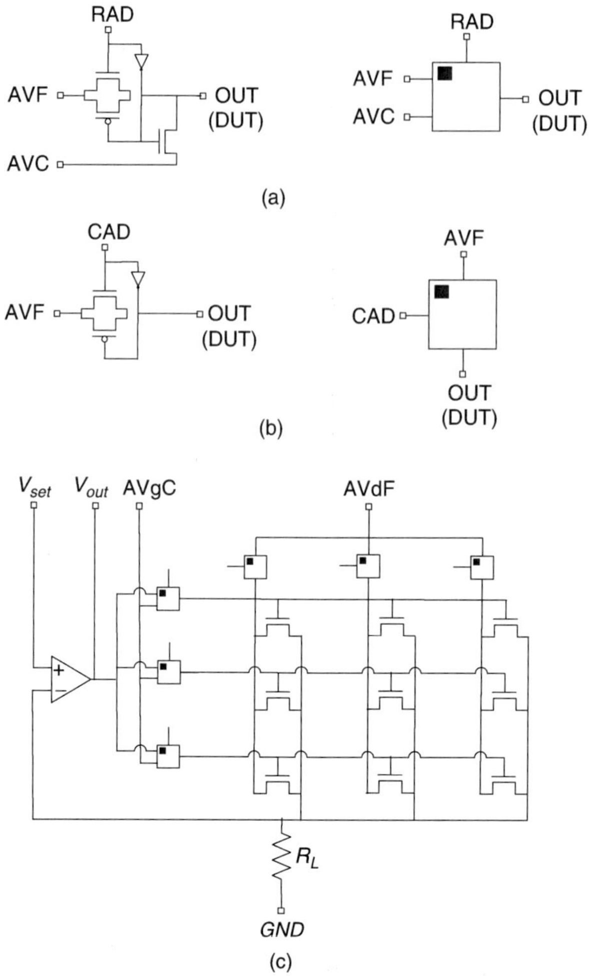

Fig. 5.22 Circuit schematic and symbol of **a** a row selection switch and **b** a column selection switch. **c** Circuit schematic of a 3×3 array of n-FETs with the op-amp

more than shown in the Fig. 5.22, with gates appropriately sized so that the voltage drop V_f–V_{set} appears primarily across the addressed n-FET, while most of the current flowing through R_L is also passing through that n-FET. Larger arrays can be similarly implemented by incorporating some of the features described in Example 4 to correct for series *IR* voltage drops and parallel current leakage paths.

The schematic of the 2D array, along with the circuitry to sequentially select the DUTs in the array and direct their V_{out} values to a common pad OUT, is shown in Fig. 5.23. An on-chip ring oscillator (RO) is enabled with a DC input EBL. The frequency of the clock signal generated by the RO is reduced to be in the 30–60 Hz range and fed to a digital counter. The output of the counter provides the

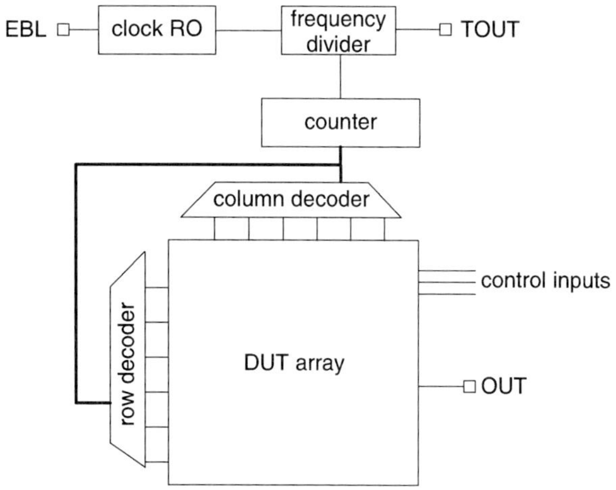

Fig. 5.23 Schematic of a 2D array with DC I/Os for collecting V_t statistics

input bits to the column and row decoders. A TOUT signal from the clock RO triggers the voltmeter, synchronizing the voltage measurements with the DUT selection sequence.

The DUTs are selected sequentially, each for a time corresponding to half the clock period. The voltage at the OUT node follows the V_t of each DUT. The output voltage recorded by the voltmeter is then a series of voltage steps. The mean and standard deviation for the array of a block of DUTs can be computed to get the V_t statistics. This approach is closely related to a ring oscillator macro design described in Section 6.2.4, with a similar DUT selection scheme and statistical analysis strategy.

References

1. Taur Y, Ning TH (2009) Fundamentals of modern VLSI devices, 2nd edn. Cambridge University Press, New York, NY
2. Sze SM (2002) Semiconductor devices: physics and technology, 2nd edn. Wiley, New York, NY
3. Schroder DK (2006) Semiconductor material and device characterization, 3rd edn. Wiley, Hoboken, NJ
4. Ortiz-Conde A, Sanchez FJG, Liou JJ, Cerdeira A, Estrada M, Yue Y (2002) A review of recent MOSFET threshold voltage extraction methods. Microelectron Reliab 42:583–596
5. Na MH, Nowak EJ, Haensch W, Cai J (2002) The effective drive current in CMOS inverters. International electron devices meeting digest, IEDM'02, pp 121–124
6. Ketchen MB, Bhushan M, Costrini G (2009) Addressable arrays implemented with one metal level for MOSFET and resistor variability characterization. Proceedings of the 2009 IEEE international conference on microelectronic test structures, 2009, pp 13–18
7. Agarwal K, Liu F, McDowell C, Nassif S, Nowka K, Palmer M et al (2006) A test structure for characterizing local device mismatches. Symposium on VLSI circuits digest of technical papers, 2006, pp 82–83

8. Agarwal K, Nassif S, Liu F, Hayes J, Nowka K (2007) Rapid characterization of threshold voltage fluctuation in MOS devices. Proceedings of the 2007 IEEE international conference on microelectronic test structures, 2007, pp 74–77
9. Drego N, Chandrakasan A, Boning D (2007) A test-structure to efficiently study threshold-voltage variation in large MOSFET arrays. Proceedings of the 8th international symposium on quality electronic design (ISQED'07), 2007, pp 281–286
10. Smith B, Arriordaz A, Kolagunta V, Schmidt J, Shroff M (2009) A novel biasing technique for addressable parametric arrays. IEEE Trans Semicond Manuf 22:134–145
11. Ji B, Pearson DJ, Lauer I, Stellari F, Frank DJ, Chang L et al (2008) Operational amplifier based test structure for transistor threshold voltage variation. Proceedings of the 2008 IEEE international conference on microelectronic test structures, Edinburgh, 2008, pp 3–7

Chapter 6
Ring Oscillators

Contents

Measurement of signal propagation delays is an important part of AC characterization of CMOS circuits. Ring oscillator based test structures are well suited for this application. The frequency of oscillation of a ring oscillator comprising a closed loop of inverting logic gates or circuit blocks is measured to obtain the signal propagation delay around the loop. If all the circuit blocks are truly identical, the delay of a single circuit block can be determined with an accuracy of <1 ps. Immunity

M. Bhushan, M.B. Ketchen, *Microelectronic Test Structures for CMOS Technology*,
DOI 10.1007/978-1-4419-9377-9_6,

of ring oscillator frequency to circuits external to the closed loop and the relative ease with which frequency measurements can be made have led to an extensive use of ring oscillators for technology and circuit characterization. The use of ring oscillator based test structures is further extended for tracking key MOSFET and parasitic parameters, enabling both AC and DC characterization of the same circuit elements. The currents drawn by a ring oscillator in its active and quiescent states, together with its frequency, provide useful information for establishing power and performance trade-offs in a CMOS product.

In Section 6.1, basic operation of a ring oscillator RO and extraction of delay parameters from frequency and power measurements are described. Examples of RO test structures, from a single stand-alone RO to complex 2D array macros are covered in Section 6.2. In Section 6.3, extraction of MOSFET and parasitic element parameters from RO measurements is described. In Section 6.4, schemes for precise measurement of circuit delays and RO designs for SRAM and analog circuits are discussed. Integration of ROs in a CMOS product as performance sort ring oscillators, PSROs, is covered in Section 6.5. A methodology of model-to-hardware correlation for ROs is discussed in Section 6.6.

There are several textbooks on CMOS circuits covering aspects of design, physical layout, and test [1–3]. For an in-depth treatment of CMOS switching characteristics and the relationship of signal propagation delay to MOSFET and parasitic parameters, readers may refer to Taur and Ning [4]. Selected publications on ring oscillator test structures are cited throughout this chapter.

6.1 Measurement of Time Delay

Measurement of the passage of time has fascinated human beings since ancient times. The diurnal motion and associated periodic phenomenon in nature provided a way to measure time intervals between events. The smallest unit of time, the second, was defined as the fraction 1/86,400 of the average length of a day. Mechanical clocks using the periodic motion of a pendulum or a balance spring to count time in units of time period of oscillation, calibrated in seconds, began to appear in medieval times. The accuracy of time measurement was further improved in the early part of the 20th century with the introduction of clocks exploiting the natural frequency of oscillations of a quartz crystal. The standard of time, with an accuracy of a few fs (10^{-15} s), is now maintained with atomic clocks, based on the resonant frequencies of atomic transitions. In all these time measurement schemes, the frequency of stable oscillations of a physical system is utilized.

Typical signal propagation delays in CMOS circuits are in the range of a few ps to a few ns. The rise and fall times of these signals in some cases may be considerably longer than the circuit switching delay. As an example, an unloaded inverter with a delay of 4 ps may have signal rise and fall times of 10 ps. A delay measurement accuracy of $\lesssim$1% requires a time resolution of the order of ~0.01 ps. Such measurements can be made with relative ease by devising a periodic structure comprising

CMOS circuits, a ring oscillator, whose time period of oscillation is a multiple of the propagation delay through a single circuit. With a large, precisely known division factor, the frequency of oscillation can be in the MHz range (time period in μs), while the circuit delay being measured may be only a few ps.

6.1.1 Ring Oscillator Operation

A ring oscillator is formed by connecting the output of a chain of inverting CMOS logic gates or circuit blocks to its input, thus closing the loop. For sustained oscillations at the fundamental frequency, the number of inverting circuit blocks in the loop must be odd. Let us consider the operation of an RO comprising five identical inverter stages as shown in Fig. 6.1a. The voltage at each of the nodes a1 to a5 varies with time in a periodic fashion, as illustrated in Fig. 6.1b. This oscillating behavior with a of period T_p is explained with the help of a timing diagram in Fig. 6.2. The voltage at each of the five nodes is plotted as a function of time, assuming equal pull-up (PU) and pull-down (PD) delays ($=T_p/10$) for all inverters. We can observe

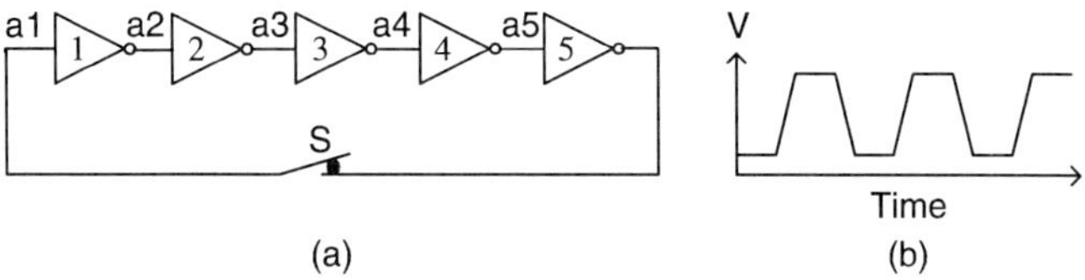

Fig. 6.1 **a** Circuit schematic of an RO comprising five identical inverters. **b** Voltage waveform at any one of the nodes a1 to a5 as a function of time

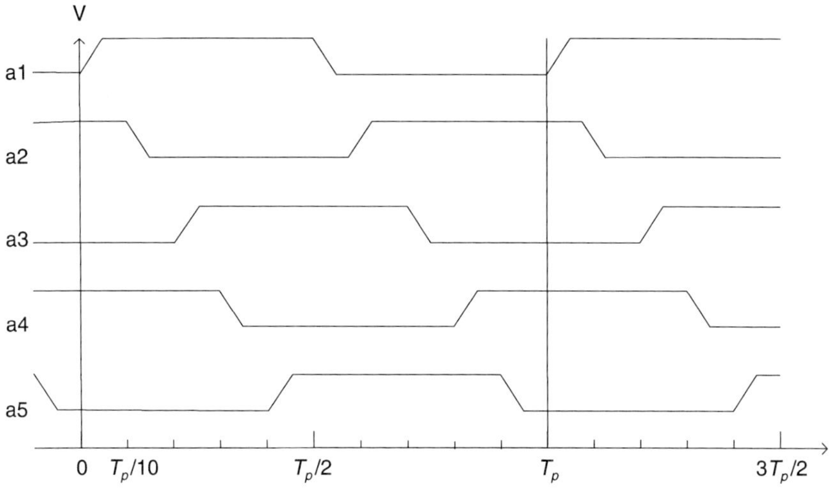

Fig. 6.2 Voltage levels at nodes a1 to a5 as a function of T_p for the RO circuit shown in Fig. 6.1a

the propagation of first rising edge at node a1, arbitrarily set at time = 0. It arrives as a falling edge at time = $T_p/10$ at a2 and propagates as a rising edge at a3, a falling edge at a4, a rising edge at a5, and as a falling edge again at a1. A cycle is completed when the second rising edge appears at node a1 at time T_p and repeated hence forth.

The RO period T_p, which is the inverse of the frequency of oscillation f is the sum of the PU delays τ_{pu} and PD delays τ_{pd} of all the inverters in the RO. If the RO has $2\alpha + 1$ inverters, where α is an integer, T_p increases with increasing α, with a corresponding decrease in f. The average of PU and PD delays per stage τ_p is given by

$$\tau_p = \frac{(\tau_{pu} + \tau_{pd})}{2} = \frac{T_p}{2(2\alpha + 1)} = \frac{1}{2(2\alpha + 1)f}. \tag{6.1}$$

In addition to RO frequency, power levels in the oscillating (active) and standby (quiescent) states are also measured. The switching power of an RO is given by

$$P_{sw} = \frac{1}{2} C_{sw} V_{DD}^2 \{2(2\alpha + 1)\} f = \frac{1}{2\tau_p} C_{sw} V_{DD}^2, \tag{6.2}$$

where C_{sw} is the average switching capacitance of a single inverter stage for PU or PD transitions, and there are $\{2\,(2\alpha + 1)\}f$ transitions per second. From Eq. (6.2), it is apparent that P_{sw} is independent of α, or the number of stages in the RO. Both τ_p and C_{sw} vary with V_{DD}.

The circuit schematic in Fig. 6.3 shows the MOSFETs in an RO comprising five inverters. The standby or leakage power of the RO in its quiescent state is measured by opening the loop and thereby suspending the oscillations and holding a1 at V_{DD} or GND. In the absence of gate oxide leakage currents, the leakage power P_{off} with node a1 held at V_{DD} is given by

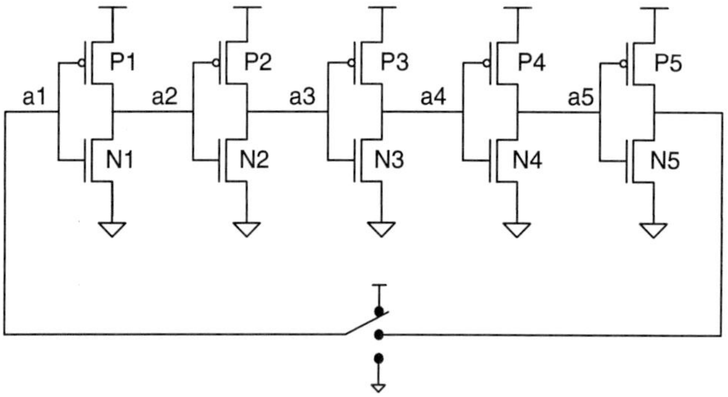

Fig. 6.3 Circuit schematic showing MOSFETs in an RO with five inverters. A switch either closes or opens the loop with node a1 connected to V_{DD} or GND

$$P_{\text{off}} = \left(3I_{\text{offp}} + 2I_{\text{offn}}\right) V_{\text{DD}}, \tag{6.3}$$

where I_{offp} and I_{offn} are the channel leakage currents of p-FETs and n-FETs respectively, and MOSFETs indicated by P1, P3, P5, N2, and N4 are in the off-state. If instead, node a1 is held at GND,

$$P_{\text{off}} = \left(2I_{\text{offp}} + 3I_{\text{offn}}\right) V_{\text{DD}}, \tag{6.4}$$

as MOSFETs indicated by P2, P4, N1, N3, and N5 are now in the off-state. In general, the magnitudes of I_{offp} and I_{offn} are different. Thus, with an odd number of stages in the RO, the value of P_{off} is dependent on the node voltages. Unlike P_{sw}, P_{off} increases as the number of stages in the RO is increased.

The circuit delay and power of a logic gate, at power supply voltage of V_{DD}, is estimated from measurements of RO frequency and currents drawn in the oscillating and quiescent states. The measured current in the oscillating state, IDDA, includes the leakage current, IDDQ, of 2α stages. This is because one out of $2\alpha + 1$ stages (or logic gates) is switching at any instant, as shown in the timing diagram in Fig. 6.2, while the remaining stages are in an off-state. Hence in terms of the measured parameters, P_{sw} is given by

$$P_{\text{sw}} = V_{\text{DD}} \left\{ \text{IDDA} - \text{IDDQ}\frac{2\alpha}{2\alpha + 1} \right\}. \tag{6.5}$$

If $\alpha >> 1$,

$$P_{\text{sw}} = V_{\text{DD}} \left(\text{IDDA} - \text{IDDQ}\right). \tag{6.6}$$

There is a small additional power dissipation because of the short-circuit current flowing between V_{DD} and GND during the switching transient, when both n-FET and p-FET are in the on-state. The fractional contribution of the short-circuit power to the total switching power increases as the ratio $V_{\text{t}}/V_{\text{DD}}$ decreases, where V_{t} indicates the threshold voltage of the MOSFETs. The origin of the short-circuit current and its impact on active power are discussed in Section 6.6.2.

From Eqs. (6.2) and (6.6), C_{sw}, which includes parasitic and wire capacitances, is expressed in terms of the measured parameters IDDQ, IDDA, and f as

$$C_{\text{sw}} = \frac{(\text{IDDA} - \text{IDDQ})}{(2\alpha + 1)\, V_{\text{DD}} f}. \tag{6.7}$$

By expressing delay as RC, the effective switching resistance of a logic gate is

$$r_{\text{sw}} = \frac{\tau_{\text{p}}}{C_{\text{sw}}}. \tag{6.8}$$

The inverse of switching resistance $1/r_{\text{sw}}$ is a measure of the current drive of the MOSFETs comprising the stage. A methodology to derive properties of MOSFETs and CMOS circuit parasitic elements from the measured C_{sw} and r_{sw} values of RO stages is discussed in detail in Section 6.3.

The frequency of an RO can be measured by coupling the voltage signal at one of its nodes to an on-chip clocked counter with its binary output read by a digital tester. Macro designs for ROs are less complex if an external instrument, such as an oscilloscope, frequency counter, or spectrum analyzer, is used to measure frequency. For τ_p in the range of 4–40 ps and α in the range of 10–1,000, the RO frequency ranges from 12.5 GHz to 12.5 MHz. For frequencies above 10 MHz, measurements using an external instrument require a high-frequency probe card, appropriate shielding of signal wires, and impedance matching to suppress reflections. A more practical approach is to include a frequency divider circuit within the macro and reduce the RO frequency to <1 MHz. In this case, a standard DC probe card may be used. The frequency can be precisely divided by a factor of two with a T flip-flop or a master–slave latch circuit. By connecting η divider units in series, the RO frequency is divided by 2^{η}.

In RO test structures the switch S, to close or open the loop in the RO circuit shown in Fig. 6.1, is included in the loop to enable or disable the oscillations. Typically a NAND2 or NOR2 logic gate in an RO serves this function. One input of this logic gate completes the loop, and an external I/O signal applied to the second input enables or disables the oscillations. The power supply of the RO itself is isolated from the peripheral circuits, such as frequency divider, buffers, and I/O driver so that current drawn only by the RO can be measured.

A major advantage of an RO test structure is that the signal propagation in the closed loop itself is isolated from the peripheral circuits external to the loop. The shapes of the waveforms in the periodic voltage pulses may be altered as they travel through the peripheral circuits to an external instrument, but the pulse period or frequency remains constant. Hence, the frequency readout circuit or test equipment may be located at some distance from the RO as long as voltage levels are maintained by correctly sizing buffers and I/O drivers.

6.2 Ring Oscillator Macro Designs

In this section, five examples of RO macro designs including physical layout strategies are described. The essential features of a single RO test structure design, layout, and test are covered in Example 1. Examples 1 and 2 with multiple ROs in a macro can both be implemented at the M1 metal level. These designs are very useful for early process monitoring in technology development or manufacturing. In Example 3, the number of ROs in a macro is increased by adding a decoding scheme to selectively enable individual ROs. In Example 4, a macro design for rapid measurements of statistical variations in RO frequencies using an internal clocking scheme is described. In these examples, all ROs in a macro share the frequency divider and output driver circuits and can be tested with standard in-line parametric ATE and an external frequency counter or oscilloscope. The macro form factors are suitable for implementation in the scribe line and accommodate 1 to ~1,000 ROs per macro. Examples 2, 3, and 4 use our standard 1×25 padset described in Appendix A. More complex test structures for testing a large number of ROs in the form of 2D-arrays are covered in Example 5. These complex test structures are better suited for testing on digital ATE.

6.2.1 Example 1: Single RO Macro Testable at M1

This first example contains all of the essential elements of an RO test structure. The test structure comprises a single RO, a switch to enable or disable the oscillations with an external voltage signal, a frequency divider circuit to lower the RO frequency, and an I/O driver to couple the output signal to off-the-shelf frequency measuring equipment. Figure 6.4 shows the components of an RO test structure circuit with 2α inverting stages and a NAND2 to enable the oscillations. The inclusion of a NAND2 in the RO brings the total number of inverting stages to an odd number, $2\alpha+1$. The oscillations may be initialized by setting the EBL signal to "1" and suspended by setting the EBL signal to "0." Inverter I1 at node a1 couples the RO output signal to the frequency divider circuit. The RO frequency is divided by a factor 2^η, where η is the number of units in the divider circuit, each unit with an output at half its input frequency. There are two power supply sectors: V_{DDE} for the RO and inverter I1 and V_{DDC} for the frequency divider and I/O driver. The test structure requires a minimum of five I/O pads, using a common GND for the V_{DDE} and V_{DDC} power supplies. The circuit and the physical layout style of this RO macro are selected to allow M1 testability.

Other schemes may be utilized to perform the enable function, such as a NOR2 or a clocked latch output. The frequency divider may be designed with clocked latches or replaced by an on-chip counter for digital readout, but in such designs it is convenient to have more than one metal layer for wiring.

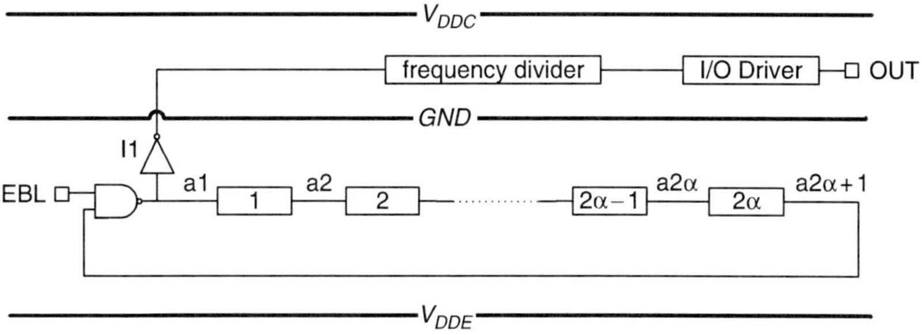

Fig. 6.4 Circuit schematic of an RO test structure with a frequency divider and I/O driver to couple the output signal to external equipment

6.2.1.1 Number of RO Stages

What is the optimum number of stages in an RO? The answer to this question depends on many factors which must be carefully considered prior to designing a macro. In order to measure the fundamental RO frequency, the number of inverting stages must be odd, but it is not required to be a prime number (see Section 6.2.1.4). Other considerations are (1) frequency limits for circuit operation and measurement, (2) macro area, (3) leakage power, (4) V_{DD} droops, (5) error in measured τ_p arising from NAND2 and peripheral circuit loading effects, and (6) error in measured τ_p arising from random and spatial variations in MOSFET properties.

The minimum number of stages in an RO including the NAND2 is three. The frequency of an RO with $\alpha = 1$, and a delay per stage of ~4 ps, is ~40 GHz which may be too high for peripheral circuit operation. With $\alpha = 50$, the RO frequency is lowered to ~1.2 GHz and is acceptable for CMOS peripheral circuits operating in the several GHz range. For some RO stages, signal fall and rise times, τ_f and τ_r, may be long. For sustained oscillations, τ_f and τ_r must be $<T_p/2$.

There are a few drawbacks in increasing the number of stages ($\alpha \gg 50$). The physical size of the RO increases in proportion to α. For very large α, macro size may need to be increased or fewer ROs accommodated in a macro with multiple ROs. Leakage current also increases with α and may become significant if a number of ROs share the same power supply sector. If only one or two metal levels are utilized for wiring, voltage droop in the power distribution grid from parasitic series resistances and background leakage current may affect the accuracy of measurements. In addition, the delay per stage may vary across the RO because of spatial variations in circuit parameters.

In the calculation of delay per stage, τ_p, using Eq. (6.1), it is assumed that all stages are identical. An error in estimating τ_p is introduced by the delay of the NAND2 which may be different than the delay of the other stages and by the additional capacitive load of inverter I1 at node a1. This error, δ_m, is expressed as

$$\delta_m = -\frac{(\tau_{ex} - \tau_p)}{(2\alpha + 1)}, \tag{6.9}$$

where τ_{ex} is the delay of the NAND2 with inverter I1 load and any additional delay error arising from the differences in waveforms on nodes a1 and a2α+1 and other nodes in the RO. The error δ_m is reduced by increasing the number of stages or α. If τ_{ex} is $1.5\tau_p$, $\delta_m = -0.10\tau_p$ for $\alpha = 2$ and $\delta_m = -0.005\tau_p$ for $\alpha = 50$. Hence the error in estimating τ_p would be 10% for 5 stages and 0.5% for a 101 stages in the RO. Another option for reducing δ_m is to keep τ_{ex} nearly the same as τ_p. This requires sizing the NAND2 and inverter I1 for each circuit stage type. This approach is not convenient when a large number of ROs for different circuit types are designed using a common macro template. For model-to-hardware correlation, measured RO parameters are compared with the values obtained from circuit simulation of the full RO circuit as discussed in Section 6.6.1 and the error term δ_m determined from the simulations. Thus the investment in additional design effort to resize the NAND2 and inverter I1 is not warranted.

The delays of individual stages in an RO may vary because of random statistical variations in MOSFET parameters such as V_t and linewidths as well as variations arising from spatial separation and local pattern densities of critical MOSFET layers. Since τ_p gives the average delay of all the stages in an RO including the NAND2, increasing the number of stages tends to smooth out the effect of these variations. If σ is the standard deviation in τ_p, the measured values of τ_p would be within $\pm\ \sigma \sqrt{}(2\alpha + 1)$ of its mean value. Hence, by increasing α, a more accurate value of τ_p is obtained. As an example, for $\sigma = 2\%$ and $\alpha = 50$, τ_p would be known within a standard deviation of $\sim 0.2\%$.

Typical values of α are in the 25–50 range. A smaller value of α (1–5) is used to capture random variations in τ_p. In this case, a large number of nominally identical ROs are measured (Section 6.2.4). Figure 6.5 shows the important components of functionality, accuracy, and physical dimensions of ROs.

6.2.1.2 Physical Layout of an RO

The physical layout of a stand-alone RO test structure, corresponding to the schematic shown in Fig. 6.4, may be carried out with either five or six I/O pads as shown in Fig. 6.6a, b. The form factors in both designs are suitable for placement in the scribe line and for testing at the M1 metal level. This compact RO design serves as a circuit delay monitor early in the technology cycle, and a maximum of five such ROs may be placed in a standard 1×25 padset. The RO is designed with 101 stages ($\alpha = 50$), a NAND2 to enable the oscillations, a frequency divide by 1,024 circuit and a large inverter buffer to serve as an I/O driver. The RO stage shown in this example is an inverter (FO = 1) but other circuits such as a NAND or a NOR with and without a load capacitance may be accommodated with M1 wiring. This same macro design may also be used for complex circuit stages requiring more than one metal level. No ESD protection is provided for routine characterization with

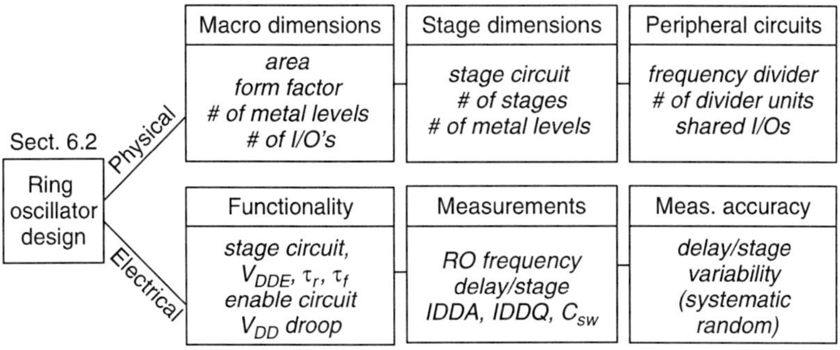

Fig. 6.5 Physical layout, design, and test considerations in an RO design

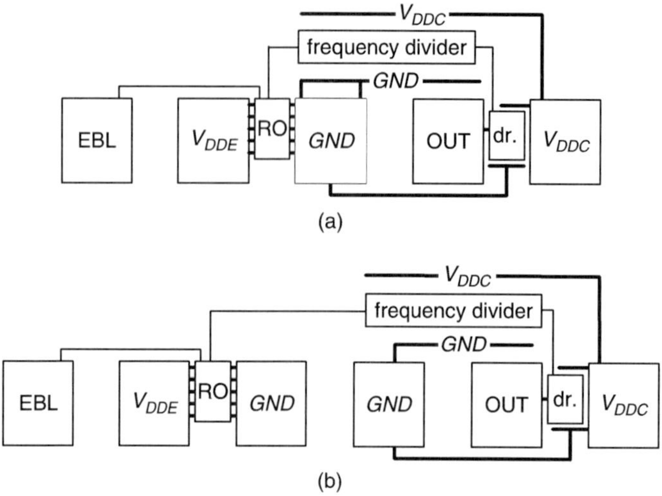

Fig. 6.6 Physical layout of a stand-alone RO macro corresponding to the schematic in Fig. 6.4: **a** with five I/O pads and **b** with six I/O pads

automated test setup. In laboratory test environment, care must be taken to prevent static discharge from damaging the thin-gate oxide (Section 2.4.5).

The RO in this design is placed between the V_{DDE} and GND pads to minimize the power bus resistance. The frequency divider and the I/O driver are powered by V_{DDC} and share a common GND pad with the RO in Fig. 6.6a. The resistance of the distributed V_{DDC} power grid to the frequency divider can be in the 50–100 Ω range as the accuracy of the voltage applied to the frequency divider is not critical and a voltage droop in V_{DDC} of the order of ~10% is acceptable. Independent GND pads may be assigned to each power supply as shown in Fig. 6.6b to eliminate the effect of GND bounce in V_{DDE} arising from periodic switching of the I/O driver. However, a common GND provides higher area efficiency, and the GND bounce can be reduced by design and by correctly setting the test voltages as discussed in Section 6.2.1.3.

A circuit physical layout of the RO template is shown in Fig. 6.7a. The 100 identical stages in the RO are arranged in a 10×10 matrix. The power grid is an interdigitated comb structure. This arrangement is useful in optimizing power grid resistance while minimizing the RO area. The stage layouts are flipped in vertically adjacent stages to enable sharing of V_{DDE} and GND connections. Alternate rows are connected in series to avoid a single long wire connecting the bottom row to the top row to close the loop. Underpasses for wires between alternate rows and crossing the power grid are drawn in DF or PS layers. The NAND2 and inverter I1 are placed at the top to facilitate connections to the EBL pad and to the frequency divider circuit respectively.

The floorplan of the RO is shown in Fig. 6.7b. The V_{DDE} and GND bus vertical spacing is adjusted to fit a standard circuit library book height of 2 μm. The stage

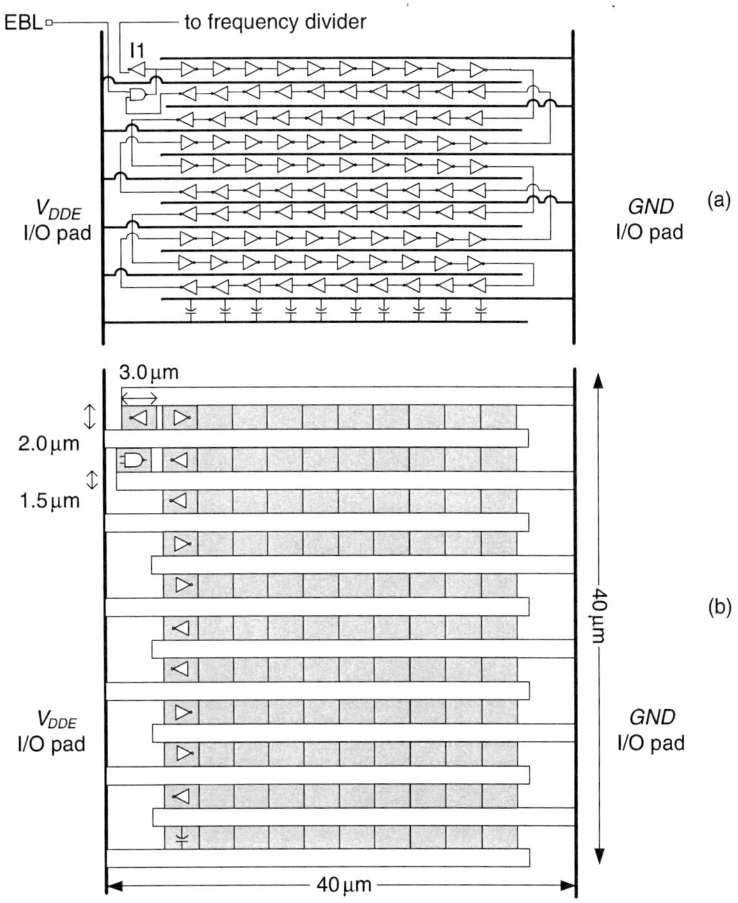

Fig. 6.7 **a** Circuit schematic of an RO template with 100 inverter stages, a NAND2 and an output inverter I1. **b** Floorplan of the RO drawn to scale

width is fixed to accommodate different logic gates and loads. A similar template may be created for wider or taller books, and the value of α is selected accordingly. The power grid bus is designed to keep the IR voltage drop to <1% of V_{DDE}. For the parameters in Appendix A, with $\tau_p = 4$ ps, $C_{sw} = 4$ fF, and $V_{DDE} = 1.0$ V, the P_{sw} from Eq. (6.2) is 0.5 mW and the corresponding IDDA value is ~0.5 mA. The resistance of each arm of V_{DDE} and GND is 6 Ω (height = 1.5 µm, width = 36 µm, $n_{sq} = 24$, $\rho_{sh} = 0.2\ \Omega/\square$, metal pattern density = 80%, and $R = 24 \times 0.2/0.8$). Adding a probe contact resistance of 1 Ω per I/O pad, the maximum voltage droop in the power supply is then ~0.4 mV. The resulting error in frequency measurements is <0.4% if f varies linearly with V_{DDE}. As V_{DDE} is lowered, IDDA and in turn the voltage droop decreases, but the sensitivity of RO frequency to variations in V_{DDE} increases.

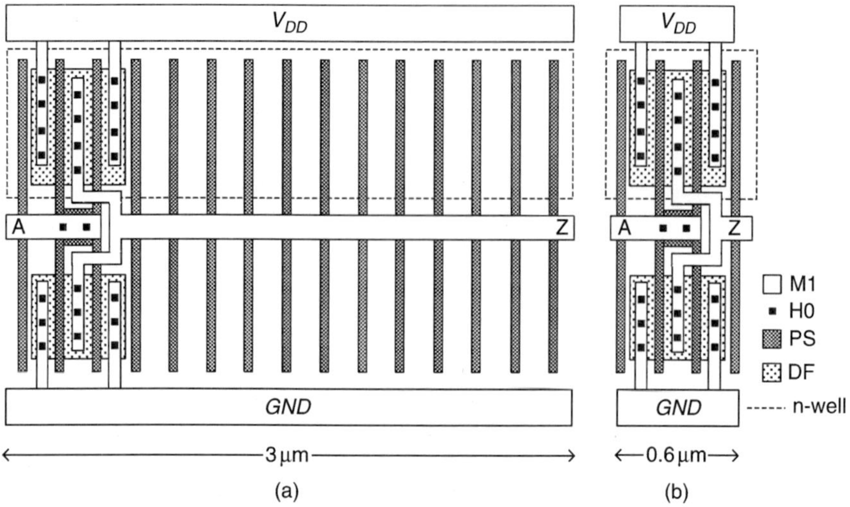

Fig. 6.8 Physical layouts of RO stages for a 2 PS finger inverter, with stage width equal to **a** 15x PS pitch, and **b** 3x PS pitch

Also shown in Fig. 6.7 are 10 DECAP cells, each placed within an RO stage area of 6 μm^2. This corresponds to a total DECAP capacitance of about 300 fF (capacitance = 5 fF/μm^2) which is on the order of the RO capacitance. As discussed in Section 2.4.6, addition of such DECAPs, while optional, can help provide immunity against sources of power supply noise external to the RO itself.

Two examples of the RO stage layout for an inverter (FO = 1) with 2 PS fingers are shown in Fig. 6.8. The input and output wire segments or the locations of pins A and Z are fixed, and the electrical connection between the stages is established by butting the stages. In Fig. 6.8a, the stage width is 3 µm, equivalent to 15 PS finger pitches. For a stage width of only 3 PS pitches or 0.6 µm as shown in Fig. 6.8b, 100 stages can be accommodated in only two rows in the RO layout shown in Fig. 6.7. It is a good practice to fix the stage width so that RO designs for different circuit types can be generated by a straightforward replacement of stages. RO templates may be created with the number of stages incremented by a factor of 2 ($\alpha = 12, 25, 50, 100$) to accommodate a large variety of circuit types.

The frequency divider function is implemented with a flip-flop unit or a master–slave latch to divide the input frequency precisely by a factor of 2. In a chain of η flip-flop units in series, the input frequency is divided by a factor of 2^{η} at the output. A circuit implementation, with a single input signal, utilizes a positive edge-triggered T-flip-flop as shown in Fig. 6.9a. It comprises four cross-coupled and-or-invert (AOI) logic gates and an inverter to generate a complementary input signal. In the subsequent blocks, complementary outputs Q and Q_n from the previous block are connected to CLK and CLK_n inputs. In Fig. 6.9b, four of these blocks are connected in series to divide the input signal frequency by 16.

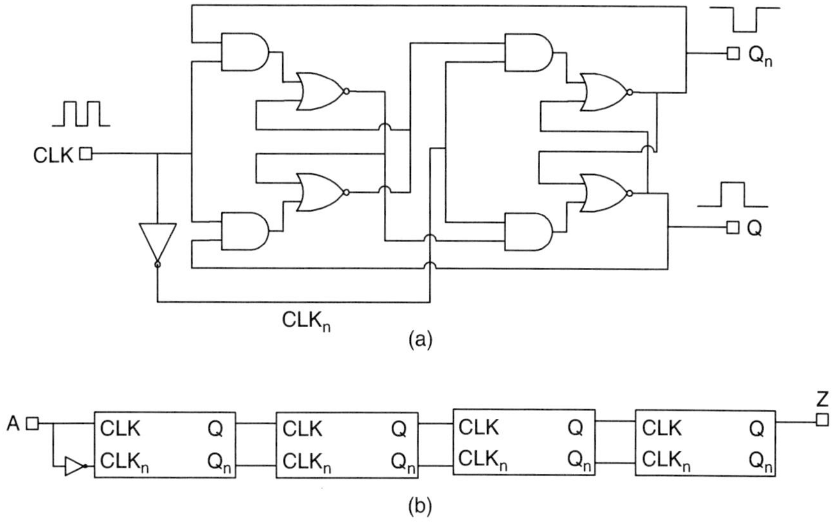

Fig. 6.9 Circuit schematic **a** of a flip-flop unit and **b** of a frequency divider with a chain of four flip-flops units

In Fig. 6.10, the circuit schematic of a flip-flop unit is reconfigured to show the physical implementation at the M1 metal level. The AOI gates are placed within the V_{DDC} and GND busses. The Q output crosses the GND bus to connect to the input CLK of the next unit. Similarly, the Q_n output crosses the V_{DDC} bus to connect to the CLK_n input of the next unit. In this way, the units are electrically connected by butting. The crossovers are aligned to use a single DF or PS rectangle to traverse multiple M1 wires. Each unit is approximately 90 PS pitches long (=18 μm) and the resistance in V_{DDC} and GND wires for a 1.0 μm wide M1 bus

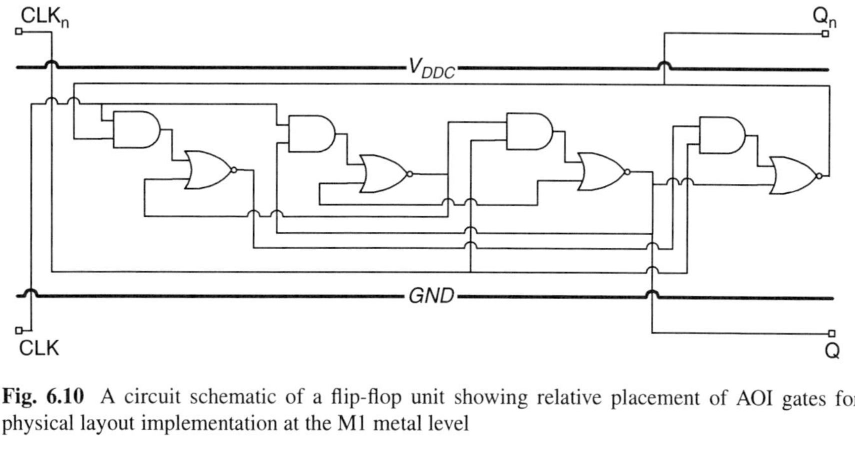

Fig. 6.10 A circuit schematic of a flip-flop unit showing relative placement of AOI gates for physical layout implementation at the M1 metal level

$(n_{sq} = 18,\ \rho_{sh} = 0.2\ \Omega/\square)$ is 3.6 Ω each. If ten units are connected in series, the net resistance is 72 Ω. If the current drawn by the circuit is <1 mA, at $V_{DDC} = 1.0$ V, there is <0.1 V drop in the power supply. As the frequency output of the divider circuit is not dependent on its power supply voltage, this voltage droop in V_{DDC} bus can be ignored as long as the circuit functionality is maintained.

The I/O driver is a two- or three-stage buffer with an output inverter of width 50 μm (Section 2.4.5). The output resistance of the driver, including any wire resistance from driver output to the OUT pad, is ~50 Ω to match a standard external load. If the external load is an off-the-shelf frequency counter with an input resistance of 50 Ω, the total resistance, driver plus load, is 100 Ω. When the RO is oscillating, the driver draws an average current of ~5 mA for V_{DDC} of 1.0 V (the OUT signal is high for only half-cycle).

The final RO design is simulated with SPICE, for verifying functionality and estimating measurement errors caused by V_{DDE} droops, with a circuit netlist extracted from the physical layout to include all parasitic R and C components. Metal M1 wires and underpasses may be strapped with M2 metal to reduce the parasitic resistances. The macro template design can then be validated in the hardware by first testing at M1 metal and again after M2 wiring is completed. The difference in measured values for a robust design should be <0.5%, i.e., within the measurement error and repeatability of the test setup.

6.2.1.3 RO Test Setup

The test equipment for an RO comprises two independently controlled constant voltage power supplies and a frequency counter, oscilloscope or spectrum analyzer. The output frequency is given by

$$f = \frac{1}{2(2\alpha + 1)\,\tau_p}\left(\frac{1}{2^\eta}\right). \tag{6.10}$$

For an RO with 101 stages ($\alpha = 50$), assuming $\tau_p = 4$ ps, and frequency divided by 1,024 ($\eta = 10$), the output frequency is ~1.2 MHz. For smaller values of α, the output frequency can be kept below ~1 MHz by adding an appropriate number of additional flip-flop units in the frequency divider. An approximate value of τ_p is obtained from SPICE simulation of the RO circuit. When a number of ROs with different α values are used in technology characterization, it is convenient to increment α by factors of ~2 ($\alpha = 12, 25, 50$).

The sequence in which the three input voltages, V_{DDC}, V_{DDE}, and EBL are applied and the quality of the EBL waveform are important factors in ensuring that a single edge circulates in the RO loop. If more than one edge propagates in the loop, the observed RO frequency is a multiple of the fundamental frequency. These higher harmonic components, which have plagued RO measurements in the manufacturing line, can be largely avoided by turning on the voltage inputs in a prescribed sequence and completely eliminated by adding a signal conditioning circuit in the RO test structure as discussed in Section 6.2.1.4. Glitches and noise in the output waveform may also corrupt frequency measurements resulting in data containing even multiples of the fundamental frequency. It is therefore important to validate the test setup by observing the output waveform on an oscilloscope.

The recommended test sequence for a stand-alone RO is as follows:

1. set all I/O signals to GND,
2. set V_{DDC} to "1,"
3. set V_{DDE} to "1" and measure IDDQ,
4. set EBL to "1," measure IDDA and *f*.

The time to measure IDDQ, IDDA, and *f* includes the settling time of each power supply and the data acquisition time. Measurements with a parametric tester and an external frequency counter can take as long as 100 ms. Measurement time can be reduced to <10 ms with a digital tester.

For model-to-hardware correlation under product application conditions, voltage dependence of RO frequency is of interest. This is obtained by varying V_{DDE} over the desired operating range of the circuit. In these measurements, it is important to maintain a constant V_{DDC} while varying V_{DDE}. The optimum range of V_{DDC} can be determined by measuring *f* at a constant V_{DDE} while varying V_{DDC}. The measured value of *f* should be independent of V_{DDC}. If V_{DDC} is too low, the output voltage signal waveform and accuracy of frequency measurements may be affected. If V_{DDC} is set too high, the current drawn by the I/O driver may reduce the V_{DDE} value on the power supply bus of the RO because of GND bounce.

When the difference in V_{DDE} and V_{DDC} voltage levels is expected to be significant, a voltage level shifter is added at the output of inverter I1 powered by the V_{DDC} power supply. If $V_{DDE} \lesssim 0.5\ V_{DDC}$, the level shifter is an inverter with a weak p-FET and a strong n-FET. Conversely, if $V_{DDE} \gtrsim 1.5\ V_{DDC}$, the voltage level shifter has a strong p-FET and a weak n-FET. The p-FET and n-FET strengths are appropriately adjusted by selecting their threshold voltages, channel lengths and widths. The level-shifter function is verified with SPICE simulations.

6.2.1.4 RO Harmonics

Integral multiples of the fundamental frequency of an RO, or harmonics, may be present if several signal edges propagate in the loop simultaneously. With an odd number of stages, only odd multiples (3*f*, 5*f*...) can be sustained. In principle, an RO can also function with even number of stages, with possible output frequencies of even multiples of the fundamental frequency. With even number of stages, two equally spaced edges propagate in the loop, one edge experiencing one set of PU and PD transitions and the second edge experiencing the complementary set of PU and PD transitions. Hence, oscillations in this case can be sustained only if the sum of the first set of PU and PD delays exactly equals the sum of the complimentary set. Otherwise the two edges will eventually annihilate each other. In practice, random variations in the MOSFET parameters and other non-uniformities in the stages prevent an RO with an even number of stages from being in a permanent oscillating state.

Odd harmonics in an RO may be generated if the RO is initialized by simultaneously applying both V_{DDE} and the input signal at EBL, by a noisy EBL signal or by the presence of other sources of noise. Three, five, or even seven edges, created by rapidly turning the RO off and on within one cycle, may propagate around the loop. The upper frequency limit or the highest harmonic in an RO is set by the condition that the voltage at a node reaches a stable value prior to the arrival of the subsequent edge.

Undesirable odd harmonics in an RO, if present, must be identified and filtered from the data. In order to prevent loss of data from higher harmonics, it is important to ensure that only the fundamental frequency is generated. By following the test sequence described in Section 6.2.1.3, the RO power supply is stabilized prior to enabling oscillations, and the occurrence of higher harmonics is reduced but not entirely eliminated. Another source of higher harmonics is a noisy EBL signal from an external power supply arriving directly at the input of NAND2. This problem can be addressed by providing an EBL signal with a sharp edge, either launched from an external pulse generator or generated within the RO macro.

A circuit to generate an EBL signal with a sharp edge within an RO macro is shown in Fig. 6.11a [5]. It utilizes a level-sensitive latch with an inverting output. A circuit schematic of the latch, which may be implemented at the M1 metal level, is shown in Fig. 6.11b. The latch is powered by the common power supply V_{DDC}. Its data input (D) is connected to V_{DDE} and the clock input (C) to EBL(I/O). With the latch powered up, $V_{DDE} = 0$ V, and EBL(I/O) at “1,” the EBL signal passed to the RO is “0,” and the RO is in its quiescent state. When the EBL(I/O) is returned to

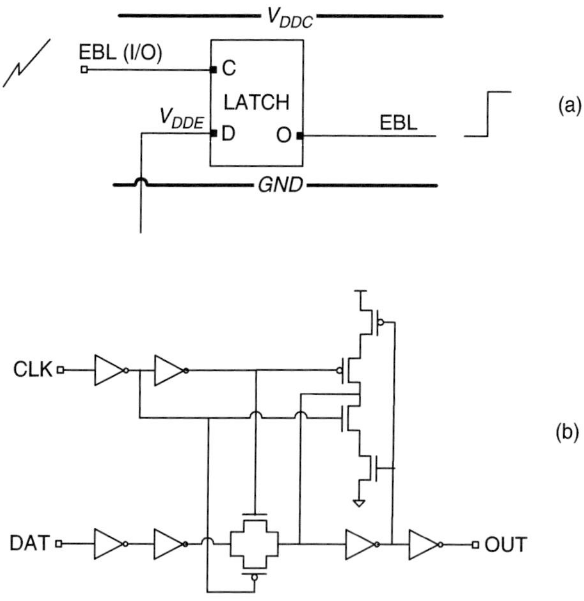

Fig. 6.11 **a** A circuit to create an EBL signal with a sharp rising edge. **b** Circuit schematic of a latch that can be implemented at the M1 metal level

"0," the latch holds a "0," and this state is maintained. Next, the V_{DDE} is turned on, setting the data input to the latch to "1." A sharp rising edge is then created at the EBL input to the RO when EBL(I/O) is raised to "1," and subsequent oscillations of the RO are harmonic free. This scheme has the advantage that it can be implemented without any additional I/O pads.

The test sequence for enabling the fundamental frequency in the RO is as follows:

1. set all I/O signals to GND,
2. set V_{DDC} to "1,"
3. set EBL(I/O) to "1" and then reset to "0,"
4. set V_{DDE} to "1" and measure IDDQ, (EBL = "0"),
5. set EBL(I/O) to "1," measure *f* and IDDA.

6.2.2 Example 2: Multiple RO Macros Testable at M1

At each CMOS technology node, several flavors of MOSFETs differing in their threshold voltages, channel lengths, and gate oxide thicknesses are offered to cover a wide range of performance, power, and applied voltages for digital, analog, and memory circuit applications. Circuit characteristics may also depend on the physical layout and local environment on silicon. It is therefore important to characterize at least one RO of each flavor at the M1 test stop for early process learning. A stand-alone RO macro in Example 1 can be utilized for one logic gate type such as an inverter, NAND, or NOR comprising one set of MOSFET flavors. A number of instances of this macro with different RO stage designs can be placed in the scribe line or test vehicle. However, silicon space utilization is improved if the I/O pads for the EBL and I/O driver are shared by multiple ROs.

In this example, the stand-alone RO test structure is extended to include multiple ROs in a macro testable at the M1 metal level. The concept is shown in Fig. 6.12, with three ROs. Each RO, with its dedicated power supply, is placed in the space between two I/O pads. An RO is selected by turning on its V_{DDE}, while the V_{DDE}s of all the unselected ROs are held at GND. The EBL signal is supplied by one I/O pad (not shown). This signal wire may be directly connected to the input of the NAND2 gate in each RO or pass through a latch to sharpen the signal edge. The

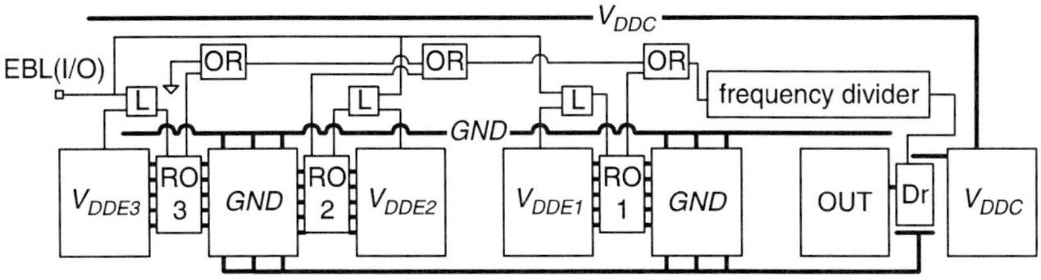

Fig. 6.12 Circuit physical layout of a test structure with three ROs sharing the frequency divider and I/O driver

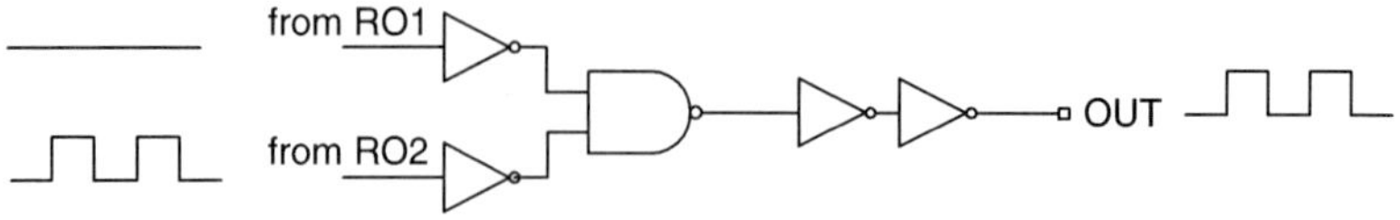

Fig. 6.13 Schematic of an OR circuit block with RO1 output at "0" and RO2 oscillating

latch wiring is included in Fig. 6.12 and can be accomplished with M1 metal and PS or DF underpasses.

All ROs in the macro share the frequency divider and I/O driver circuits. Only the one RO, with its V_{DDE} turned on, oscillates while the output signals of all the other ROs are at "0." As only one RO output signal at a time must travel to the frequency divider, an OR circuit block shown in Fig. 6.13 is added to the output of each RO. One of three RO power supplies, for example, V_{DDE2}, is switched on and the EBL signal set to "1." The output of RO2 is now oscillating and the output of RO1 and RO3 are at "0." With the top input of the NAND2 in Fig. 6.13 at "1," the output of RO2 is propagated through the "OR" circuit. A two-stage buffer may be added to the output of each OR circuit block to drive a long wire ($\sim$100 μm) to the next RO while maintaining the logic level. Any number of ROs may be added to the left of RO3 in Fig. 6.12. Each pair of ROs requires three additional I/O pads, with two ROs sharing a GND pad. With this arrangement, a 25 I/O pad macro can accommodate 13 ROs. The resistance of the V_{DDC} and GND wires for the peripheral circuits, traveling across the macro, is reduced with common GND pads distributed across the macro and two V_{DDC} I/O pads, one on each end of the macro.

The test sequence for each RO in this macro is the same as in Section 6.2.1.4, and only one RO can be measured at a time on a parametric tester. The test time for each RO includes the settling times of the power supplies and the time to measure the frequency output. If digital ATE is used, the test time is substantially reduced by introducing a design for parallel test, and a test time reduction of $\sim 1/N$ over the serial approach is obtained, where N is the number of ROs in the macro. Digital ATE with an internal clock can perform a frequency measurement in $\sim$1 ms, further reducing the test time. A circuit schematic for implementing the parallel test scheme is shown in Fig. 6.14a. The RO macro design is modified to allow parallel measurements of IDDQ and IDDA of all ROs by turning on all the V_{DDE} power supplies simultaneously. The measurement time of a single RO is dominated by settling time, and here all V_{DDE} power supplies settle in parallel. With all ROs in the oscillating state, the frequency measurement is carried out sequentially, immediately following the IDDA measurement, in a time short compared with the settling time. The EBL signal when supplied by a digital channel has a sharp edge, and the latch implementation described above is not needed. Instead, a scan chain of master–slave latches to steer the output of only one RO to the frequency divider at a time is included. The timing diagrams for CLK, DAT, and SEL signals to select the RO to be measured are shown in Fig. 6.14b. The measured frequency corresponds to the output of the RO whose SEL is at "1." The CLK period is dependent on the minimum RO frequency and the number of cycles over which the frequency measurement is carried out.

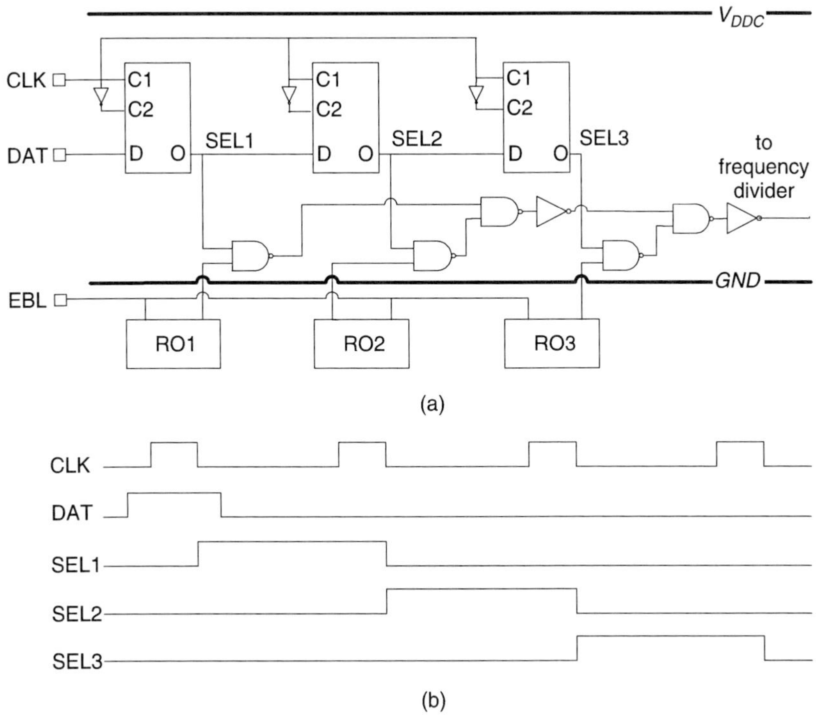

Fig. 6.14 **a** Schematic of a circuit with a chain of MS latches to select an RO output. **b** Timing diagram for scan chain operation

The macro for parallel test requires two additional I/O pads for CLK and DAT and only 12 ROs may be accommodated in a standard 1×25 padset macro. The wiring can still be carried out with only M1 metal together with PS or DF underpasses. The macro design preserves the feature of independent V_{DDE} power supplies, and ROs operating at different V_{DDE} values can still be tested in parallel.

6.2.3 Example 3: Multiple RO Macros Testable at M4

RO stages consisting of complex logic circuit blocks or memory elements generally require more than one metal layer for wiring. Measurements of ROs comprising such circuit blocks are made at a test stop later than M1. With three or more metal layers, more ROs can be packed into a standard macro template, while still maintaining a form factor suitable for placement in the scribe line. Here, we describe two such macro configurations. The first one, with independent V_{DDE} supplies, is an extension of Example 2. A decoder circuit is included in the second configuration to increase the RO density, and several ROs share a V_{DDE} power supply. These macros are of medium design complexity and can be tested with parametric ATE equipped with a frequency counter.

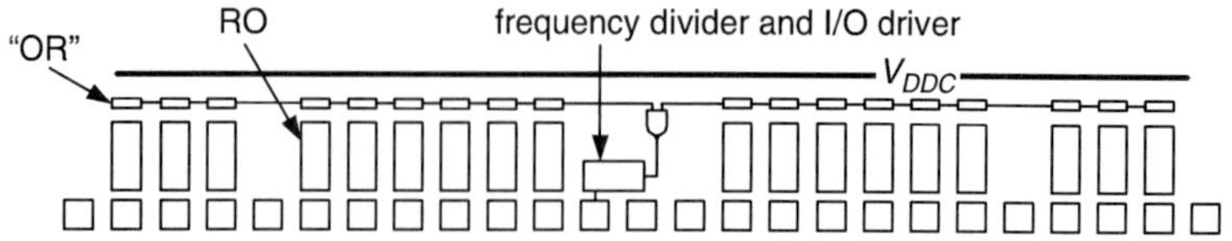

(a)

1	2	3	4	5	6	7	8	9	10	11	12	13	14	15	16	17	18	19	20	21	22	23	24	25
EBL	VE1	VE2	VE3	GND	VE4	VE5	VE6	VE7	VE8	VE9	OUT	GND	VDDC	VE10	VE11	VE12	VE13	VE14	VE15	GND	VE16	VE17	VE18	WS

(b)

Fig. 6.15 **a** Floorplan of an 18 RO macro with independent V_{DDE} supplies (VE1 to VE18). **b** Macro I/O pad assignments

A modified version of the RO macro in Example 2 is shown Fig. 6.15a. The ROs are placed above the pads, providing a greater flexibility in stage dimensions and wiring. This design preserves the isolated V_{DDE} feature, allowing mixing of ROs of different MOSFET flavors in the same macro with independent measurements of IDDA and IDDQ. The number of ROs in a 1×25 padset macro is increased from 13 in Example 2 to 18 by reducing the number of GND pads. Low resistance power distribution grid of the type shown in Fig. 2.36b is on orthogonal metal wires. The I/O pad assignments for the macro are shown in Fig. 6.15b.

Silicon space utilization is further improved by adding a decoder to enable one RO at a time from a set of ROs sharing a common V_{DDE} pad. An example floorplan of a macro with 42 ROs is shown in Fig. 6.16a, and the I/O pad assignments are shown in Fig. 6.16b. There are six RO sectors, each with a 3-bit decoder to enable

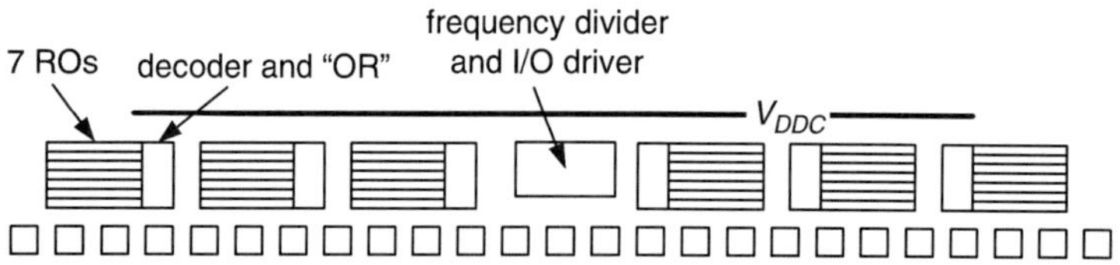

(a)

1	2	3	4	5	6	7	8	9	10	11	12	13	14	15	16	17	18	19	20	21	22	23	24	25
OS	GND	VE1	GND	GND	VDDC	VE2	GND	A2	VE3	GND	OUT	GND	VDDC	GND	VE4	A1	GND	VE5	A0	GND	GND	VE6	GND	WS

(b)

Fig. 6.16 **a** Floorplan of a 42 RO macro with six power supply islands (VE1 to VE6). **b** Macro I/O pad assignments any one or none of the seven ROs in that sector. There are three common GND pads for the seven power supplies including V_{DDC}. An optional sub-circuit is included to select either the divided or the undivided frequency signal from the ROs by setting the OS input bit. Pad 25 labeled WS is available for making contact to the substrate or used as an additional GND pad.

The circuit schematic of a three-input decoder with seven decode outputs to select one of seven ROs is shown in Fig. 6.17a. The decoder input bit "000" corresponds to disabling all seven ROs for measurement of their combined IDDQ. The decoder input bits A0, A1, and A2 are shared by the six sectors and travel on parallel wires across the macro for a distance of ~1.5 mm. These wires carry DC signals and do not require buffer insertions. In Fig. 6.17b, a circuit schematic of an RO sector is shown. The power grid arrangement for the V_{DDE} and V_{DDC} power supplies, with common GND, is partitioned to place the seven ROs on the V_{DDE} power supply and the decoder and the "OR" circuits on the V_{DDC} power supply.

Each RO occupies two vertical power grid pitches. The stages can be flipped in the horizontal direction in the two rows for sharing of V_{DDE} and GND busses. Alternatively, with the availability of more metal layers and independent V_{DDE} and GND wires for each row, the stage orientation can be the same in both rows. With 101 stages and a stage width of 20 PS pitches (4 μm), each row spans ~200 μm and may be subject to spatial variations. The number of stages in any RO or the stage width in any RO may be altered with very minor alterations in the template. The design can also be easily scaled with 2-, 3-, or 4-bit decoders, accommodating 3, 7, and 15 ROs per sector respectively. An upper limit on the number of ROs in a sector is set by the voltage drop due to background leakage current.

The test setup for this macro is similar to that for the macro in Example 2. One power supply sector is turned on at a time, and the decoder bits enable a single RO in this sector. The other 41 ROs remain in the quiescent state and the output of the selected RO propagates through the "OR" circuit to the frequency divider. The switching capacitance and resistance estimates of an RO can be accurately made by subtracting the combined IDDQ value of all seven ROs in the sector from the measured IDDA value. Other schemes such as adding a footer circuit to disconnect unselected ROs from the power supply can be used to measure the IDDQ values of individual ROs.

An additional circuit may be included in the macro to select either the divided or the undivided frequency signal as shown in Fig. 6.18. This feature is useful for validating the frequency divider function. DC input signal OS is set at "1" to enable the frequency divider function and is set to "0" to measure the undivided RO frequency at the OUT pad. The OUT signal pad has adjacent GND pads, providing an arrangement suitable for measuring an output signal in the GHz range with an appropriately designed high-speed probe card.

The macro can be reconfigured for parallel test on a digital ATE by adding a scan chain to select the output, as described in Example 2. The six V_{DDE} power supplies are turned on simultaneously, and the IDDQ values of ROs in each sector are measured in parallel by setting the decoder input bits to "000." The other decoder input bits select one RO in each sector, and their respective IDDAs are measured

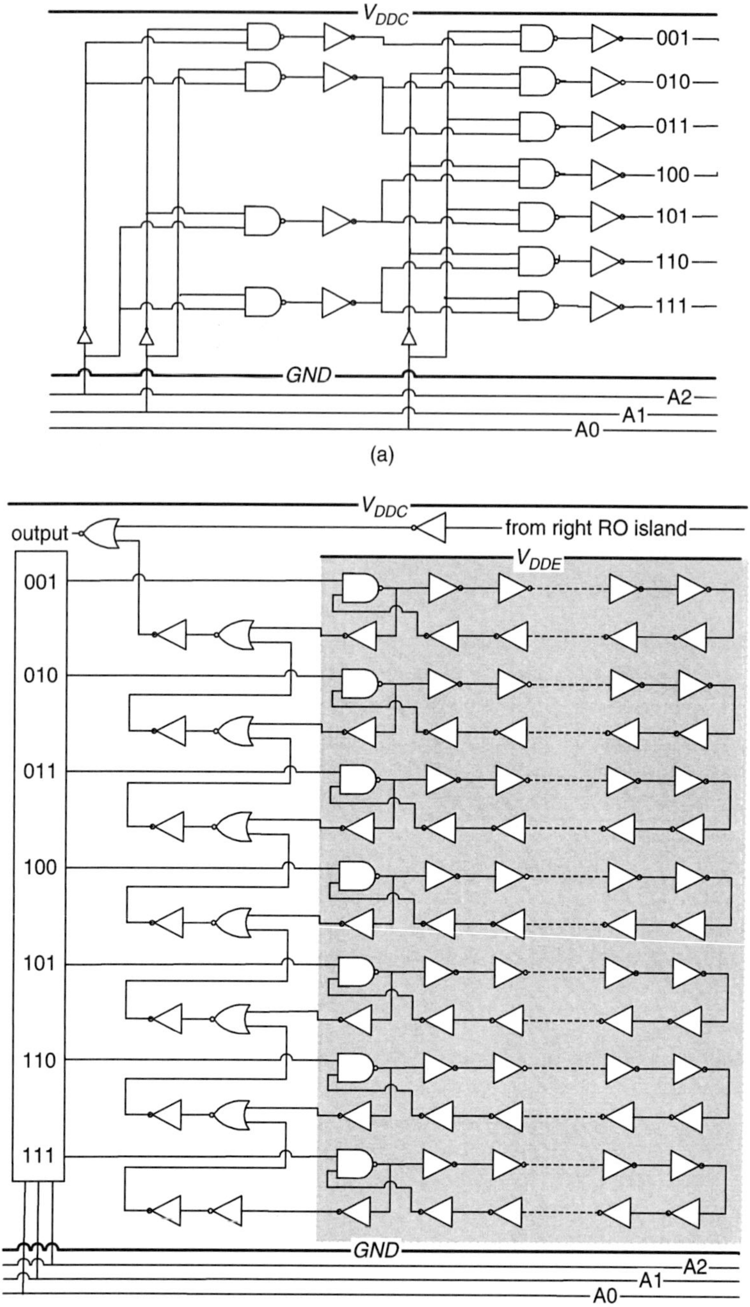

Fig. 6.17 **a** Schematic of a three-input decoder circuit. **b** One RO segment including seven ROs on the V_{DDE} power supply island (*shaded area*) and a decoder and OR circuit blocks on the V_{DDC} power supply

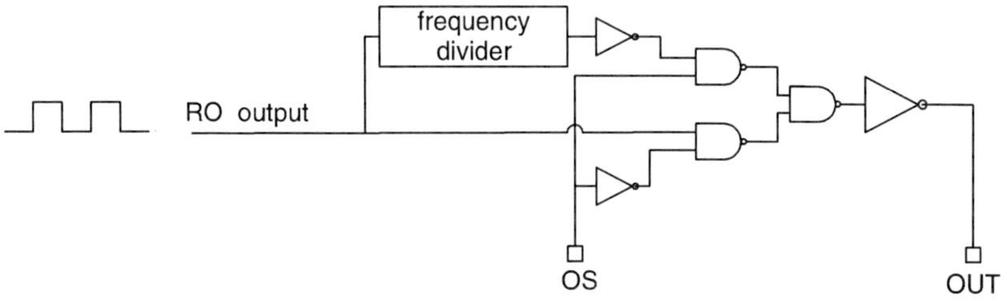

Fig. 6.18 Circuit schematic to select frequency divided or undivided RO output signal

in parallel also. With two additional scan chain inputs (CLK and DAT) for the digital clock on the tester, the RO frequencies are measured sequentially. This can be accomplished by eliminating the undivided output feature and using pads 1 and 25 (OS and WS) in Fig. 6.16b for CLK and DAT input signals.

6.2.4 Example 4: Macro for RO Variability Statistics

ROs may be designed to measure circuit delays of logic gates with different loads, device sizes, and physical layouts. Typically, a sub-set of these ROs are placed in different locations on a reticle field to map systematic spatial variations in circuit delays. Measurement of variations in circuit delays arising from random statistical variations in the constituent circuit elements (MOSFETs, parasitic R and C) requires an RO design with a small number of stages (3–9) and placement of ~10–50 nominally identical ROs of each design type in close physical proximity. A large number of ROs are hence needed to capture both systematic and random variability.

In this example, a macro for measuring random variability in circuit delays by rapidly collecting data on a large number of ROs is described [6]. The number of ROs in the macro template, with a standard 1×25 padset, can be >500. The design is of medium complexity and testable with parametric ATE equipped with a frequency counter or oscilloscope or with digital ATE. Efficiency in test and data analysis is obtained with additional circuitry in the macro.

The basic idea in this design is to use an internal or external clock to drive a counter which provides the input bits to a decoder to sequentially enable a single RO in an array of nominally identical ROs. The outputs of the ROs are muxed to provide a common low frequency output as in previous examples. This output signal follows the frequency of the selected RO in the array during a time interval equal to half the clock period. The frequency statistics of the output signal, giving the mean frequency of all the ROs in the array and its standard deviation, is directly obtained by utilizing the statistical functions in the frequency counter or oscilloscope. With only DC or low frequency I/Os, the macro can be tested with parametric ATE. Important features in this example are (1) on-chip clock generation, (2) external clock option, (3) a circuit to select the clock signal, (4) a circuit to trigger a frequency counter

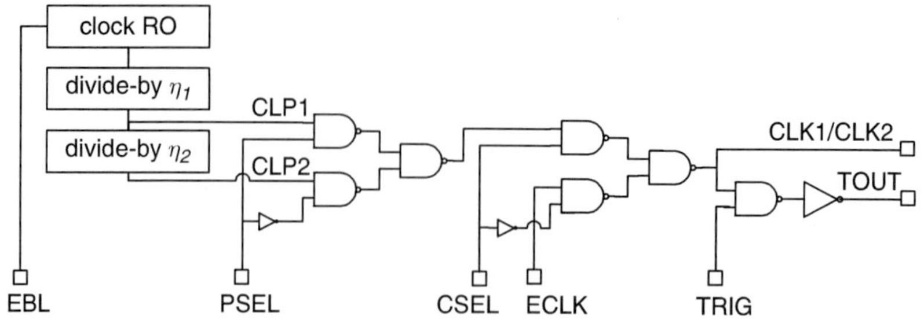

Fig. 6.19 A circuit scheme for selecting an internal or external clock to initialize an RO array and for generating a synchronized trigger signal

either synchronously or asynchronously with the clock, (5) a counter coupled to a decoder to sequentially enable the ROs in a selected power supply sector, and (6) direct measurement of frequency statistics.

The circuit scheme to generate a clock with two different output periods is shown in Fig. 6.19. The clock RO is enabled with EBL set to "1." The clock RO output is fed to a frequency divider configured to output two different frequencies at nodes CLP1 and CLP2, corresponding to clock periods T_{c1} and T_{c2}, with $T_{c1} < T_{c2}$. The PSEL voltage level is toggled between "1" and "0" to select either clock period, T_{c1} or T_{c2}. An option to use an external clock signal ECLK instead of the internally generated clock, with a control signal CSEL, is included. Input signal TRIG enables a trigger output TOUT to synchronize the internal or external clock with the frequency measurement equipment.

A circuit schematic of an RO array block is shown in Fig. 6.20a. It comprises 32 nominally identical ROs with a common output, a four-stage frequency divider circuit configured as a counter coupled to two 4-bit decoders to enable the ROs. Both the true and the complementary outputs Q and Q_n of each unit of the frequency divider circuit shown in Fig. 6.9b serve as inputs to the two 4-bit decoders. The ROs are sequentially enabled for time durations equal to half the selected clock period ($T_{c1}/2$ or $T_{c2}/2$), and their outputs are muxed together via a distributed OR function. To capture the effect of random variations on circuit delays, the number of stages in the RO is kept small, typically five, as shown in Fig. 6.20b. The circuit stages may be appropriately designed to emphasize variations in the parameter of interest. A two-stage frequency divider circuit may be placed at the output of the array to lower the frequency of the signal traveling across the macro to an acceptable value.

The output voltage and the frequency as a function of time are shown in Fig. 6.21. The frequency in a time interval equal to half the internal clock ($T_{c1}/2$) corresponds to a single RO in the array. The total time to cycle through all the ROs is $NT_{c1}/2$, where N is the number of ROs in the array, and the cycle is repeated as long as the clock is running. This time period should be smaller than the maximum time over which the frequency counter can process the frequency statistics. A suitable clock time period may be selected with PSEL input signal. Frequencies of individual

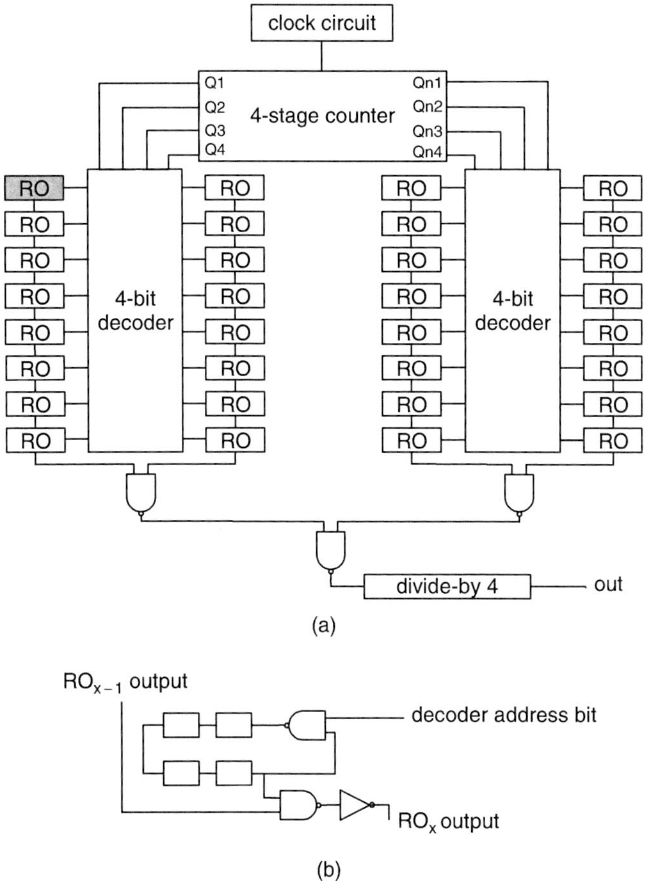

Fig. 6.20 **a** Circuit scheme for automated sequential enabling of 32 ROs. **b** Schematic of a five-stage RO circuit including an output OR function

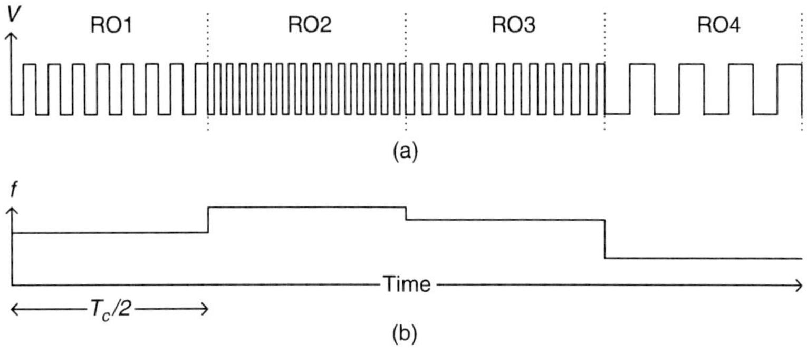

Fig. 6.21 **a** Voltage output of four sequentially enabled ROs in an RO array. **b** Corresponding output frequency as a function of time

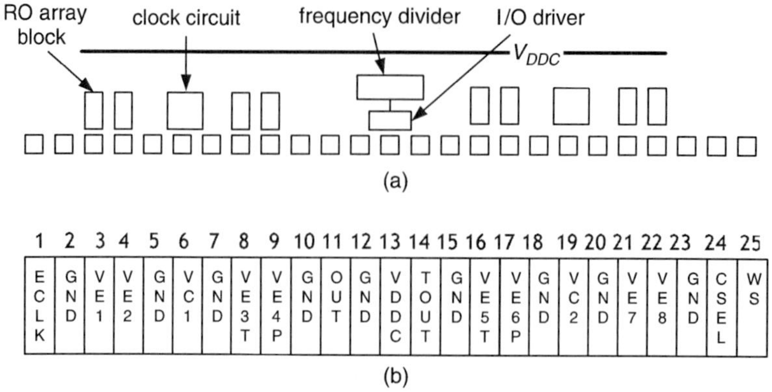

Fig. 6.22 **a** Floorplan of a macro with eight RO array blocks. **b** Macro I/O pad assignments with eight V_{DDE}s (VE1 to VE8)

ROs may be measured by using the TOUT signal in Fig. 6.19 to synchronize the frequency measurements with the enablement of each RO. If 1 of the 32 ROs in the array, indicated as a shaded rectangle in Fig. 6.20a, is omitted from the design, the frequency output corresponding to that position is zero. In the absence of a resettable counter, this serves as an RO position marker in the output frequency stream.

The macro template with eight RO array blocks is shown in Fig. 6.22. Four RO array blocks share a common clock control circuit. The ROs in each block are powered by an independent V_{DDE} power supply. Only one V_{DDE} out of eight is turned on at a time to select the RO array block to be tested. The V_{DDE} pads for RO blocks 3 and 4 on the left side of the macro are shared with the PSEL and TRIG pads for the right side of the macro. Similarly, V_{DDE} pads for RO blocks 5 and 6 on the right side of the macro are shared with the PSEL and TRIG pads for the left side. The frequency divider at the output and the I/O driver are shared by all RO array blocks. These together with the clock RO are powered by a common V_{DDC} power supply. By increasing the number of units in the counter to five and using a 5-bit decoder, the number of ROs in an array block is increased to 128. Hence, a total of 1,024 ROs in a 1×25 padset macro may be accommodated.

6.2.5 Example 5: 2D RO Array Macro

In the previous four examples, RO macro designs comprising isolated or linear arrays of ROs can accommodate from one to a few hundred ROs. For DFM applications and for evaluating systematic and random variations in all the circuit blocks in a product circuit design library, a large number of ROs (>>1,000) need to be measured. In this case, a 2D array approach is more suitable. RO macro designs for 2D arrays tend to be more complex and require four or more metal levels for a robust power grid and wiring channels along with a non-standard I/O pad configuration. It is preferable to use digital ATE for rapid data acquisition from such a large number of ROs.

The basic concept of a 2D array described in Section 2.5.4 is applicable to RO macros. The ROs are arranged in a $N_c \times N_r$ matrix, where N_c and N_r are the number of columns and rows in the array. A multiplexer or a wide OR circuit is used to feed the outputs of all ROs to a frequency measurement circuit on-chip or to external frequency measurement equipment following a frequency divider circuit. The ROs may be enabled sequentially by adding appropriate selection circuitry to individual ROs. Alternatively, the ROs may be enabled simultaneously and the selection circuit placed at the RO output.

All ROs and peripheral circuits may share a common power distribution grid in which case only the RO frequencies are measured. If the number of available metal levels is limited, a robust power distribution is ensured by creating smaller power supply islands for groups of ring oscillators. In this case, the IDDQ and IDDA values can be measured as described in Example 3.

With increasing design complexity, variations in number of available metal levels, silicon area, and test equipment specifications, a number of different macro designs options may be exercised [7, 8]. Many of these options are extensions of circuit ideas described in previous examples. Here, we describe one implementation of a 2D RO array macro. In this example shown in Fig. 6.23, there is an option to place each column of ROs on an isolated power supply island to reduce the background leakage current. A column decoder activates switches to connect the islands to V_{DD} and GND pads, while the row decoder is used to sequentially enable the ROs. Voltage sense wires may be added for accurate measurements of voltage applied to an RO.

In Fig. 6.23a, a NAND2 and an inverter are included in the RO circuit for column and row address inputs. An RO is enabled when both CAD and RAD inputs are high. A circuit scheme to apply power only to a selected column is shown in Fig. 6.23b. The V_{DD} and GND wires of each column are isolated and all the unselected columns (CAD = 0) are disconnected from the power supply, with a clamp voltage $V_C = 0$ V applied instead to ensure full isolation. A multiplexing scheme as described in previous examples is used to feed the RO signals to a common output in the 2D array shown in Fig. 6.23c. The RO frequencies are measured sequentially using an external or on-chip frequency counter. The decoder inputs may have dedicated I/O pads, or scan chains may be used to reduce the number of I/O pads.

In general, different macro templates for 2D arrays may be needed for measuring systematic and random variability in circuit delays. A rule of thumb is to keep the number of stages $\geq$25 to measure systematic variations in RO frequencies and $\leq$9 for investigating random variations, as discussed in Section 6.2.1.1. The optimum number of stages may depend on the circuit topology and CMOS technology node as random variations increase with reduction in feature size. The design resources may be minimized by adopting a floorplan such that the number of ROs, the number of stages in an RO, and stage dimensions can be easily altered for different applications.

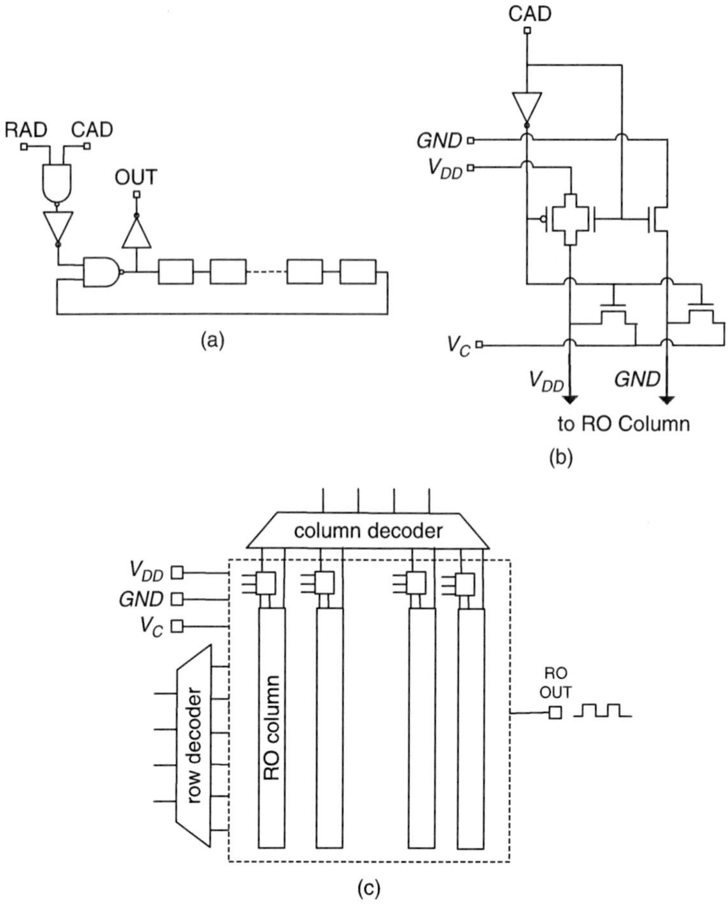

Fig. 6.23 **a** Circuit schematic of an RO with row and column selectors. **b** A selection circuit to connect V_{DD} and GND busses of unselected columns to $V_C = 0$. **c** A 2D array macro design with column and row decoders

6.3 MOSFET and Parasitic Parameter Extraction from ROs

In Chapters 3, 4, and 5, test structure designs for characterization of discrete resistors, capacitors, and MOSFETs are described. The non-linear *I–V* and *C–V* characteristics of MOSFETs are represented by a few parameters defined at fixed DC voltage bias points, V_{gs} and V_{ds}, or at a fixed drain-to-source current, I_{ds}. In a CMOS logic gate, V_{gs} and V_{ds}, and as a result I_{ds} values of each MOSFET may vary continuously during a switching transient. It is therefore convenient to define an effective resistance and capacitance, r_{sw} and C_{sw}, of each logic gate corresponding to a specific switching configuration, such as top, bottom, or both inputs switching simultaneously in a NAND2 (Section 2.4.2). With $\tau_p = r_{sw}C_{sw} = R_{sw}C_{sw}/W$, these delay parameters (τ_p, R_{sw}, and C_{sw}) coupled with the IDDQ of the logic gate are useful aids for circuit sizing and design optimization [9]. These delay parameters can be derived from circuit models and correlated with the values obtained from ring oscillators using the procedure described below.

The delay parameters and their relationships to MOSFET properties are introduced in Section 2.4.2 and covered in depth by Taur and Ning [4]. Here, we describe RO designs and a methodology for self-consistent determination of key MOSFET and interconnect properties from high-frequency operation of logic gates [10–12]. This approach is useful for validation of device models derived from DC characterization, for relating CMOS product performance to the DC properties of circuit elements measured in the manufacturing line and for AC performance optimization of silicon technology.

In the following discussion, RO stage designs for R and C parameter extraction are described. It is assumed that all stages in an RO are nominally identical and the number of stages in the RO is large (>50) so that any deviation caused by the random variations in MOSFETs and by inclusion of a NAND2 stage in the RO may be neglected. Average delay parameters of a single stage are then correctly determined from the measured time period (or frequency) of the RO.

An equivalent circuit schematic of two successive inverter RO stages is shown in Fig. 6.24. The shaded area in Fig. 6.24, indicates the inverter as a current source, charging or discharging the capacitances C_{in} and C_{out} each time the logic level applied to its input changes state from "1" to "0" or from "0" to "1." The signal propagation delay, τ_p, following Eq. (2.4) is expressed as

$$\tau_p = r_{sw}C_{sw} = R_{sw}\,(C_{in} + C_{out}), \tag{6.11}$$

where C_{in} and C_{out} are the input and output capacitances per unit width W of the inverter and R_{sw}/W is its effective switching resistance. In terms of measured RO

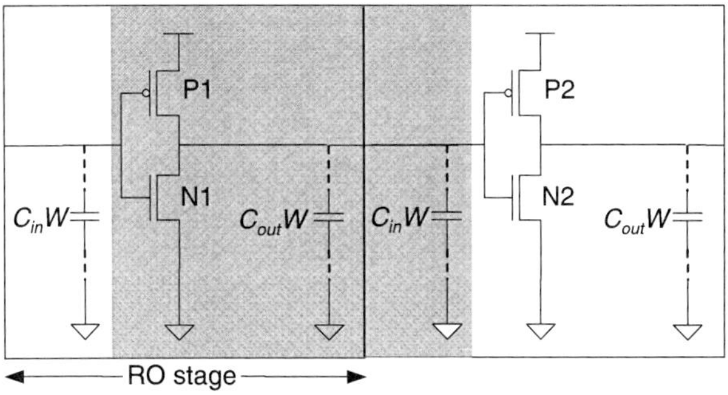

Fig. 6.24 A circuit schematic of two inverter RO stages delineated by *dashed lines*. The *shaded area* indicates the capacitance components in C_{sw}

parameters, C_{sw} is expressed in Eq. (6.7) and restated here:

$$C_{sw} = (C_{in} + C_{out})\,W = 2\tau_p \frac{(\text{IDDA} - \text{IDDQ})}{V_{DD}}. \tag{6.12}$$

The individual C_{in} and C_{out} components are determined from the measured C_{sw} of a pair of appropriately designed ROs. A load capacitance may be added to each stage, and its value determined by subtracting the capacitance of an unloaded stage, treated as a reference. With this differential measurement scheme, the capacitance values of different components in CMOS circuits may be obtained.

The r_{sw} of an inverter is a measure of the inverse of the average current drive strengths of the n-FETs and p-FETs during switching. The r_{sw} value for different logic gates is dominated by different regions of the I_{ds}–V_{ds} characteristics of the n-FETs or p-FETs or both. A judicious selection of logic gate designs allows us to probe different regions of the I_{ds}–V_{ds} curves and relate the differences in measured r_{sw} values to specific MOSFET properties such as V_t and source-drain resistance.

A key advantage of this methodology is that the physical layouts and local environment of MOSFETs and parasitic elements in a CMOS circuit are representative of circuits on a product chip, and their properties are derived under high-speed switching conditions. With the number of nominally identical stages in an RO > 50, the delay parameters are averaged over a large number of MOSFETs, and the influence of random local variations is negligible. This results in a significant reduction in test time, as measurements on a pair of ROs can be performed in less time than the time required for full DC characterization of >50 MOSFET pairs.

There are, however, some limitations in the methodology of extracting MOSFET properties from the delay parameters of a logic gate. In Fig. 6.25, voltage waveforms at the input and output nodes of an inverter, the I_{ds} values of the n-FET and the p-FET (I_{dsn} and I_{dsp}), and their gate capacitances (C_{gsn} and C_{gsp}), which contribute to both C_{in} and C_{out}, are plotted as functions of time. The transient current is dominated by the p-FET current drive during a PU transition and by the n-FET current drive during a PD transition. The gate capacitances also vary during a transition as these MOSFETs change their state. The R_{sw}, C_{in}, and C_{out} values represent time averaged currents and capacitances during transitions, and the measured τ_p is the average of the PU and PD delays. Hence, the n-FET and p-FET properties cannot be easily separated. In addition, except in the case of an inverter, there are more than one n-FET and one p-FET in each stage contributing to the propagation delay.

In general, the ring oscillator approach is suited for tracking relative shifts in parameters with silicon process or physical layout changes. We describe two classes of RO test structure designs for AC parameter extraction. In the first, covered in Sections 6.3.1 and 6.3.2, the R and C components are obtained from a differential pair of ROs. In the second, an additional input is connected to each RO stage to apply an independent voltage bias to the gate of a MOSFET. The RO stages are configured to either map discrete bias points in the C_{gs}–V_{gs} curve or determine relative changes in the average V_t value of a large number of MOSFETs. This second class of designs is covered in Sections 6.3.3 and 6.3.4.

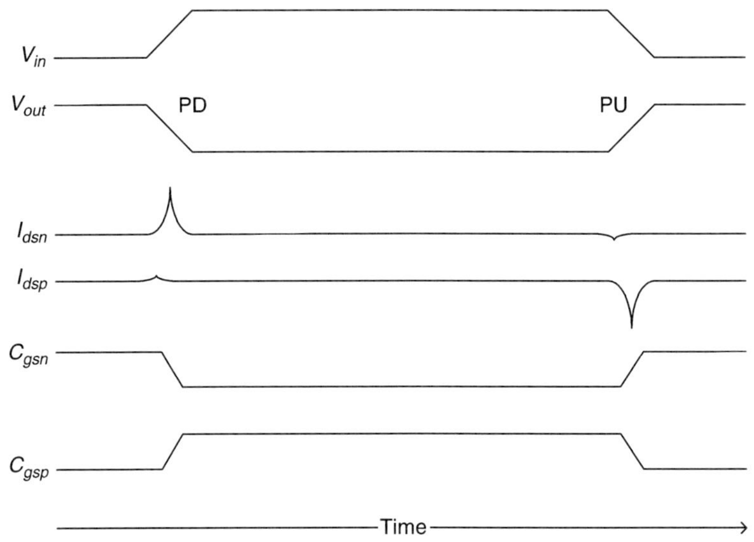

Fig. 6.25 Input and output voltages, I_{dsn}, I_{dsp}, C_{gsn}, and C_{gsp} as a function of time for a PD and a PU transition of an inverter

6.3.1 Capacitance Extraction

A set of three RO stage circuit schematics for capacitance extraction is shown in Fig. 6.26. The design and physical layout of the driving inverter in each RO stage circuit is identical. The RO stage in Fig. 6.26a comprising an unloaded inverter (FO = 1) serves as a reference for the RO stages in Fig. 6.26b, c and all other RO stage designs discussed here. Its R_{sw} is derived from the measured delay/stage, τ_{pa}, C_{swa}, and design value of W using Eqs. (6.11) and (6.12). The RO stage in Fig. 6.26b is that of an inverter driving two identical inverters. The third inverter load is that of the following RO stage (FO = 3). Each of the two load inverters is in turn loaded with a capacitance C_l equivalent to FO = 3, so that the switching characteristics of the three load inverters are nearly identical. The load capacitances may be constructed from MOSFET gate loads described in more detail later. The C_{sw} of this FO = 3 stage includes the $C_{in}W$ and $C_{out}W$ of the three inverters and $2C_l$. The delay/stage values of the two RO stages, τ_{pa} and τ_{pb}, are

$$\tau_{pa} = R_{sw}\left(C_{in} + C_{out}\right) + \frac{R_{sw}C_p}{W} \tag{6.13}$$

and

$$\tau_{pb} = R_{sw}\left(3C_{in} + C_{out}\right) + \frac{R_{sw}C_p}{W}. \tag{6.14}$$

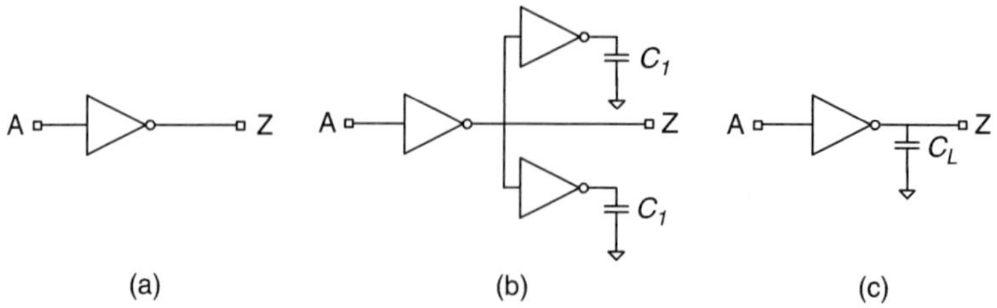

Fig. 6.26 Circuit schematics of inverter RO stages **a** with FO = 1, **b** with FO = 3, and **c** with a capacitive load C_L

Note that a capacitance C_p has been included in Eqs. (6.16) and (6.17). This capacitance, representing wire interconnections in each stage, is eliminated by subtraction as long as the physical layouts of all stages follow a template with a standard width and height, as shown in Fig. 6.8a. Assuming R_{sw} of the FO = 3 stage to be identical to that previously determined for the reference stage, C_{in} is obtained by subtracting Eq. (6.13) from Eq. (6.14):

$$C_{in} = \frac{\tau_{pb} - \tau_{pa}}{2R_{sw}}. \tag{6.15}$$

It is apparent from Eq. (6.14) that the delay/stage varies linearly with FO (=3 in this case). Knowing the delay τ_{pa} of an unloaded inverter stage and its C_{in}, the delay/stage can be calculated for any FO value. This relationship is useful for circuit sizing as discussed in Section 2.4.2. A direct extraction of C_{out} is not feasible as the parasitic capacitance C_p is an unknown. However, variations in C_{out} may be tracked in RO stages comprising different logic gates, provided C_p is maintained approximately constant by design.

In Fig. 6.26c, a capacitive load C_L, which may be a MOSFET or a metal wire, has been added. The switching capacitances of the stages in Fig. 6.26a, c, C_{swa} and C_{swc} respectively, are determined from Eq. (6.12) and expressed as

$$C_{swa} = C_{in} + C_{out} + C_p \tag{6.16}$$

and

$$C_{swc} = C_{in} + C_{out} + C_L + C_p. \tag{6.17}$$

The load capacitance C_L is determined by subtraction of the switching capacitance of the reference stage C_{swa} from the switching capacitance C_{swc} of the loaded RO stage.

The circuit schematics of an inverter, a NAND2 and a NOR2, are shown in Fig. 6.27. With the top input A switching, and the bottom input B tied to V_{DD} in the NAND2 and to GND in the NOR2, the switching behavior of these gates is similar to that of an inverter with a series n-FET, N2, in the NAND2 and a series p-FET,

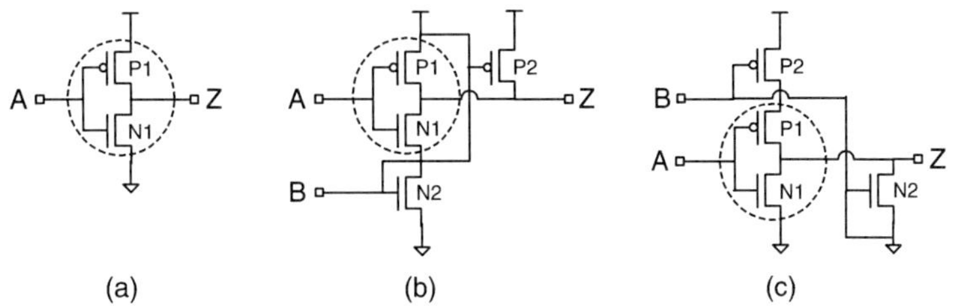

Fig. 6.27 RO stage designs with **a** an inverter, **b** a NAND2, and **c** a NOR2. In each case, the n-FET and the p-FET participating in the switching transition with signal input at node A are *circled*

P2, in the NOR2. If the n-FET and p-FET widths, W_p and W_n, and gate lengths, L_p, are selected to be the same in all three logic gates, then with input A switching, the C_{in} of all three gates is approximately the same. The C_{out} of the NAND2 and NOR2 is incremented by the parasitic drain-to-substrate junction capacitance C_j of N2 and P2 respectively. With increase in the number of NAND or NOR gate inputs (e.g., NAND3, NAND4), the increase in C_{out} is proportional to the number of series n-FETs or p-FETs in the stack.

The main n-FET and p-FET capacitance contributors to C_{in} and C_{out} are gate oxide capacitance C_{gT} which includes gate-to-source and gate-to-drain overlap capacitances C_{ov} and diffusion area, and junction capacitance C_j. The contribution to C_{in} of the gate-to-drain side C_{ov} is doubled by the Miller effect which comes into play as the voltages on the gate and drain terminals (A and Z) vary with time during switching events [4]. These individual capacitance components can be experimentally determined with RO stage designs shown in Fig. 6.28.

The output of the inverter in Fig. 6.28a has an n-FET and a p-FET gate-capacitor load. The source (S) and drain (D) terminals are tied to V_{DD} for the p-FET and GND for the n-FET. A physical layout of this stage with a gate load equivalent of FO = 2 is shown in Fig. 6.29. This layout, using the standard inverter design in Appendix A, can be done with just M1 metal level unlike the FO = 3 stage in Fig. 6.26b which requires two metal layers. Note that these MOSFET gate capacitors are not subject to the Miller effect as only the voltage at the gate (G) terminal varies with time. Hence, the gate capacitance obtained from this design has a smaller value than C_{in} of an equivalent inverter.

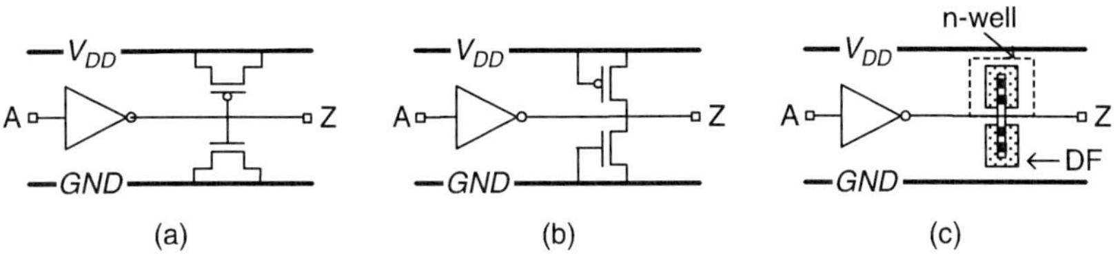

Fig. 6.28 RO stage circuit schematics for an inverter with n-FET and p-FET, and load of **a** gate capacitance, **b** overlap capacitance, and **c** silicon diffusion capacitance (physical abstraction)

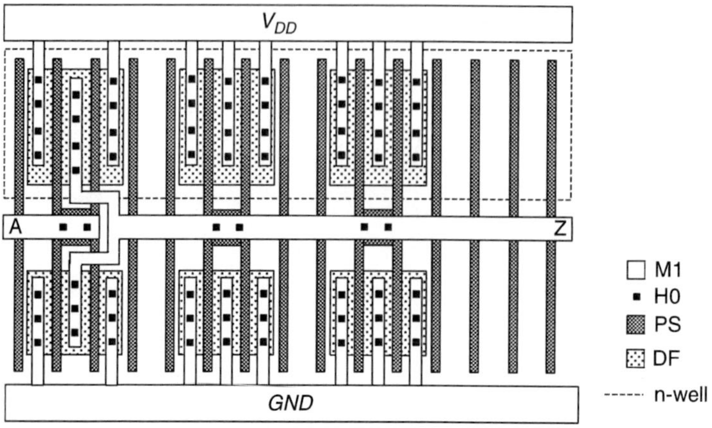

Fig. 6.29 Physical layout of an RO stage corresponding to the circuit schematic in Fig. 6.28a

The IDDQ of the stage in Fig. 6.28a includes gate oxide leakage current, a component of which is proportional to the gate width and length of the load MOSFETs. Gate oxide leakage current can be significant in the 90, 65, and 45 nm technology nodes. Its value may be obtained by subtracting IDDQ of the reference RO stage. For this purpose, the gate area of the load MOSFETs is increased by increasing the number of fingers or channel length.

The C_{ov} component can be characterized with an RO stage design shown in Fig. 6.28b. The source or drain is connected to the signal line, and the other two terminals of the MOSFET are tied to the GND (n-FET) or V_{DD} (p-FET) bus. An RO stage for measuring C_j which includes a physical representation of the DF island is shown in Fig. 6.28c. Contact to the diffusion area is made through M1 metal which is connected to the signal wire, and the contact resistance is minimized by using an adequate number of interconnect H0 vias. The measured capacitance includes both area and perimeter components of the diffusion capacitance.

Interconnect wire capacitances are measured primarily at a test stop later than M1. The capacitance of a wire is the sum of capacitances to its vertical and horizontal neighbors (Fig. 2.30). The values of the vertical capacitance components C_{up} and C_{down} vary with the inter-level dielectric thicknesses, effective dielectric constants, and metal linewidths. The lateral capacitance components C_{left} and C_{right} are functions of metal height and spacing and the dielectric properties of material separating the wires. Each of these capacitance components provides information on the wire geometry or the effective dielectric constant of the insulating layers. An RO stage design for extracting vertical capacitances is shown in Fig. 6.30, and another for extracting the total capacitance is shown in Fig. 6.31. In both cases, the signal wire is on M2, and orthogonal M1 and M3 wires (not shown) serve as GND planes above and below it. The wire resistance is minimized by creating a comb structure. In Fig. 6.31, the nearest neighbor M2 wires are also held at GND. The signal

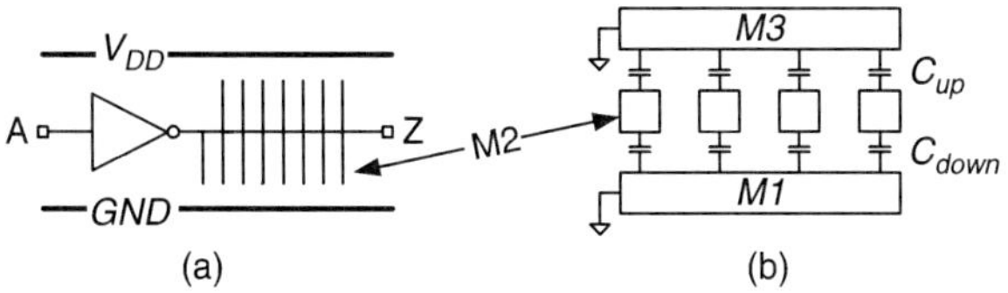

Fig. 6.30 **a** An RO stage design to extract inter-level dielectric properties. **b** Schematic cross section of M2 capacitor, with M3 and M1 GND planes

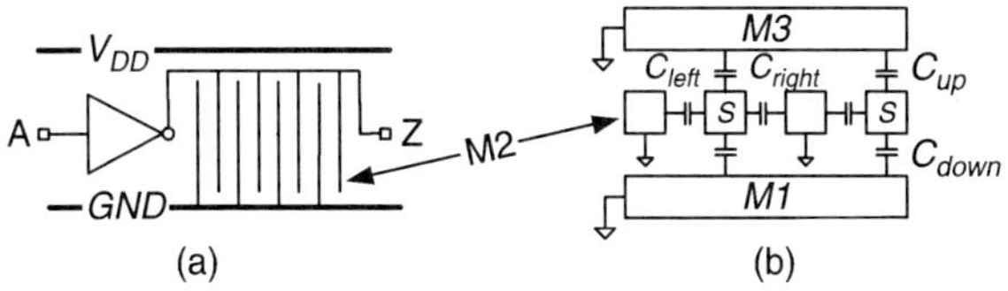

Fig. 6.31 **a** An RO stage design to extract total M2 wire capacitance. **b** Schematic cross section of M2 capacitor, showing M2 signal lines (S) with M2 neighbors and M3 and M1 GND planes

wire resistance in these configurations is negligible compared to the inverter driver resistance of a few hundred ohms. The wire capacitance is set equivalent to at least FO = 1, which is ~2–5 fF and corresponds to a wire length of ~10–25 μm. As the metal pitch increases for upper layers, the stage width and number of stages have to be adjusted to fit the ROs in a standard area in an RO macro template. Hence, this technique is more suitable for narrow wires in lower metal layers.

6.3.2 Resistance Extraction

The RO stage designs shown in Fig. 6.27 also provide R_{sw} values of inverters, NANDs, and NORs. It is instructive to examine the I_{ds}–V_{ds} trajectories of MOSFETs during switching to interpret the r_{sw} values of the logic gates. The trajectories of n-FETs in the inverter and the NAND2, overlaying the DC I_{ds}–V_{ds} characteristics, are shown in Fig. 6.32. Let us consider the trajectory of an n-FET in an inverter undergoing a PD transition shown in Fig. 6.32a. Initially, the inverter input voltage or the V_{gs} of the n-FET is "0," and the inverter output voltage or the V_{ds} of the n-FET is "1." The n-FET is in the off-state, with its $I_{ds} = I_{off}$. As the input voltage at node A transitions from "0" to "1," the n-FET V_{gs} and I_{ds} increase while V_{ds} is decreasing. When $V_{ds} \approx V_{DD}/2$, the inverter in the following stage begins its PU transition, and the signal has propagated through this inverter. As the n-FET passes through the saturation region, its I_{ds} begins to decrease with V_{ds} ultimately going to zero as the transition is completed. The instantaneous resistance of N1, at time t, $V_{ds}(t)/I_{ds}(t)$, averaged over the switching time is equivalent to the r_{sw} of the inverter.

In the case of a NAND2, with the top input A switching and the bottom input B, tied to V_{DD}, n-FET N1 at the top of the stack follows a similar $I_{ds} - V_{ds}$ trajectory, shown in Fig. 6.32b, as the n-FET in the inverter. With identical p-FET

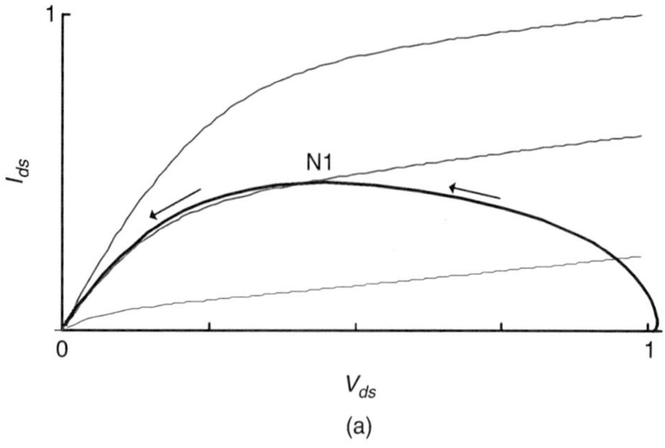

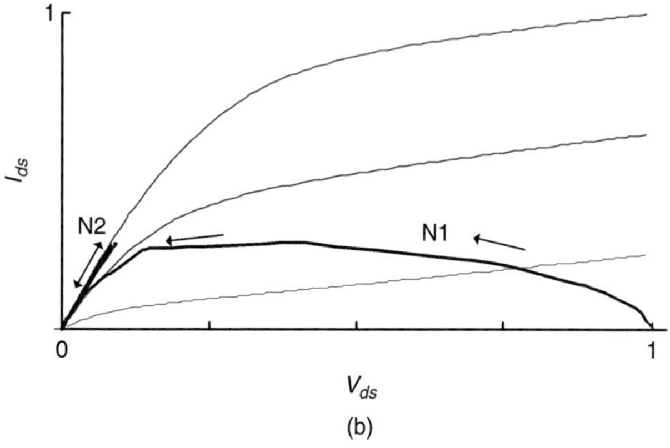

Fig. 6.32 $I_{ds} - V_{ds}$ trajectories, overlaying DC $I_{ds} - V_{ds}$ curves, of n-FETs during a PD transition of **a** an inverter and **b** a NAND2

and n-FET widths, W_p and W_n, in the inverter and NAND2, the contribution of N1 in the inverter and NAND2 to r_{sw} are comparable. There is an additional contribution to the r_{sw} of the NAND2 from the n-FET N2. Its I_{ds}–V_{ds} trajectory with $V_{gs} = V_{DD}$ and $V_{ds} < V_{DD}$ lies in the linear region, adding a constant resistance value to the r_{sw} of the inverter. The p-FET P2 meanwhile remains in the off-state.

The difference in the switching resistances of the NAND2 and the inverter is then the resistance of N2, which provides a measure of the source-drain resistance R_{ds} of the n-FET. In the same way, information on the R_{ds} of the p-FET is obtained from the NOR2 stage, with the inverter RO as a reference. The sensitivity of r_{sw} to R_{ds} is increased by increasing the stack height. Alternatively, the sensitivity to R_{ds} may be increased by reducing the widths of N2 or P2 in a NAND or NOR gate respectively, as R_{ds} is inversely proportional to the device width.

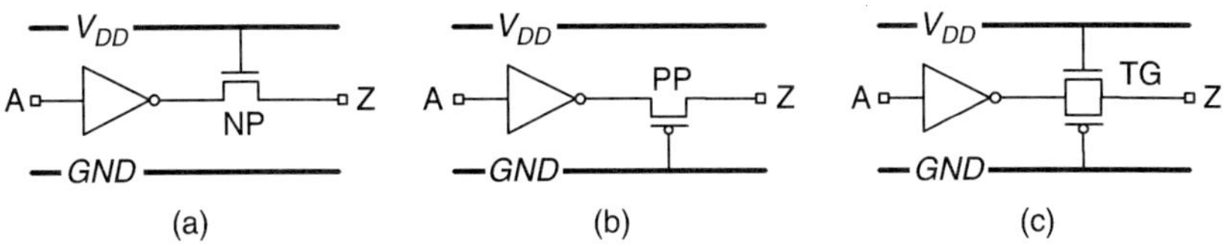

Fig. 6.33 Inverter driving **a** an n-passgate, **b** a p-passgate, and **c** a transmission gate

RO stages to characterize passgate or transmission gate circuits provide information on a different region of $I_{ds} - V_{ds}$ trajectory than a static logic gate such as an inverter, NAND, or NOR. In Fig. 6.33, RO stage designs of an inverter driving an n-passgate (NP), a p-passgate (PP), and a transmission gate (TG) are shown. The gate of the NP is tied to V_{DD} and it is always in the on-state. Let us consider the RO stage comprising an inverter driving an n-passgate, as shown in Fig. 6.33a. The $I_{ds} - V_{ds}$ trajectory of the n-passgate NP for an inverter PU transition is shown in Fig. 6.34. Initially, the output of the inverter is "0," the V_{gs} of the NP is equal to V_{DD}, and its $V_{ds} = 0$ as it passes a "0" to node Z. During the transition, as the output voltage of the inverter begins to increase, the V_{ds} of NP increases. At the end of the transition, its $V_{gs} = 0$ and $V_{ds} \sim V_t$. The n-FET traverses the low $V_{gs} - V_{ds}$ region, and the effective switching resistance of the NP is sensitive to its V_{tlin}. The RO delay sensitivity to V_{tlin} is increased by choosing the NP width to be much smaller than the widths of the MOSFETs in the inverter, as this increases the resistance contribution of the NP to the measured r_{sw}. In a similar fashion, the delays of the RO stages in Fig. 6.33b, c are sensitive to the V_{tlin} of the p-FET and a combined effect of n-FET and p-FET respectively. The capacitance increase in the stages in Fig. 6.33 compared to a reference inverter stage is related to the overlap capacitances of the passgate transistors. The IDDQ values of the all three ROs are that of an inverter and any additional gate oxide leakage currents in the passgates.

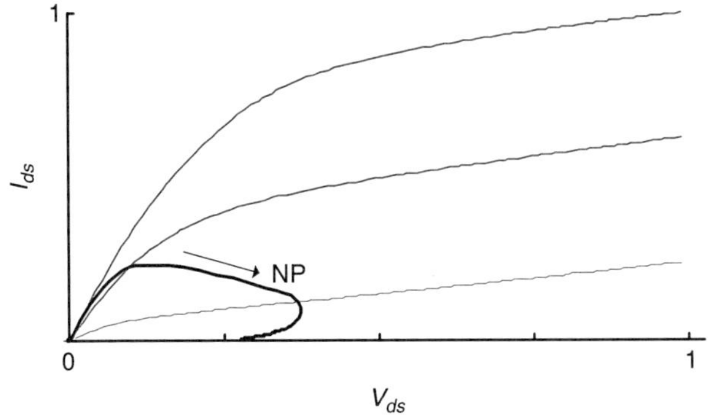

Fig. 6.34 I_{ds}–V_{ds} trajectory of n-passgate NP for a PU transition overlaid on the DC I_{ds}–V_{ds} plots

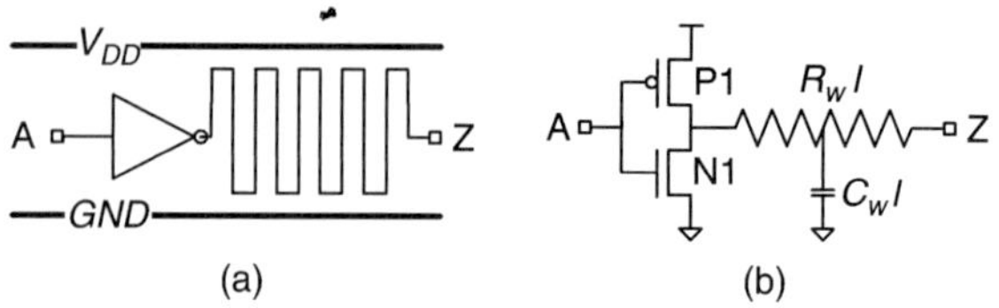

Fig. 6.35 **a** Schematic representation of inverter driving a wire load. **b** Circuit elements of the stage

An RO stage design for tracking of interconnect wire resistance is shown in Fig. 6.35. For an inverter driving a wire load of length l, the delay per stage (Eq. 2.12) is given by

$$\tau_{\mathrm{p}} = R_{\mathrm{sw}}\left(C_{\mathrm{in}} + C_{\mathrm{out}}\right) + \frac{R_{\mathrm{sw}} C_{\mathrm{w}} l}{W} + R_{\mathrm{w}} l C_{\mathrm{in}} W + \frac{R_{\mathrm{w}} C_{\mathrm{w}} l^2}{2}, \tag{6.18}$$

where C_{w} and R_{w} are the wire capacitance and resistance per unit length. The RO stage delay is more sensitive to changes in wire resistance if $R_{\mathrm{sw}}/\mathrm{W} \lesssim R_{\mathrm{w}} l$. Hence, a wide inverter is used to drive a long signal wire with the upper and lower metal layers serving as ground planes. In this case, an RO stage with an unloaded inverter of the same design serves as the reference stage.

Although the I–V and C–V characteristics of MOSFETs are non-linear, the delay/stage varies linearly with load capacitance or equivalent FO as shown in Fig. 6.36a, b. The logic gate capacitances (C_{in} and C_{out}) and R_{sw}, which are estimated from measurements of RO frequency and power, vary in a linear fashion with stack height as graphically illustrated in Fig. 6.36c, d. Hence, the average properties of a large number of MOSFETs over a switching cycle of a logic gate can be derived from a limited number of DC and low-frequency measurements on ROs.

If only RO frequencies are measured because of test time constraints, or RO power supplies are not isolated, ratios of frequencies or stage delays of loaded ROs with reference ROs can be used for tracking process variations [12]. Data analysis and data visualization techniques for tracking variation in process parameters from measured RO frequencies are described in Chapter 10.

6.3.3 MOSFET C–V Characterization

An RO can be configured to extract C–V characteristics of MOSFET gate capacitance by adding an independently controlled DC input signal to each stage. One example of an RO stage with this arrangement is shown in Fig. 6.37a in which an inverter drives an n-FET (or p-FET) gate load and a DC input signal V_{CG} is applied to its gate terminal. This RO design is used for mapping the $C_{\mathrm{gT}} - V_{\mathrm{gs}}$ characteristics of the n-FET load [13]. The RO is operated at a low V_{DDE} value (<0.4 V). The inverter output voltage provides a small signal excitation for measuring C_{gT} at

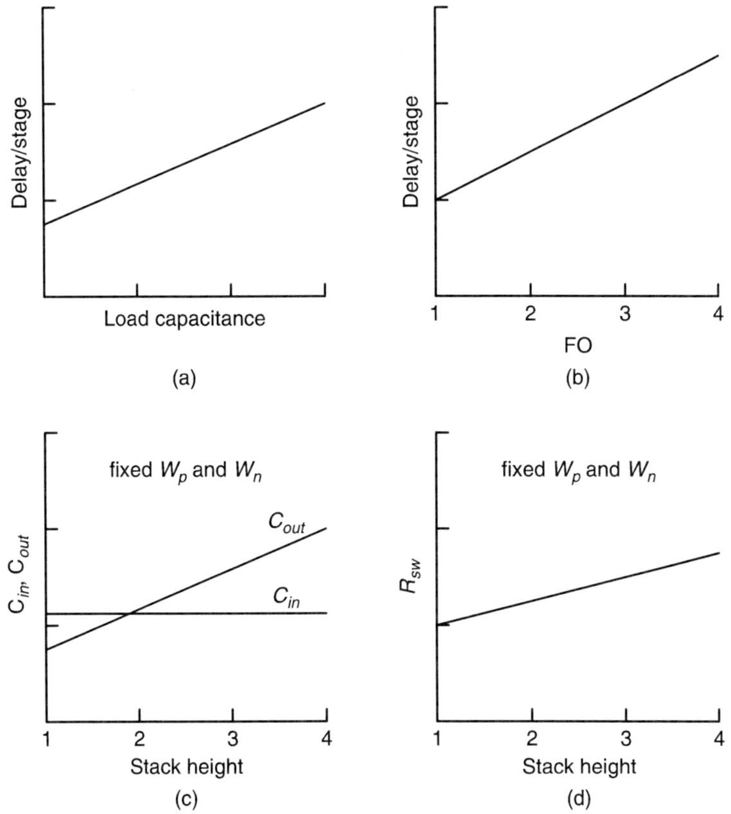

Fig. 6.36 Delay/stage as a function of **a** load capacitance and **b** FO for a static logic gate. **c** C_{in} and C_{out} and **d** R_{sw} as a function of stack height for fixed W_p and W_n in an inverter, NANDs, and NORs

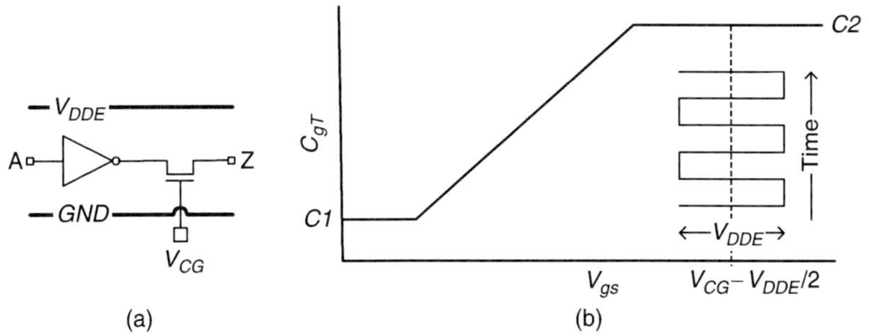

Fig. 6.37 **a** An RO stage with independently controlled V_{CG} bias. **b** $C_{gT}-V_{gs}$ characteristics of an n-FET with a small signal excitation voltage swing of V_{DDE} supplied by the inverter at a DC bias point of $V_{gs} = (V_{CG} - V_{DDE}/2)$

different discrete bias points on the C–V curve. The n-FET gate capacitance C_{gT} as function of its V_{gs} value is shown in Fig. 6.37b. The gate-bias voltage is given by

$$V_{gs} = V_{CG} - \frac{V_{DDE}}{2}. \tag{6.19}$$

An unloaded inverter RO serves as a reference, and the gate capacitance of the n-FET is obtained by the subtraction technique discussed in Section 6.3.1.

The MOSFET C_{gT}–V_{gs} characteristics obtained from this RO design are averaged over a large number of MOSFETs (>50). The capacitance measurements are carried out at a very high frequency (~1 GHz), which is equal to the frequency of operation of the RO while only DC or low-frequency I/O signals are used. The gate capacitance can be measured in the presence of significant gate oxide leakage current, as is the case for the 90 nm technology node and beyond, prior to the introduction of high-k dielectric (HK) materials. The capacitance values $C1$ and $C2$ in Fig. 6.37b are that of the n-FET in depletion and inversion modes respectively. The difference ($C2$–$C1$) is a measure of the effective electrical channel length of the n-FET. The voltage bias at $C_{gT} = (C2 - C1)/2$ or at any selected point in the transition region is a measure of the V_t of the n-FET.

The simulated $C_{gT} - V_{gs}$ plots obtained from this type of RO design for n-FETs of three different V_t values are shown in Fig. 6.38a and for three different channel lengths (L_p) in Fig. 6.38b. With a common reference RO, a total of seven ROs are used. Test time may become significant for detailed mapping of $C_{gT} - V_{gs}$ characteristics, as one frequency measurement is required for each V_{gs} bias point. Measurement at three V_{gs} bias points for each RO, shown as dark circles in Fig. 6.38, may be sufficient for rapid tests and to track relative differences in V_t and L_p values of the n-FETs arising from design or process variations. In addition, the gate oxide thickness t_{ox} may be determined from the C_{gT} of a long channel n-FET load. Note that this technique has some errors associated with systematic and random variations in inverters between the reference RO and ROs for $C_{gT} - V_{gs}$ characterization. However, it provides a self-consistent determination of all the essential AC and DC properties even in the presence of significant gate oxide leakage current.

The RO design with the V_{CG} bias lead in each stage is shown in Fig. 6.39a. This design may be implemented at the M1 metal level by modifying the RO macro design in Example 2. The physical layout of the RO for M1 implementation is shown in Fig. 6.39b. The GND busses are split to allow the V_{CG} leads traversing the physical width of the RO, to be connected to all the stages. The underpasses for the V_{CG} wire are in DF or PS levels. Care must be taken to ensure that the IR drop in the V_{CG} wire arising from the gate oxide leakage current flowing through it is small. The configuration of such ROs in an M1 test structure is shown in Fig. 6.40. Two ROs may share a V_{CG} lead and a maximum of five such RO pairs can be accommodated in a standard 1×25 padset macro.

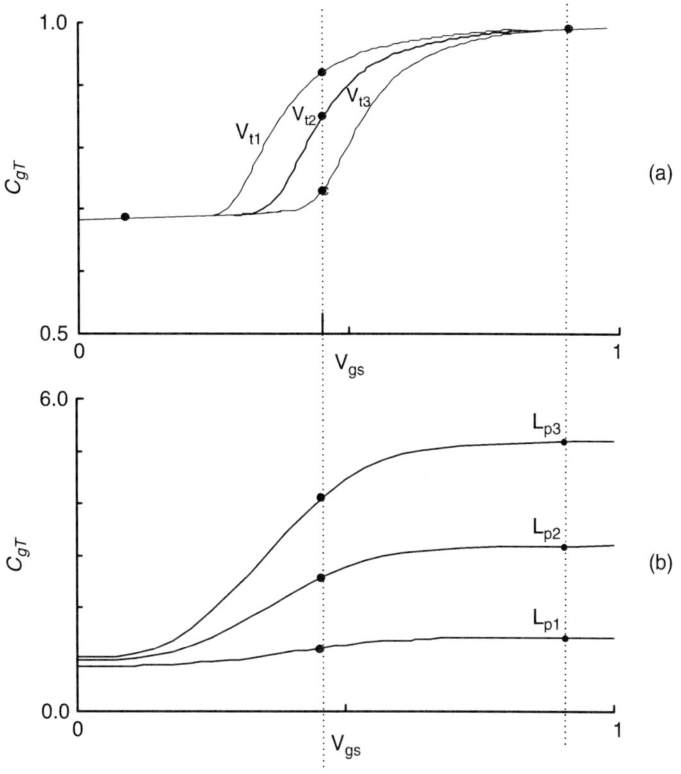

Fig. 6.38 C_{gT}–V_{gs} curves obtained from ROs for n-FETs of **a** three different V_t values and **b** three different L_p values. Reproduced from [11], with permission, © 2008 IEEE

6.3.4 ΔV_t Extraction

A second example of an RO design with an independent V_{CG} input, as described in Section 6.3.3, can be used for measurement of small differences in average V_t of two MOSFETs types. In this case the V_{CG} bias is applied to the gate of n-passgate or a p-passgate driven by an inverter. The schematic of such an RO stage with an n-passgate is shown in Fig. 6.41a. Note that a small p-FET keeper is added to assist the PU transition. If the RO stage has a p-passgate, an n-FET keeper is used to assist the PD transition. The effective resistance of the n-passgate and hence the RO delay/stage is modulated by the V_{CG} bias, in a similar way as it would be modulated by a change in its V_t. In Fig. 6.41b the simulated change in delay/stage, $\Delta\tau_p$, is plotted as a function of change in V_{CG}, $\Delta\ V_{CG}$, or a change in V_t, $-\Delta V_t$, of the passgate. The response of $\Delta\tau_p$ to ΔV_{CG} and $-\Delta V_t$ is identical, at least for swings of $<\pm 40$ mV. The passgate operates in the low V_{ds} region as is apparent from the $I_{ds} - V_{ds}$ trajectory of an n-passgate is shown in Fig. 6.34. Hence the measured $-\Delta V_t$ is related to the change in V_{tlin}.

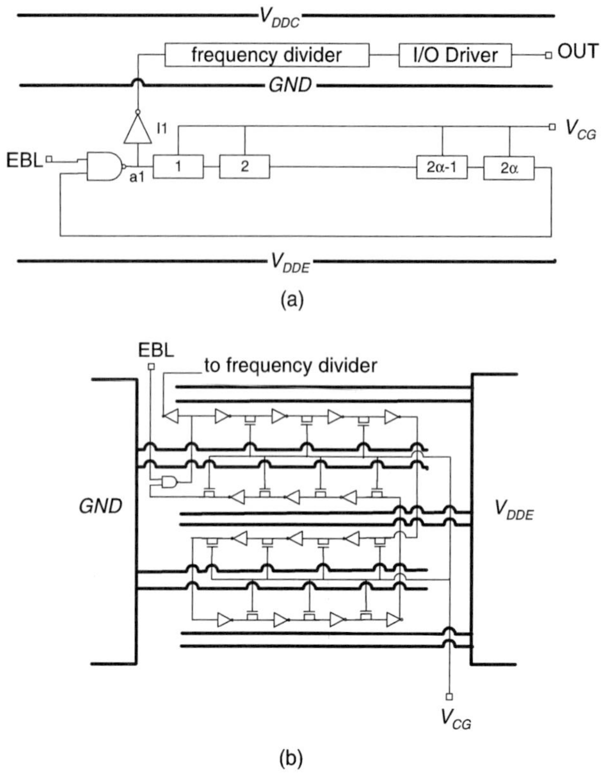

Fig. 6.39 **a** Schematic of an RO circuit with V_{CG} function. **b** Physical layout of a part of RO with V_{CG} leads placed in the split GND bus for implementation at the M1 metal level

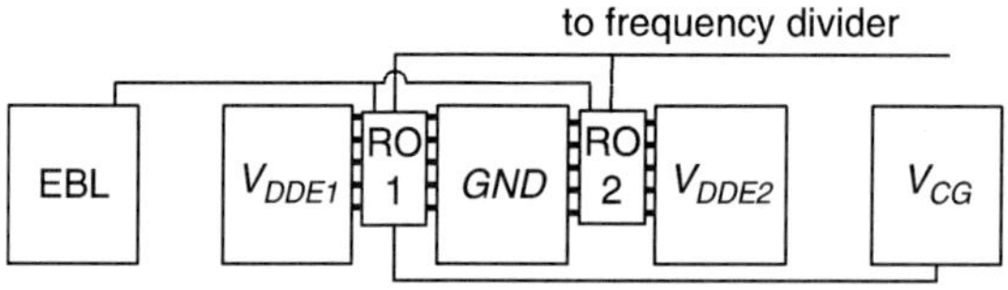

Fig. 6.40 Physical layout of RO test structure with a common V_{CG} input for two adjacent ROs. The ROs, testable at the M1 metal level, can be accommodated in a macro design similar to that described in Example 2

A pair of ROs is designed with their stages differing only by the V_t of the n-passgate (p-passgate), and the number of stages in each RO is large ($\alpha \geqslant 50$). The difference in the average V_t values of the passgates in the RO pair can be obtained within an accuracy of a few mV. One of the ROs serve as the reference and several points on the $\Delta\tau_p$ vs. ΔV_{CG} plot are measured to get a calibration curve shown in Fig. 6.41b. Using this calibration curve, the difference in the average V_t values of the passgates ΔV_t is determined from the difference in τ_p of the two ROs at

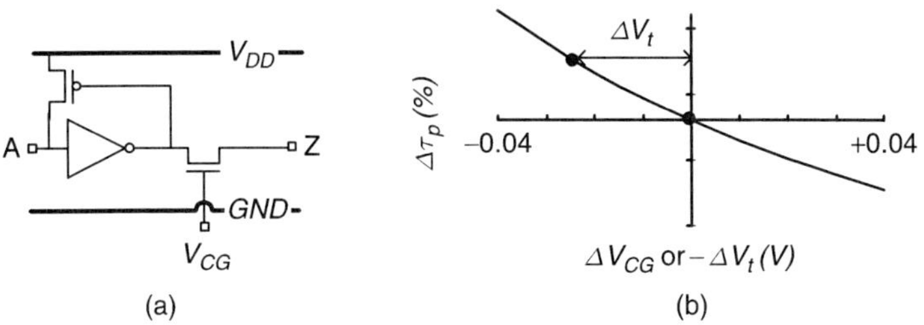

Fig. 6.41 **a** Schematic of an RO stage circuit with V_{CG} wire for measuring ΔV_t. **b** A simulated plot of $\Delta\tau_p$ as a function of ΔV_{CG} or $-\Delta V_t$

$V_{CG} = 0$, $\Delta\tau_p$. The advantage of this technique is that an average difference in V_t of a 100 FETs is obtained with a few RO frequency measurements and the test time is significantly reduced.

This technique is particularly useful for measuring V_t degradation in p-FETs with aging [14]. For a p-FET in the on-state, an increase in its V_t is observed with time from the NBTI effect. This effect is also present in n-FETs (PBTI) in 45 nm technology nodes and beyond where an HK gate-dielectric is used. The V_t degradation is accelerated with increase in V_{gs} and temperature. An average V_t shift is obtained by first obtaining a $\Delta\tau_p$ vs. ΔV_{CG} ($V_{CG} = -V_{DDE}$ for p-FETs) calibration curve for the RO. The RO V_{DDE} is turned off and with the source and drain of the p-FET pass-gates at GND, a negative V_{CG} (positive V_{CG} for the n-FET) is applied for a specified time. In this configuration, the MOSFETs in the inverters are in the off-state and do not undergo any degradation. Following the stress, the V_{DDE} is turned on and with $V_{CG} = -V_{DDE}$, τ_p is measured again. The shift in V_t after stress is determined from the measured $\Delta\tau_p$ and using the calibration of $\Delta\tau_p$ vs. ΔV_{CG}. Here the RO under pre-stress conditions serves as the reference.

Alternatively, the offset in V_{CG} from V_{DDE}, necessary to return the RO frequency to its original pre-stress value (with initial $V_{CG} = -V_{DDE}$ for p-FETs) is a direct measure of ΔV_t. In accelerated stress tests, a pair of ROs may be used and only one RO is stressed. The unstressed RO is measured along with the stressed RO each time to ensure that its delay/stage remains unchanged and that the test conditions are identical for all measurements. This RO design may be implemented at the M1 metal level.

6.4 Special RO Applications

In this section several additional applications of ring oscillators are described. The applications cover precise measurement of circuit delays and variability and model-to-hardware correlation of memory and analog circuits. Some of the RO designs described here may require new macro template designs to include additional control I/O signals.

6.4.1 Precise Measurements of Circuit Delays

The RO test structures described so far include a NAND gate or similar scheme in the loop to enable/disable the oscillations and a load inverter to couple the RO output to the peripheral circuits such as multiplexers or a frequency divider. The error introduced in measured circuit delay τ_p from these perturbations can be reduced by either increasing the number of stages $(2\alpha + 1)$ or obtaining the error correction from SPICE simulations. An alternative is to design two ROs with different number of stages and obtain a precise measurement of τ_p

$$\tau_p = \frac{(f_1 - f_2)}{2f_1f_2(\alpha_1 - \alpha_2)}, \tag{6.20}$$

where f_1 and f_2 are the frequencies of the two ROs with $(2\alpha_1 + 1)$ and $(2\alpha_2 + 1)$ stages respectively. The effect of random statistical variations in MOSFET parameters on τ_p may be reduced with $(\alpha_1 - \alpha_2) > 25$. This technique can also be used to validate a macro template design and to ensure that the effects of the enable circuit and inverter load are within the desired precision limit.

With additional control input signals, a single RO with switches to select the number of stages in the loop may be used as shown in Fig. 6.42 [15]. Control signals for the switches S1, S2, and S3 with true and complementary voltage levels Q and Q_n are provided. In this case, by exercising switches S1 and S2, the average delay of two stages labeled 1 and 2 is measured.

A similar approach can be used for measuring the delay of a single large circuit block. A non-inverting circuit block is inserted in one RO of the pair and its delay determined from the measured frequency difference between the ROs. A more precise delay determination is obtained with the arrangement shown in Fig. 6.42 where stages 1 and 2 are replaced by a non-inverting circuit block of interest. A number of such circuit blocks, varying in design, may be inserted in a single RO with a decoder scheme to select a circuit block. Care should be taken to keep the input and output loads of the circuit the same as the other stages in the RO. This technique is area efficient when designing ROs with large circuit blocks and when control signals are readily available, as for example, on a product chip.

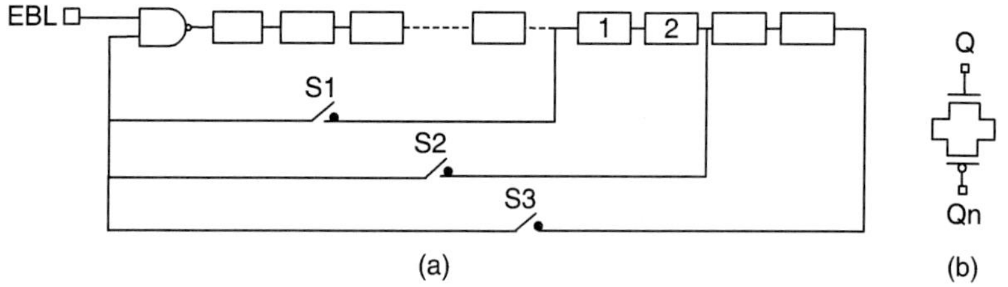

Fig. 6.42 **a** An RO design with switches S1, S2, and S3 to select the number of stages. **b** A transmission gate switch configuration for S1, S2, and S3

6.4.2 Matched RO Pairs

In Example 4 of this chapter, an RO macro design for measuring random variability in circuit delays is described. Here, we cover several other techniques for measuring random and systematic circuit delay variations based on a pair of ROs. For random variability, a number of nominally identical RO pairs are measured within a chip or within a wafer and the standard deviation of delay of a single stage σ is derived from the standard deviation of the measured difference in the average delay per stage of each of the RO pairs σ_r where

$$\sigma = \sigma_\mathrm{r}\sqrt{\frac{(2\alpha + 1)}{2}}. \tag{6.21}$$

Special care must be taken to minimize any systematic variations in the RO pair. The physical layout of the ROs may be carried out with interleaving stages as shown in Fig. 6.43. In this arrangement, any variation arising from spatial separation and local physical environment between ROs is minimized.

A single RO, with two independent control signal inputs for each stage, may be used for comparing systematic differences in stage delays incorporating two different MOSFET layouts. A circuit schematic of such an RO stage for comparing differences in p-FET layouts is shown in Fig. 6.44a. By setting CT1 = "0"

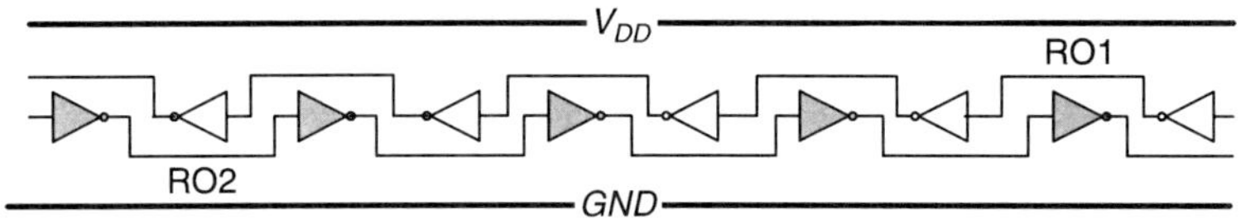

Fig. 6.43 A section of a matched pair of ROs, RO1 and RO2, with interleaved stages. The inverters in RO2 are shaded in gray

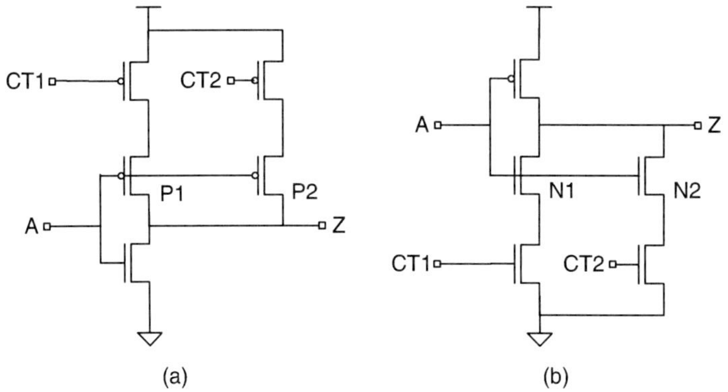

Fig. 6.44 Circuit schematics of an inverter stage with control inputs CT1 and CT2 for selecting **a** p-FET P1 or P2 and **b** n-FET N1 or N2

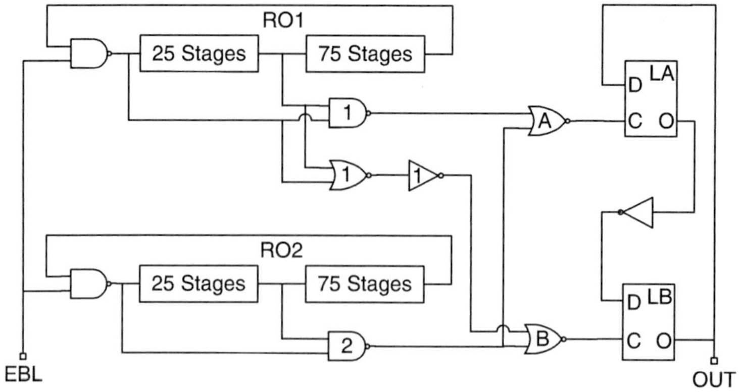

Fig. 6.45 Circuit to measure difference (BEAT) frequency of two 101 stage ROs, RO1 and RO2. Reproduced from [14], with permission, © 2007 IEEE

and CT2 = "1," P1 is included in the inverter circuit. With complementary inputs, CT1 = "1" and CT2 = "0," P2 is included instead. A similar design for selecting any one of two n-FETs is shown in Fig. 6.44b.

Accurate determination of the difference in frequency values for a pair of ROs requires that this difference be much larger than the error or uncertainty in a single measurement. Measurement accuracy is improved by directly measuring the difference in frequencies of two ROs. A circuit to measure the frequency difference of a pair of ROs or the beat frequency is shown in Fig. 6.45 [14]. The output of this circuit is at half of the frequency difference between RO1 and RO2. This circuit is particularly useful for measuring frequency shifts in reliability tests for BTI and hot-electron effects. By improving the measurement accuracy, stress conditions and time may be significantly reduced. In such experiments, only one RO in the pair is subjected to accelerated stress conditions, and the beat frequency is measured periodically to obtain the frequency degradation. Additional logic may be incorporated to measure the individual frequencies of RO1 and RO2 as well as their beat frequency. RO stage designs described in Section 6.3.4, with higher sensitivity to changes in the MOSFET V_t, further enhance the frequency shift signal. Stress times as low as few seconds may be sufficient, and the reliability tests may be carried out in the manufacturing line without a significant impact on test time.

6.4.3 SRAM ROs

Performance of memory and analog circuits can also be measured in a ring oscillator configuration. The circuit schematic of a 6T SRAM cell is shown in Fig. 6.46a, with word line ports W1 and W2 and bit-line ports B1 and B2. A number of different RO

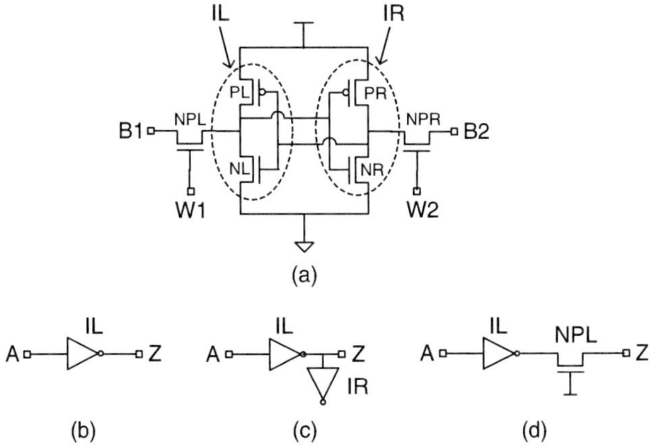

Fig. 6.46 **a** Circuit schematic of a 6T SRAM cell. Circuit schematics of RO stages with **b** inverter IL, **c** inverter IL having load IR, and **d** inverter IL having passgate load NPL

stage designs may be composed from the inverters, IL and IR, and the n-passgates, NPL and NPR. In Fig. 6.46b, d, RO stages for the inverter IL, IL with load IR, and IL with an n-passgate (NPL) load are shown. Similar stages may be constructed for the IR inverter with IL inverter and NPR loads. If the physical layout of the cell is not perfectly symmetric, comparison of RO stage delays for the IL and IR inverters provides useful information on any unintentional asymmetry introduced by fabrication process recipes.

An RO stage configured to measure the write delay of an SRAM cell is shown in Fig. 6.47a [16]. This circuit is drawn with inverters IR and IL in Fig. 6.47b. The word line inputs W1 and W2 (in Fig. 6.46a) are connected to the inputs of the cross-coupled inverters in the latch. Signal propagation takes place from Z1 and Z2 of one stage to A2 and A1 respectively of the following stage, as shown in Fig. 6.47c. With EBL = "0," the RO is initialized with node a1 at GND. The oscillations are enabled by setting EBL = "1." There are an odd number of SRAM stages in the RO as the standard NAND2 enable scheme is not used. The RO output is coupled to a frequency divider circuit which may be shared with other ROs in a standard RO macro template.

In high-density SRAM memories, MOSFET widths are smaller than in logic circuits and customized OPC for each SRAM cell design is needed to faithfully reproduce the design dimensions of the circuit elements. It is therefore important that in the physical layout of an SRAM ring oscillator, the local environment of the modified SRAM cells is closely matched with that of an SRAM memory block. Filler SRAM cells, as described in Section 5.3, may surround each RO stage. Capacitances of interconnect wires and of the inverter at the output should be kept at a minimum because of the relatively small current drive strengths of the MOSFETs in the SRAM cell.

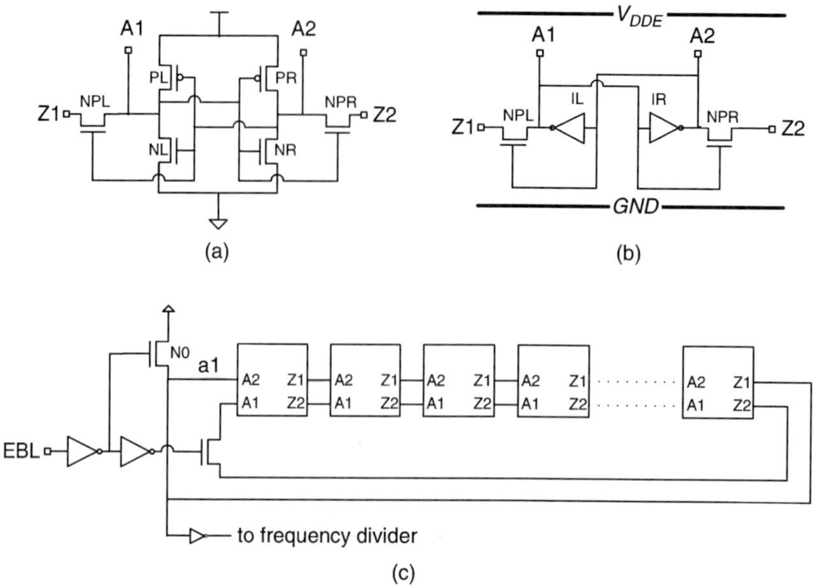

Fig. 6.47 **a** and **b** Circuit schematics of an SRAM stage configured to operate in the "write" mode. **c** Ring oscillator comprising an odd number of SRAM cell stages

6.4.4 Voltage Controlled Oscillators

In CMOS products, a phase locked loop (PLL) circuit is employed for generating a high-frequency internal clock in synchronization with an external clock [1, 2]. An essential component of a PLL is a voltage controlled oscillator (VCO) whose output frequency is a function of the magnitude of an input control voltage. The VCO design in a PLL may be independently characterized in a standard RO macro template with provision for an additional input to supply the control voltage.

One example of a VCO circuit is shown in Fig. 6.48. It comprises an odd number of inverter stages, each with an additional series p-FET and n-FET. The input voltage V_r along with a current mirror circuit comprising Pm and Nm, sets a limit on the maximum current drive of the MOSFETs in the inverter, P1 and N1, which can be consequently starved for current. The output frequency of the VCO is thus controlled by V_r. This current starved inverter configuration in Fig. 6.48 may also be used for varying the ratio of PU and PD delays such that the RO frequency is dominated by either p-FET or n-FET. Instead of input voltage V_r, independent voltage bias control signals can be provided for P2 and N2. The maximum current through P1 and N1 can then be controlled independently, with the stage delay dominated by PU delay for a p-FET starved inverter and by PD delay for an n-FET starved inverter.

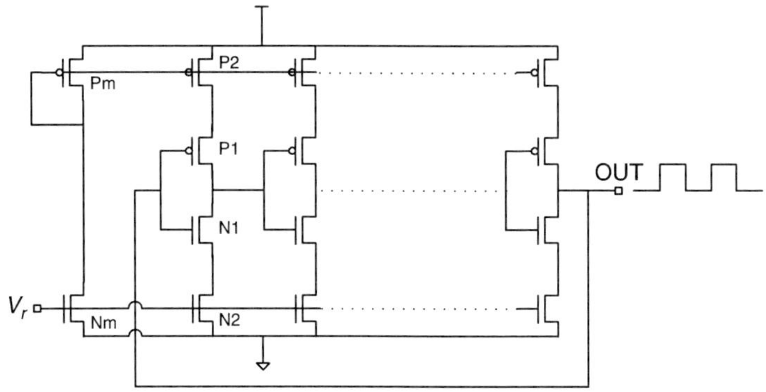

Fig. 6.48 Circuit schematic of a current starved voltage controlled oscillator

6.5 On-Product ROs

The RO test structure examples in Sections 6.2, 6.3 and 6.4 are situated in an independent padset or pad cage, suitable for placement in the scribe line or on a test chip. These types of designs are useful for monitoring the process, understanding process–circuit interactions and building models for circuit simulations. The test structures in the scribe line, however, may not correctly reflect the circuits on the product chip because of across reticle field variation and differences in circuit topologies and physical layouts. It is highly desirable to integrate ROs in the CMOS product itself, to more directly monitor circuit performance within the product, and to facilitate correlation of the product performance with that of the ROs placed in the scribe line. By distributing the ROs across a product chip, the across-chip and across-wafer variations in MOSFET parameters can be monitored and the process tuned to minimize these variations. On-product ROs are also utilized to measure the impact of process changes and to monitor circuit performance over the life-time of the product [17].

An example of the placement of spatially distributed ROs is shown in Fig. 6.49a. There are four product chips, separated by scribe lines, on each reticle field of the wafer. Here, RO locations are on a uniform grid to map across-chip variations. The RO locations are generally subject to available space and wiring tracks on the product. ROs may also be strategically placed near critical chip logic and memory functional blocks. Each location may have a set of ROs to monitor functional blocks in its vicinity, such as logic, memory or PLL, or variations in parameters such as L_{p} and V_{t}.

If ROs are placed at the edges of the chip and have dedicated I/Os, both frequency and power measurements on individual ROs can be made. It is generally preferable to distribute the ROs across the chip with the ROs sharing the chip power grid and I/Os. In this case, RO enable and output selection circuitry is integrated with the chip logic. The RO frequency may be read out by an on-chip frequency counter circuit and no special external test equipment for frequency measurement is required.

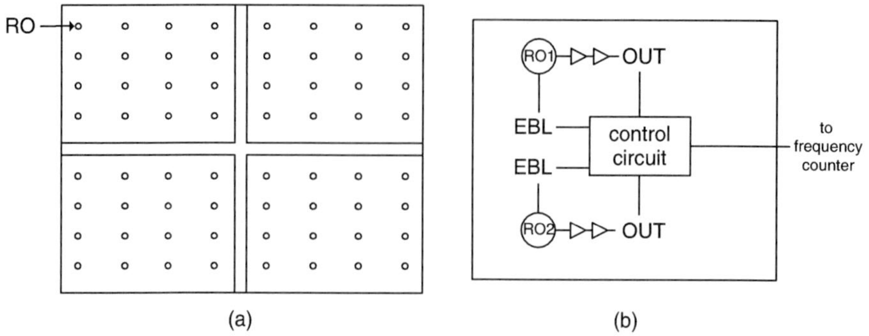

Fig. 6.49 **a** Placement of ROs on a reticle field with four product chips. **b** Integration of spatially separated ROs with a common control unit

In high-performance chips such as a microprocessor, there may be variations in temperature and V_{DD} across the chip when the chip clock is running, which in turn may influence the RO frequencies. In order to determine systematic across-chip variations arising from silicon processing, the RO frequencies must be measured without enabling the chip clock. A macro with an RO selection scheme and a common frequency divider circuit of the type described in Example 3 may be included on the chip and the frequency measurements carried out with an off-chip frequency counter. Buffers are placed in the RO output signal path to the control macro as indicated in Fig. 6.49b.

Because of systematic variations in circuit properties across chip and local random variations in circuit elements as well as drift in process parameters, there is a spread in RO frequencies measured on chips from wafer lots processed over a period of time. This is illustrated in Fig. 6.50 with measured frequencies of RO1 plotted

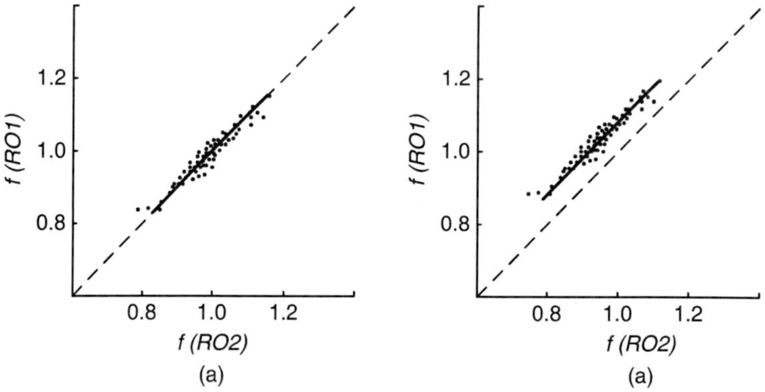

Fig. 6.50 Normalized measured frequencies of RO1 vs. RO2 from many chips on many wafers, with each *dot* representing one chip: **a** with 1:1 correspondence and **b** with a systematic offset

against measured frequencies of RO2 on each chip where the frequencies are normalized to the target values. The solid line is a linear fit to the data and the dashed line indicates a 1:1 correspondence. The frequencies are spread over 0.8 and 1.2 on different chips because of process variations across wafers, with most of the data points clustered around 1.0, as is generally the case. In Fig. 6.50a, the frequencies of RO1 and RO2 are matched, and in Fig. 6.50b, RO1 has a higher frequency than RO2 arising from a systematic across-chip variation. Data analysis techniques for tracking systematic variations in distributed ROs and for correlating RO and product performance are covered in more detail in Chapter 10.

The on-product RO frequencies typically correlate with the maximum frequency of operation of a product f_{max} defined at a fixed V_{DD} [18]. Measurement of product f_{max} takes up considerable test time, whereas individual RO frequencies can be measured rapidly. An RO design that most closely follows f_{max} is selected for sorting product chips into different frequency bins. Such ROs, strategically located on the products, form a group of performance port ring oscillators (PSROs).

Product chips are tested at the wafer level after completion of processing or after dicing and packaging. Correlation of measured frequencies of on-product PSROs with identical RO designs placed in the scribe line is carried out routinely to facilitate prediction of product f_{max} from RO measurements at the M1 metal level. This provides feedback for process centering and tuning in the manufacturing line to the desired f_{max} range early in the process. The relationships of f_{max} and scribe-line PSRO frequency to on-product PSRO frequency f(RO) is validated with a linear fit of the data to provide calibration curves as shown in Fig. 6.51a, b.

Idealized RO leakage power and active power as a function of frequency are plotted in Fig. 6.51c, d and have behaviors similar to product standby and active power respectively. Thus, RO measurement in the scribe line along with appropriate calibration curves can be used to predict product power as a function of operating frequency and also for process tuning for power-performance optimization.

The local temperature and V_{DD} on a product chip in its fully functional state may vary across the chip due to variation in local density of switching circuits. The difference in on-product RO frequencies, with and without functional operation, is an indication of localized switching activity or V_{DD} droops. RO designs of the type shown in Fig. 6.48 have less sensitivity to V_{DD} variations and may be used as on-product temperatures sensors.

RO frequencies for determining systematic circuit delay variations across a product chip, arising from silicon process variations, are measured after completion of full process and availability of I/O contacts such as C4s or wire bonds. It is often desirable to measure the variability early in the process to allow timely corrective actions. Special masks for one or more metal layers may be designed to delineate I/O pads and wiring specifically for measuring the on-product ROs. These masks can be used to process a small number of wafers which are removed from the standard product process flow.

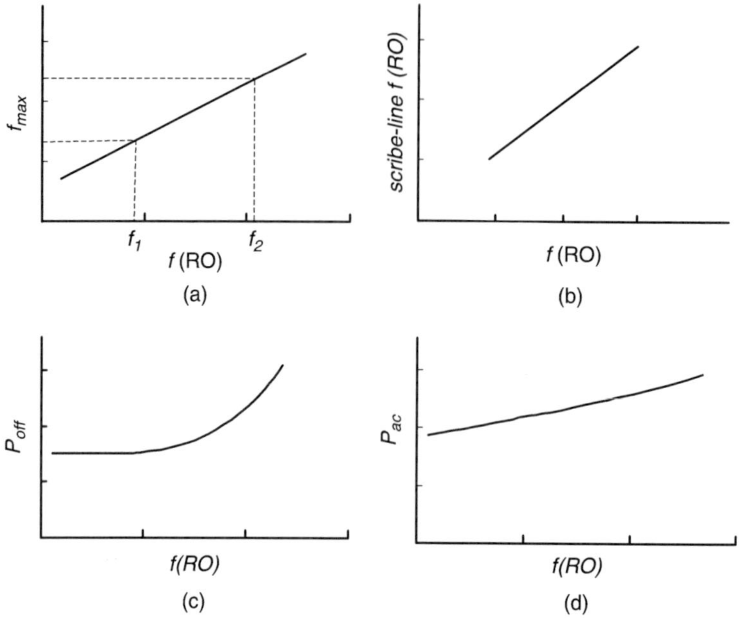

Fig. 6.51 Correlation plots for **a** product $f_{\max}$ as a function of on-product RO frequency f(RO) and **b** scribe-line RO frequency as a function of f(RO). **c** P_{off} as a function of f(RO). **d** P_{ac} as a function of f(RO)

A non-contact technique for enabling and measuring RO frequencies is particularly useful for monitoring systematic variations across product chips in the manufacturing line. One such RO design includes a silicon diode activated by an incident laser beam to deliver power and an antenna to couple the output signal to a frequency measurement system [19]. This pad-less, non-intrusive contacting scheme is very attractive from a process development perspective; however, it requires a special test setup which adds to the cost and test time for routine monitoring.

6.6 Model-to-Hardware Correlation

RO measurements form an important link between products designed with circuit timing models and SPICE simulations and the measured AC performance of circuits in silicon hardware. Hence, correlation of circuit models to measured circuit delays in the hardware plays an important role in product debug. Prior to model-to-hardware correlation, RO frequency and power measurements must be properly validated. In general, errors may originate from design, test equipment, test conditions, and data manipulation. Each of these items must be addressed for the data to be reliable.

6.6.1 RO Circuit Simulations

Models used for simulating circuit behavior in a product design environment contain a large number of variables such as parameters of n-FETs and p-FETs and resistances and capacitances of interconnects and parasitic elements. Statistical distributions of key parameters are included to cover their ranges in the hardware. Some of these variables may also depend on physical layout of the circuits. Model-to-hardware correlation is complicated by the fact that any one or all of these circuit parameters for silicon wafers processed over a period of time may fall anywhere within their distributions. A number of measurements sampling many silicon wafers and Monte Carlo simulations using algorithms for random sampling of model input parameters are needed to cover the full range of parameter distributions. Here we give guidelines for generating the distributions of measured RO parameters from circuit simulations.

Ideally, simulation of an RO macro may be carried out with a circuit netlist obtained by full parasitic extraction of the physical layout and with input excitation levels and timing matching the conditions during testing. Monte Carlo simulations are then carried out to obtain the mean and standard deviation of RO frequency and power for the parameter distribution range observed in the hardware. Although this procedure gives a true representation of the physical system under test, it can be cumbersome and time consuming even for a single V_{DDE} value for each RO, especially for large macros with many ROs and complex control circuits. If the frequency divider is included in the circuit, the simulation time may become too long for the simulator to handle. A practical approach is to use a simplified procedure for simulating delay/stage without any significant loss of accuracy.

An RO forms a closed system and its frequency is not influenced by the control or output circuitry. Hence, a circuit netlist may be generated for each RO in a macro and simulated independently to obtain the IDDQ values of ROs. The simulation time with the RO in its oscillating state is reduced by considering only a few sequential stages. This scheme is shown in Fig. 6.52. A chain of at least nine RO stages is stimulated with a single-pulse input. The PU and PD delays across the fifth stage are obtained. The stages before and after the fifth stage are for conditioning the input and output signals, to match an infinitely long chain or a loop, by eliminating the influence of discontinuity at the ends of the chain. The average of PU and PD delay of the fifth stage is compared with the delay/stage obtained from full RO circuit simulation. If the difference between the delays/stage obtained by these two methods is negligible, the delay chain may be used for conducting a large number of simulations to study the impact of systematic variability in MOSFET and other circuit parameters. Monte Carlo simulations may be carried out using statistical distributions of all parameters. Early in the technology cycle, only a few key parameters, such as L_p and V_t, may be varied to get the RO stage delay sensitivity to individual parameters.

This simulation methodology works well for bulk silicon CMOS technology. In PD-SOI technology, the circuit delays are a function of its switching history and the delay/stage in a ring oscillator configuration with periodic switching may

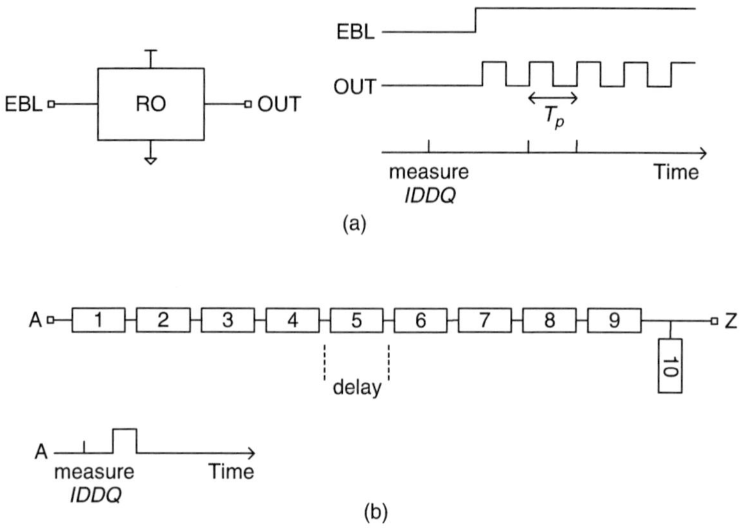

Fig. 6.52 Circuit schematic for simulations and signal waveforms for **a** full RO circuit simulation and **b** delay chain comprising nine RO stages

be different than the delay obtained with a single-pulse excitation. The simulation methodology for PD-SOI technology is covered in Section 8.4.

6.6.2 Sources of Error

There are several sources of error in determining circuit delays, leakage, and switching capacitances from RO test structures. These arise from intrinsic and design dependent behavior. Corrections can be applied based on circuit simulations. Some of the sources of error and a design methodology for error minimization are included in Section 6.2.1.1. These sources of error and possible corrective actions are summarized in Table 6.1.

Table 6.1 Sources of error in estimating circuit delays and power from ROs and possible solutions for minimizing errors

Parameter	Source of error	Comments
Delay/stage	Random variability	$\alpha \geqslant 50$
Delay/stage	NAND2 stage	$\alpha \geqslant 50$
Delay/stage	C_{in} of inverter I1	Minimum MOSFET widths
IDDQ/stage	NAND2 and Inverter I1	
IDDQ/stage	Peripheral circuits	V_{DDC} isolation
Delay/stage	V_{DD} droop	Robust power grid, placement of DECAPs I/O GND isolation

Test equipment and measurement related errors include presence of harmonics, excess leakage from peripheral circuits, power supply droops and GND bounce, noise from impedance mismatch in cables and connectors, and short-circuit power.

Harmonics oscillations in ROs may be eliminated by ensuring that the RO power supply is stable before enabling the oscillations with a sharp signal edge as described in Section 6.2.1.4. Reducing the length of cable connections to the test equipment and impedance matching measures are used to avoid noise glitches in the output voltage waveform. Typical frequency measurements are made by counting the voltage crossings within set voltage threshold levels. The levels are set near the mid-point of the voltage output to avoid switching noise and voltage spikes (Section 9.2.4).

The third set of errors arises from the conditions under which measurements are carried out. At a very low V_{DDE}, the RO may stop functioning. Typically an RO comprising static gates can operate at a lower voltage than an RO with a small number of mixed circuit types with large unequal loads such that the rise and fall times become comparable to the RO period. At high V_{DDE}, the voltage droops in the power supply grid may become significant. When varying V_{DDE}, it is recommended to maintain the power supply for the I/O driver and peripheral circuits V_{DDC} as discussed in Section 6.2.1.3.

Capacitance determination from measured frequency and power of ROs is valid only at low V_{DDE}, when the short-circuit current is small compared to the charging current. The short-circuit current flows directly between V_{DDE} and GND when both n-FET and p-FET are momentarily partially on during switching. This introduces an error in the estimation of C_{sw} from Eq. (6.7). The short-circuit power P_{sc} of an RO is given by

$$P_{sc} = \frac{\beta}{12} (V_{DDE} - 2V_t)^3 \frac{\tau_{rf}}{\tau_p}, \tag{6.22}$$

where τ_{rf} is the rise or fall time of the signal at the input or output of a stage ($\tau_r = \tau_f$) and β is the MOSFET gain factor [1]. The magnitude of P_{sc} relative to P_{sw}, and hence the error in estimated C_{sw} increases with increase in V_{DDE}. As an example, in an inverter RO stage (FO = 1) with $V_t \sim 0.2$ V for both n-FET and p-FET, there is an increase in estimated C_{sw} of ~2% when V_{DDE} is raised from 1.0 to 1.1 V. The increase is <1% for an inverter with a capacitive load (FO ~ 3). Circuit simulations are carried out to determine maximum V_{DDE} for estimating C_{sw} within a desired accuracy. The sources of errors resulting from test setup and suggested solutions are summarized in Table 6.2.

6.6.3 Macro Design Validation

Macro designs are typically verified for full functionality by circuit simulations using a netlist extracted from the physical layout. The design is validated in silicon to detect any accidental errors created during data preparation for optical mask

Table 6.2 Sources of error in RO test and possible solutions

Parameter	Source of error	Comments
Frequency	Harmonic content	Harmonic free design
Frequency	Output voltage glitches	Short output cables, impedance matching
Frequency	GND noise	Higher voltage threshold on frequency counter
Frequency	GND bounce	Low I/O driver V_{DDC}
Capacitance	Short-circuit power	Lower V_{DDE}

generation or by systematic defects during processing. Errors in documentation, test code, and test setup may also produce erroneous results. Some of the key steps in RO design and test validation are listed below:

- verification of electrical isolation of each power supply sector;
- comparison of measured IDDQ of each power supply sector, for at least one V_{DDE}, with simulated values;
- comparison of frequency and IDDA values of ROs at different values of V_{DDC} to select optimum V_{DDC} value;
- determination of measurement accuracy for each parameter.

References

1. Weste NHE, Eshraghian K, Smith MJS (2000) Principles of CMOS VLSI design, 2nd edn. Addison Wesley, Reading, MA
2. Baker RJ (2007) CMOS circuit design, layout and simulation, 2nd edn. Wiley, IEEE Press
3. Uyemura JP (2001) CMOS logic circuit design. Kluwer Academic, Norwell, MA
4. Taur Y, Ning TH (2009) Fundamentals of modern VLSI devices, 2nd edn. Cambridge University Press, New York, NY
5. Bhushan M, Ketchen MB (2010) Generation, elimination and utilization of harmonics in ring oscillators. Proceedings of the 2010 IEEE international conference on microelectronic test structures, 2010, pp 108–113
6. Bhushan M, Ketchen MB, Polonsky S, Gattiker A (2006) Ring oscillator based technique for measuring variability statistics. Proceedings of the 2006 IEEE international conference on microelectronic test structures, 2006, pp 87–93
7. Fuketa H, Hashimoto M, Mitsuyama Y, Onoye T (2010) Transistor variability modeling and its validation with ring-oscillation frequencies for body-biased subthreshold circuits. IEEE Trans Very Large Scale Integration (VLSI) Syst 18:1118–1129
8. Drego N, Chandrakasan A, Boning D (2010) All-digital circuits for measurement of spatial variation in digital circuits. IEEE J Solid-State Circuits 45:640–651
9. Sutherland I, Sproull B, Harris D (1999) Logical effort. Academic Press, New York, NY
10. Ketchen M, Bhushan M, Pearson D (2005) High speed test structures for in-line process monitoring and model calibration, Proceedings of the 2005 IEEE international conference on microelectronic test structures, 2005, pp 33–38
11. Ketchen M, Bhushan M (2006) Product-representative "at speed" test structures for CMOS characterization. IBM J Res Dev 50:451–468
12. Bhushan M, Gattiker A, Ketchen MB, Das KK (2006) Ring oscillators for CMOS process tuning and variability control. IEEE Trans Semicond Manuf 19:10–18

13. Bhushan M, Ketchen MB, Cai M, Kim C (2008) Ring oscillator technique for MOSFET CV characterization. IEEE Trans Semicond Manuf 21:180–185
14. Ketchen MB, Bhushan M, Bolam R (2007) Ring oscillator based test structure for NBTI analysis. Proceedings of the 2007 IEEE international conference on microelectronic test structures, 2007, pp 43–47
15. Zhou B, Khouas A (2005) Measurement of delay mismatch due to process variations by means of modified ring oscillators. IEEE international symposium on circuits and systems, ISCAS 2005, pp 5246–5249
16. Chan YH, Srinivasan U (2006) SRAM ring oscillator. US Patent 7,142,064,B2
17. Kuhn KJ (2007) Reducing variation in advanced logic technologies: approaches to process and design for manufacturability of nanoscale CMOS. IEEE international electron devices meeting, IEDM 2007, pp 471–474
18. Gattiker A, Bhushan M, Ketchen MB (2006) Data analysis techniques for CMOS technology characterization and product impact assessment. IEEE international test conference ITC'06, pp 1–10
19. Steinbrueck G, Vickers JS, Babazadeh M, Pelella MM, Pakdaman N (2009) Non-contact padless measurement technology and test structures for characterization of cross-wafer and in-die product variability. Proceedings of the 2009 IEEE international conference on microelectronic test structures, 2009, pp 91–95

Chapter 7
High-Speed Characterization

Contents

Discrete circuit element and ring oscillator-based test structures described in Chapters 3, 4, 5, and 6 are essential for CMOS technology characterization. However, these test structures do not capture all of the transient behavior exhibited by circuits and devices during product operation. High-speed tests on large logic and memory blocks are very useful for validating product functionality and predicting product yield but provide only limited information on CMOS technology process and circuit models. Rapid debug of product chips and technology characterization in general are assisted by another class of high-speed test structures devoted to model-to-hardware correlation for circuit functionality, timing, and noise under different operating conditions. In these test structures, measurements are carried out either at high frequencies (10 MHz to several GHz), or with signal rise and fall times of <100 ps, or both.

One approach to high-speed test structure designs used for detailed characterization is to create a macro template with high-speed signal I/Os, DC control inputs, multiple power supply sectors, circuitry to modulate input signal waveform shapes, and I/O drivers for the output signals. A variety of high-speed experiments can then be plugged into the macro template and all macros of this type tested with the same custom probe card and laboratory bench test equipment. Such tests are generally carried out on a limited number of chips for model build and product debug. A second approach is to design macros with only DC I/Os in which all the high-speed

M. Bhushan, M.B. Ketchen, *Microelectronic Test Structures for CMOS Technology*, DOI 10.1007/978-1-4419-9377-9_7, action takes place within the macro. These types of macros can be tested with parametric ATE in the silicon manufacturing line and are suitable for routine monitoring of circuit behavior at high operating speeds. Examples of test structure designs for bulk CMOS technology based on these two approaches are described in this chapter and for PD-SOI CMOS technology in Chapter 8.

An introduction to high-speed measurement techniques is given in Section 7.1. Design and test of a high-speed macro template for measuring time delays with sub-ps precision is described in Section 7.2 along with the test setup in Section 7.3. Five examples of high-speed experiments are covered in Section 7.4. Examples 1, 2, and 3 include measurements of PU and PD delays, signal wire cross talk and latch delay, and metastability. Example 4 is a simplified design of a high-speed macro template that is testable at the M1 metal level. In Example 5 a macro design for pulse *I–V* characterization of MOSFETs using only DC I/Os is described.

Circuit blocks used in this chapter are introduced in Section 2.4.8 and covered in standard CMOS circuit textbooks [1–3]. Specific citations for high-speed test structures are included in the "References" section at the end of this chapter.

7.1 High-Speed Measurements

Measurement of time by utilizing a periodic phenomenon forms the basis of pendulum clocks developed in the medieval times as well as the very sophisticated atomic clocks used for defining the standard of time today. Measurement of elapsed time for a single shot event requires stopwatch-like instrumentation in which the measurement accuracy is dependent on the relative precision of start and stop times and the time resolution of the instrument itself.

In Chapter 6, ring oscillator (clock)-based test structures for measurement of circuit delays are described. The ring oscillator output frequency may be lowered to ~1 MHz or less with a frequency divider circuit, enabling sub-ps measurement precision with only sub-μs resolution measurement equipment. A ring oscillator is a closed system with each stage having fixed inputs and outputs and signal rise and fall times. The measured circuit delay represents only one of the many possible switching scenarios experienced by a logic gate in a complex functional block. Furthermore, the measured circuit delay for these fixed conditions is an average of pull-up (PU) and pull-down (PD) delays of all the stages in the ring oscillator.

Differential time measurement techniques, analogous to a stopwatch, are used for precise determination of differences in signal propagation delays through two DUTs under a variety of switching scenarios. The output signal waveforms of the DUTs, driven by a common input, are identical in shape with a time displacement corresponding to their delay difference. The time measurement is referenced to the exact same point on the each of the waveforms as shown in Fig. 7.1. The time difference between these two points can be measured with the precision of the measurement equipment, limited only by the jitter-induced noise. With a high degree of common mode rejection by design, many of the potential sources of errors are eliminated by subtraction.

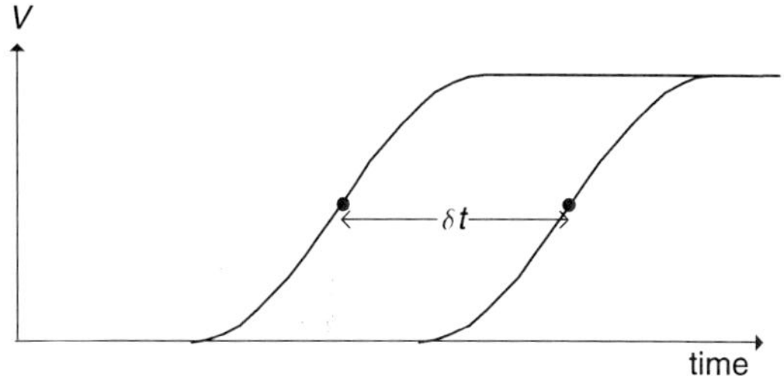

Fig. 7.1 Output signal waveforms for a differential time measurement scheme. *Dark circles* indicate arbitrary but identical reference locations on the waveforms

In some applications, pulse-based high-speed characterization is conducted using only DC I/Os. One or more narrow pulses or a signal edge with adjustable rise or fall time can be generated using DC inputs. In one scheme, circuit operation at high speeds is then characterized by converting an AC output to a DC current. In another scheme, a signal voltage level generated by a timed event is captured in a latch, and the DC output voltage of the latch is recorded. In both cases, high-speed actions take place within the test structure and measurements are made using DC I/Os, thereby simplifying the test setup.

These basic ideas of high-speed test structure designs described in this chapter and Chapter 8 are depicted in Fig. 7.2. In Fig. 7.2a, high-speed inputs A, B, and C conditioned by Ckt_F drive two DUTs, differing only in circuit properties of interest, which results in different circuit delays. The relative timing of the high-speed input signals is controlled externally and may be further adjusted with analog DC control voltages (AJ) for the circuitry within Ckt_F. Digital input S, in Ckt_S, directs the output of either DUT1 or DUT2 to the OUT terminal. The arrival times of the output signals from DUT1 and DUT2 are recorded with a sampling oscilloscope. With the path delays being common through Ckt_F and Ckt_S, the time difference in the two cases gives the signal propagation delay difference δt between DUT1 and DUT2. The DUTs can be further characterized by measuring δt with different values of V_{DD} and temperature.

In Fig. 7.2b, the test structure is operated with only DC or low-frequency I/Os. A clock signal to activate the DUT is generated internally by applying a slow rising input voltage (rise time $\sim$ μs to ms) at the EBL node of a ring oscillator. The output of the DUT is a DC current whose value is related to, for example, the circuit capacitance or the leakage current. This arrangement is suitable for characterizing MOSFETs in a pulse mode, as described in Section 7.4.5.

In Fig. 7.2c, a pulse with rise and fall times as low as $\sim$20 ps or less is generated within the test structure with a slow rising signal voltage applied at the EBL input. This high-speed pulse initiates an event within a DUT whose output is captured in a latch as a "1" or a "0." The latch output voltage can be read out at any time after the completion of the event. Digital input signals, AL1 and AL2, are used to steer the pulse and also as control inputs to decoders and multiplexers. High-speed test structures based on this idea are described in Chapter 8 for PD-SOI-specific applications.

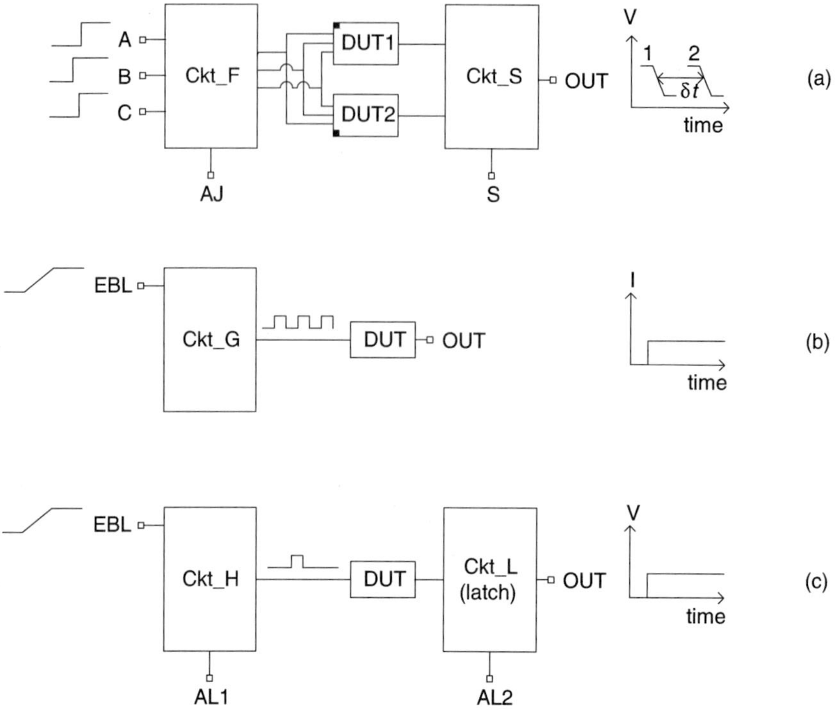

Fig. 7.2 High-speed test structure designs with **a** high-speed I/Os using differential time measurement, **b** DC inputs and a constant current output, and **c** DC inputs for pulsed mode activation and a constant voltage output captured in a latch

7.2 Differential High-Speed Macro Template

In high-speed macro designs, careful consideration must be given to signal integrity, power supply noise, and cross talk. It is therefore convenient to create a macro template for high-speed differential measurements, following the scheme shown in Fig. 7.2a, such that a variety of experiments can be plugged into a single macro [4, 5]. The I/O pad assignments are fixed, and a single probe card design serves for all macros of this type. The description of a hierarchical and modular design of such a macro template is given below. This macro template requires a minimum of three metal layers for wiring. A simplified M1 testable version is described in Section 7.4.4.

The macro comprises eight experiments (EXPTs), a decoder to select any one at a given time, and a common high-speed output for all EXPTs. The EXPTs share five high-speed input signals. A template is created for an EXPT circuit block into which different DUTs, the lowest modules in the hierarchy, can be plugged. I/O pin locations in the physical layout of the EXPT block are also fixed for easy replacement or interchanging of EXPT blocks within or among macros. Minor changes in the

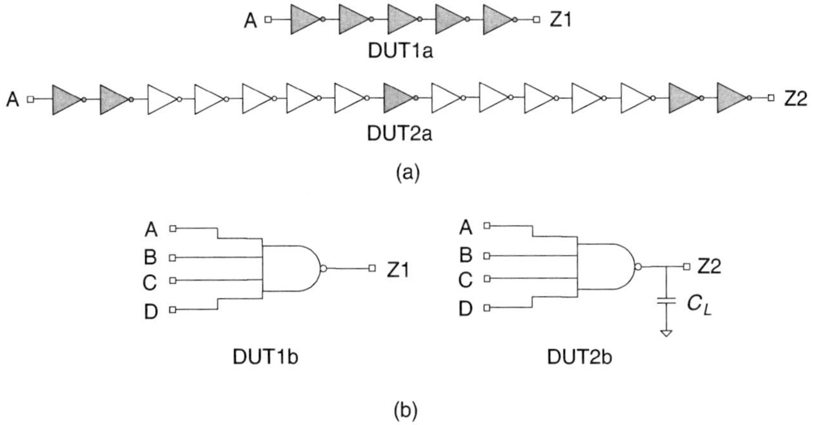

Fig. 7.3 DUT configurations of **a** a chain experiment to determine average of PU and PD delays, τ_p, and **b** a multiple input logic gate (NAND4) to determine individual PU and PD delays

internal design of Ckt_F and Ckt_S blocks may be made to accommodate different numbers of DUTs or different DUT types in an EXPT.

Two different configurations of the DUTs for differential time measurements are shown in Fig. 7.3. In Fig. 7.3a, DUT1a and DUT2a comprise a chain of 5 and 15 nominally identical inverting logic gates. The difference in signal propagation delay through the two DUTs gives the average PU and PD delay of 10 inverting stages. For macro design validation, the average delay/stage τ_p obtained with this arrangement can be compared with the τ_p value determined from a ring oscillator comprising the same logic gates.

Any number of gates may be used in the chains in the configuration shown in Fig. 7.3a but the number should be either even or odd in both chains to maintain the same output waveform shape. Random statistical variation is averaged over all the gates. Hence the difference in the number of gates should be small (<10) for characterizing random variations and large (>30) for systematic variations.

The DUT configuration in Fig. 7.3b is a four-input NAND (NAND4) gate with four high-speed signal inputs. DUT2b drives a capacitor load C_L and hence has a longer signal delay than does DUT1b. The relative timing and rise and fall times of the input signals are individually controlled by Ckt_F block. Any one or all the NAND4 input signals may be switching and they may be rising or falling.

The physical layouts of the individual logic gates in DUT1 and DUT2 are designed to be identical to maintain the same parasitics. The differences in delay through DUT1 and DUT2, for the two cases shown in Fig. 7.3, are then

$$\delta t = \delta n \tau_p \text{ for case (a)} \tag{7.1}$$

and

$$\delta t = r_{swp} C_L \text{ or } \delta t = r_{swn} C_L \text{ for case (b),} \tag{7.2}$$

where δn is the difference in the number of gates in the chains (= 10) and r_{swp} and r_{swn} are the switching resistances of the NAND4 gates for PU and PD transitions (Section 2.4.2), respectively. In the configuration shown in Fig. 7.3b, the delays and r_{sw} values for PU and PD transitions of a logic gate or a circuit can be characterized independently.

Estimated values of δt are obtained from circuit simulations in the macro design phase. The δt values under different measurement conditions (V_{DD}, temperature, and process variations) should be >10× the time resolution of the test setup.

Circuit schematics for two different EXPT configurations are shown in Fig. 7.4a, b. Both EXPTs comprise Ckt_F, DUTs, and Ckt_S. Ckt_F has four primary high-speed input signals, A, B, C, and D enabled by an additional high-speed input signal F that is generally held at "1." The rise and fall times of the four signals that drive the DUTs are controlled by current-starved inverters with analog inputs AJP and AJN. The DUT configurations can be independently customized in each EXPT.

In Fig. 7.4a, a DUT configuration with four input signals is shown. As an example, a NAND4 gate can have a single-input switching, with the other three inputs held at "1." Alternatively with inputs A = B = C = D = "1," simultaneous inputs are applied to the NAND4 by toggling input F. The DUT outputs are selected by setting the DC voltage level of input S. With S = "1," the signal path through DUT1a is selected and with S = "0," signal path through DUT2a is selected.

Two or more DUTs with fewer input signals can be situated in an EXPT as shown, for example, in Fig. 7.4b. Here, one set of DUTs has one input signal (A) and the second set has three input signals (B, C, and D). If DUT1b and DUT2b are NAND3 logic gates, then with signal levels B, C, and D set at "0," and S = "1" or "0," a signal propagates from input A through DUT1a or DUT2a, respectively, to the OUT terminal of Ckt_S. Similarly, if DUT1a and DUT2a are inverters, with input A set at "0," input signals B, C, and D can be propagated through DUT1b and DUT2b to the common OUT terminal.

Input Z is the output of a decoder element to select only one EXPT in the macro with Z set to a "1." This DC signal also enables the propagating signal to reach the OUT terminal. With Z = "0," the OUT signal of an unselected EXPT is a "0." This is a useful feature when the outputs of a number of EXPTs are "OR"ed within a macro, as it prevents any malfunctioning EXPTs (design or process error) from disabling other EXPTs.

The circuit schematics shown in Fig. 7.4a, b are two of the many different DUT configurations that can be accommodated in an EXPT, without altering its inputs and outputs. The DUT configurations may have any combination of inputs with the constraint that the total number of high-speed inputs in an EXPT placed in the macro template is limited to five. The Ckt_S block design is customized to the number of DUTs and the block template is sized to accommodate up to four pairs of DUTs.

The signal waveforms at the input and output nodes of each circuit block are shown in Fig. 7.4c. With the DUT being an inverting stage and with even numbers of inverting stages in Ckt_F and Ckt_S, the I/O signals of the EXPT have the same

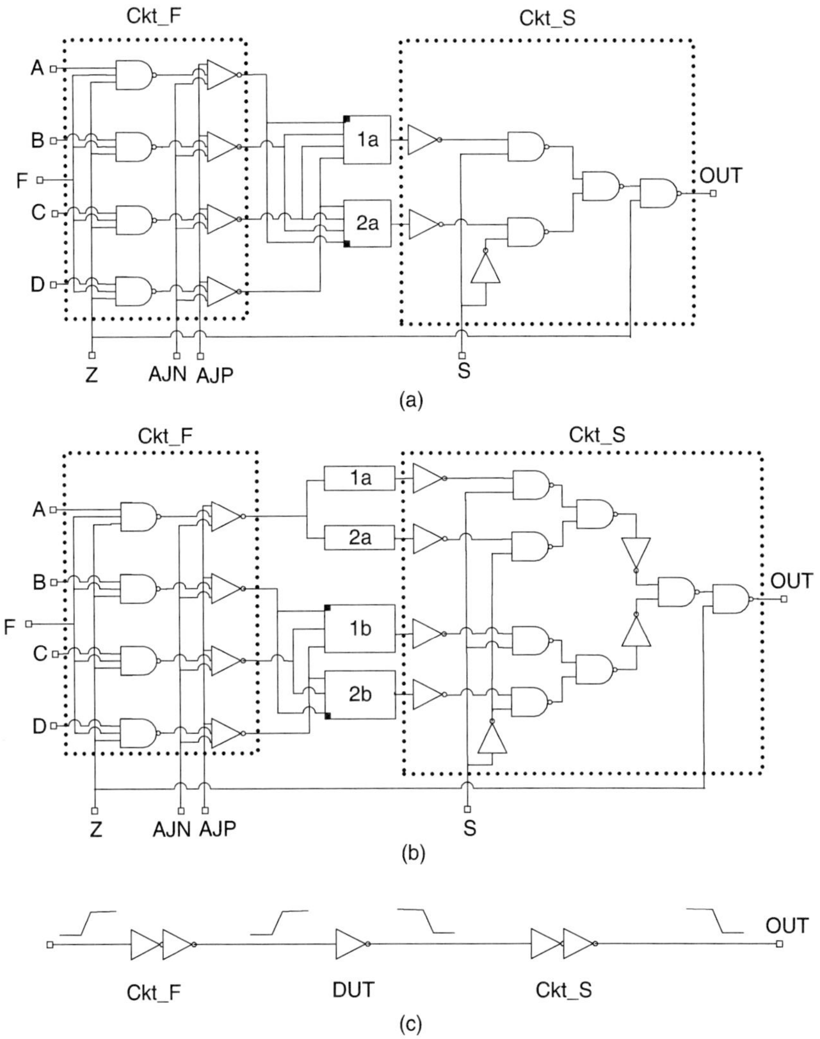

Fig. 7.4 Circuit schematics of EXPTs with **a** one four-input DUT configuration, **b** one single-input and one three-input DUT configurations. **c** Signal waveforms at the inputs and outputs of each circuit sub-block

polarities as for the DUT itself. Maintaining this convention throughout the entire macro proves to be very helpful in design, debug, and characterization as the test signal polarities match those of the DUT itself.

The floorplan of a high-speed macro template with a 1×25 padset (Appendix A) is shown in Fig. 7.5a and the I/O pad assignments are shown in Fig. 7.5b. There are two power supply sectors, V_{DDE} and V_{DDC}, with a common GND. All EXPTs are on the V_{DDE} power supply and the I/O driver, buffers, decoder, and OR circuit are on the V_{DDC} power supply. The five high-speed inputs and one high-speed output

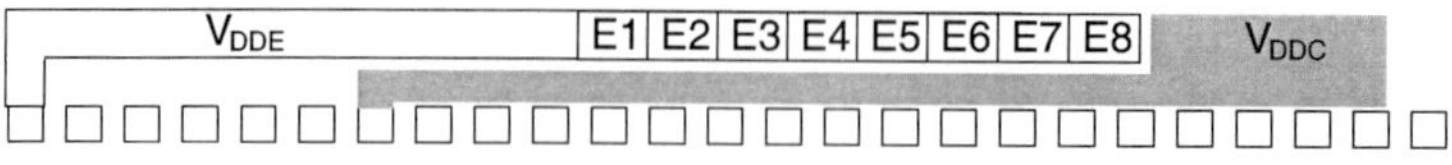

(a)

1	2	3	4	5	6	7	8	9	10	11	12	13	14	15	16	17	18	19	20	21	22	23	24	25
VDDE	GND	A0	A1	A2	GND	VDDC	GND	OUT	GND	A	GND	B	GND	F	GND	C	GND	D	GND	S	GND	AJN	AJP	GND

(b)

Fig. 7.5 **a** Physical layout of a high-speed macro template with power supply sectors and location of EXPTs, E1–E8. **b** I/O pad assignments

I/O pads are arranged in the G–S–G (GND–signal–GND) configuration. The analog inputs AJN and AJP, to control the DUT input signal waveform shapes, can be set anywhere between "1" and "0." For default waveform shapes, AJN = "1" and AJP = "0." DC input S is toggled between "1" and "0" to alternately select the output from DUT1 or DUT2. DC signal lines A0, A1, and A2 are decoder input bits to select any one of the eight EXPTs.

The wiring scheme of the macro for two of the eight EXPTs is shown in Fig. 7.6. The high-speed and DC inputs travel in the horizontal direction. Buffers are inserted in the high-speed wires to maintain signal integrity. Guidelines for buffer sizing and spacing are provided in Section 2.4.4. Decoder input bits A0, A1, and A2 enable

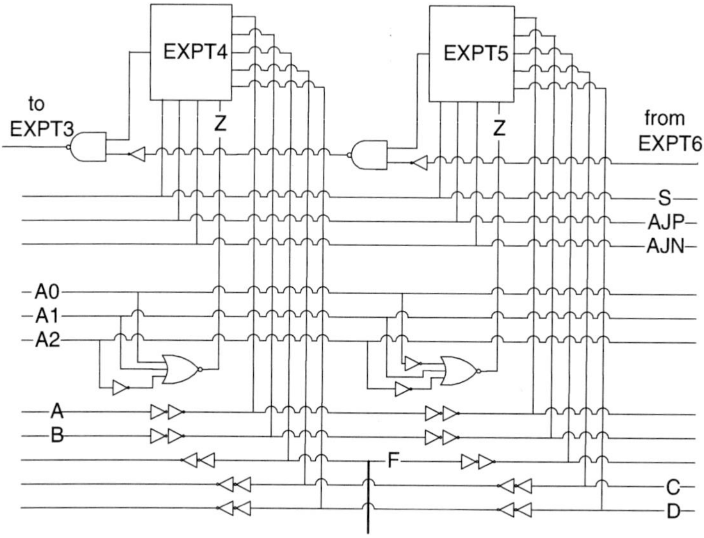

Fig. 7.6 A wiring scheme for the high-speed macro template showing two out of eight EXPTs located in the center of the macro

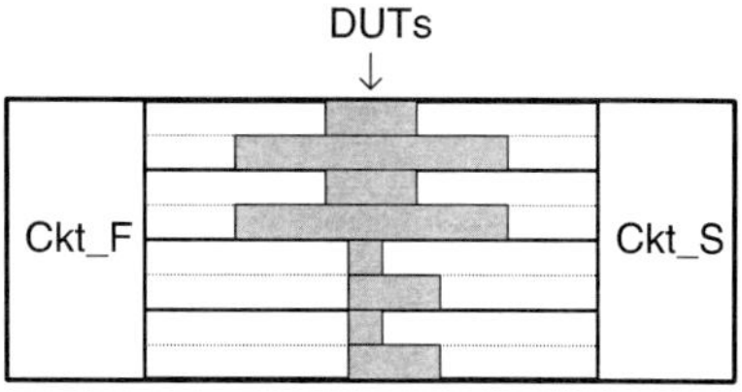

Fig. 7.7 Floorplan of an EXPT block to accommodate four pairs of single-input DUT designs

high-speed signals, with input Z = "1" for only one out of eight EXPTs. The outputs from all the EXPTs go through an OR circuit to a single I/O driver whose output is connected to the high-speed I/O pad, OUT.

An example physical floorplan of the EXPT block is shown in Fig. 7.7. The height of the EXPT block is equivalent to 8× the standard cell library book height (power grid pitch) to accommodate a maximum of eight DUTs (four pairs). The width of the block is approximately equal to the I/O pad width, sufficient to accommodate most DUTs of interest. The wire lengths, connecting the DUT inputs to CKT_F and outputs to Ckt_S, for the DUTs within a pair, are closely matched. This ensures that the signal path lengths and parasitic delays external to the DUTs are identical.

The power consumed by the high-speed macro is dominated by that of the I/O driver connected to the OUT pad. The output of the I/O driver passes through a high-speed probe, followed by a 50 Ω cable and on to the 50 Ω input of a conventional sampling oscilloscope. To minimize V_{DD} droop in the macro, the output voltage is lowered, subject to the constraint that the measurement of δt not be compromised. This implies an output voltage of $\sim V_{DD}/3$, where V_{DD} is the nominal technology voltage, corresponding to the I/O driver internal resistance of ~100 Ω.

The integrity of the input and output signals passing through the macro is a major concern, and issues of jitter and V_{DD} droop must be considered carefully. Two-stage buffers, with an output stage of ~3× the standard logic gate width, are inserted in long wires to limit the signal rise and fall times (Section 2.4.4). Even when varying V_{DDE} for an EXPT, V_{DDC} is always maintained close to nominal technology V_{DD}. The jitter of this type of macro template has been experimentally determined to be $\lesssim$1 ps at nominal V_{DD} [6].

Measurement conditions may vary from a situation approaching isolated high-speed pulse inputs to square wave inputs at frequencies >1 GHz. It is important that there be sufficient decoupling capacitance (Section 2.4.6) included to allow only minimal V_{DD} droop from the time of input signal arrival until the I/O driver switches. With typically 10–20 logic gates between macro input and output, along with significant on-macro wiring, sufficient decoupling to maintain voltage droop of <1% can be incorporated within the macro. When the I/O driver with a 50 Ω output load switches its state, the current it draws at V_{DD} of 1.0 V changes by ~7 mA. However, by then the waveforms for δt measurements have already been recorded. The V_{DD} level recovers before the arrival of the next isolated input pulse.

The design of the dedicated probe card includes substantial decoupling capacitance between V_{DDE} and GND and V_{DDC} and GND, which further limits and smoothes out power supply voltage excursions. At the high-frequency extreme, the system is essentially in steady state with a constant current draw smoothed out by the presence of the decoupling capacitors, a situation resembling that of a ring oscillator. At intermediate frequencies, the situation is more complicated and there can be some periodic droop that exceeds 1%. The amplitude of this voltage droop scales with the current drawn by the I/O driver, which is the reason behind maintaining a low-voltage level at the OUT pad of the macro. The decoupling capacitors in the probe card also help to limit voltage excursions at intermediate frequencies.

The differential nature of the measurement technique provides immunity to the residual periodic voltage droop, as long as the variation in V_{DDE} is small (<1%) during the longest measured δt interval. This is typically the case for δt in the range of 1–100 ps, which covers most measurements of interest.

7.3 High-Speed Test Setup

A representative test setup for high-speed measurements is shown in Fig. 7.8. Pulse generators with one or more synchronized channels provide the high-speed signal inputs, their relative timing adjusted by the internal controls in the pulse generators and additionally with external variable delay lines. SMUs provide the necessary digital and analog inputs. DC and microwave switch matrices route inputs through coaxial cables to the appropriate locations on the probe card. The output voltage transitions are measured with a conventional sampling oscilloscope and the time shift δt determined. With programmable switch matrices and appropriate computer interfaces in the SMUs, pulse generators, delay lines, and sampling oscilloscope, the measurement system can be fully automated, although for macro debug and limited data collection, manual operation is often preferred.

A custom probe card is an absolute necessity for this experiment. For the macro under discussion, there are a total of six high-speed I/Os, with five inputs and one output. These six I/Os are physically adjacent to each other as indicated in

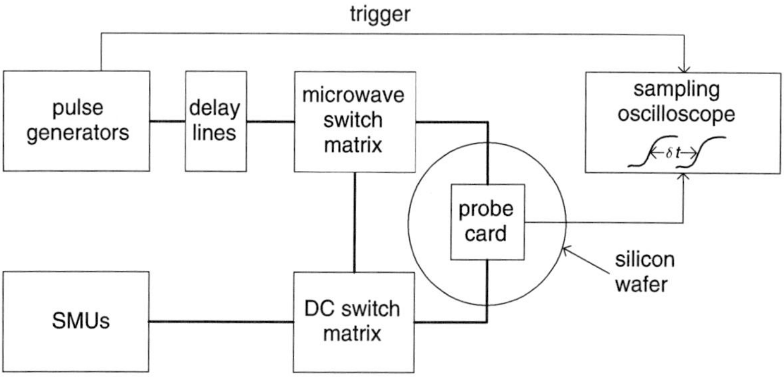

Fig. 7.8 A schematic of a test setup for high-speed differential delay measurements

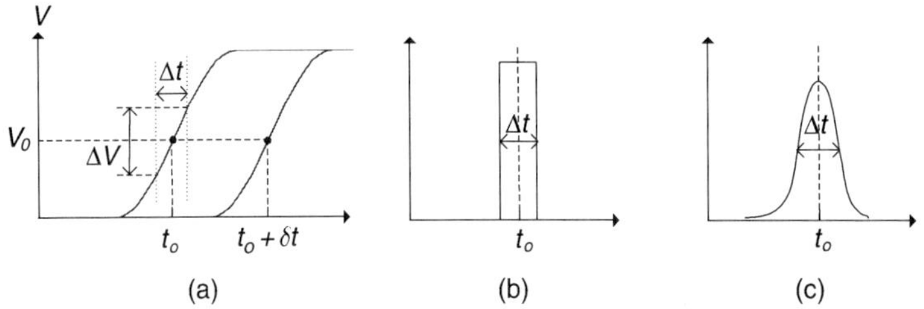

Fig. 7.9 **a** Signal output waveforms for δt measurements with a sampling oscilloscope. **b** Jitter-free histogram of arrival times within a voltage window ΔV. **c** Histogram of arrival times in the presence of jitter

Fig. 7.5b, with nearest neighbors sharing intermediate GND pads. On the probe card, each high-speed I/O is in a G–S–G configuration. The V_{DDE} and V_{DDC} inputs are provided with significant decoupling capacitance to GND, as mentioned in Section 2.4.6, both out on the probe needle tips (~100 pF) and on the card (~10 μF).

Measurements are carried out in a sampling mode and repeated many times for each data point. A representative rising output waveform is shown in Fig. 7.9a. It is the movement of this waveform in time, as the path through the macro is shifted between DUT1 and DUT2 in a selected EXPT, that is measured. This is accomplished by determining the movement of the location on the waveform at some specified voltage V_0, as input S is switched from "1" to "0." In practice the histogram of arrival times within a voltage window ΔV centered on V_0 is measured. In the absence of jitter, this histogram will have a rectangular shape as shown in Fig. 7.9b, with a width of $\Delta t = \Delta V / (\mathrm{d}V/\mathrm{d}t)$. Jitter from various sources smears out the histogram, as shown in Fig. 7.9c.

Jitter arises primarily from two sources: from the sampling oscilloscope including the timing of its trigger signal with respect to that of the macro input and from intrinsic statistical switching time variations of the circuits comprising the paths through the macro. The time to make the measurement scales with ΔV, so it is of advantage to increase the measurement window until it significantly widens the measured histogram. In practice, with ΔV set at nearly zero, 1σ (σ = standard deviation) of the measured histogram can be optimized to a value of <1 ps, suggesting that the net jitter from all sources is of that order. With dV/dt of about 10 mV/ps, ΔV can be increased to ~15 mV, to give an effective 1σ value in the range of 1.5–2 ps. In this configuration, a number of points (typically 100) are averaged to determine the time, t_0. Input S is then switched from "1" to "0" and the measurements are repeated to obtain $t_0 + \delta t$, and thus δt. In this manner it is possible to reproducibly determine δt with an accuracy of better than 1 ps and frequently better than 0.6 ps. The lower limit of δt measurement appears to be set by slow drifts in the measurement electronics, or within the macro itself and not by the statistical nature of the measurements.

The pulse generator(s) is used to produce one or more trains of synchronized pulses of fixed period and width. To measure a PU or a PD transition, a

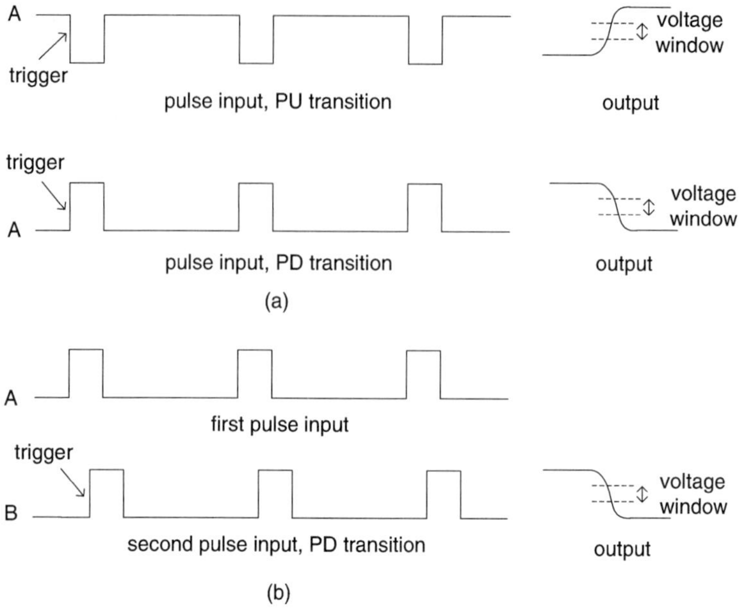

Fig. 7.10 **a** Periodic pulse input signal A for a PU and a PD transition. **b** Periodic pulse input signals A and B for a PD transition induced by B

single-pulse train may be used as shown, for example, in Fig. 7.10a. The output waveforms shown at the right are collections of points measured with the sampling oscilloscope, each point corresponding to a different discrete measurement with the oscilloscope trigger timed by the same edge that induces the measured transition in the DUT. In the case of multiple pulse trains, one pulse train may be displaced from another by some adjustable amount as shown in Fig. 7.10b. To obtain low jitter performance, it is essential that for each point, the sampling oscilloscope trigger signal and the signal to induce the desired transition in the DUT originate at the same pulse generator edge. This is accomplished by using a splitter at the primary pulse generator output or using its low relative jitter complement output, if available, to drive the sampling oscilloscope.

With eight EXPTs in a macro and as many as four DUT pairs in an experiment, it is essential to carefully manage all of the I/Os to ensure that the intended experiments are actually being conducted. Documenting the signal and DC voltages for each test as shown in Table 7.1 is useful for debug and automated test program generation.

As in the case of ring oscillators, a stringent test can be done to detect if the voltage droop associated with the I/O circuitry is introducing an error. If the measured result does not change as V_{DDC} or the input voltage signal levels are varied, then the residual voltage droop is not significant.

Table 7.1 Decoder inputs to select an EXPT and input voltages for inverter chain (input A), inverter (input B) and NOR2 (inputs C and D) DUTs in EXPT4, and NAND4 (inputs A, B, C, and D) DUTs in EXPT5. PL denotes a high-speed pulse

		Decoder inputs			Signal inputs					
EXPT	Description	A2	A1	A0	A	B	C	D	F	S
4	5 Inverter chain	1	0	0	PL	0	0	0	1	1
4	15 Inverter chain	1	0	0	PL	0	0	0	1	0
4	Inverter	1	0	0	0	PL	0	0	1	1
4	Inverter + C_L	1	0	0	0	PL	0	0	1	0
4	NOR2	1	0	0	0	0	PL	PL	1	1
4	NOR2+ C_L	1	0	0	0	0	PL	PL	1	0
5	NAND4	1	0	1	PL	PL	PL	PL	1	1
5	NAND4+ C_L	1	0	1	PL	PL	PL	PL	1	0

7.4 High-Speed Macro Designs

The macro template for differential high-speed measurements described in Section 7.2 may be utilized for a variety of measurements requiring high-speed I/Os. In Examples 1, 2, and 3, we describe EXPT designs for measurement of circuit delays, signal cross talk, and latch metastability. All of these designs require more than one metal layer for wiring. In Example 4, a simplified version of the high-speed macro template for measuring PU and PD delays and matched pairs of logic gates is implemented at the M1 metal level. In Example 5, a macro design for measuring pulse *I–V* characteristics of MOSFETs using only DC I/Os is described.

7.4.1 Example 1: Macro for PU and PD Delay Measurements

The differential time measurement scheme described in Section 7.1 is well suited for independent measurements of changes in PU and PD delays with different circuit excitation schemes as well as average of PU and PD delays of a small number of gates. Quantities that can be investigated include the following:

- switching resistance r_{sw} for PU and PD transitions
- changes in PU and PD delays with relative timings of multiple inputs
- changes in PU and PD delays with input fall and rise times
- PU and PD delays of matched pairs of logic gates for variability study
- change in r_{sw} with time (BTI and hot-e degradation)
- change in r_{sw} with switching history in PD-SOI technology (Section 8.3.3)

DUT configurations for measurement of PU and PD delays of logic gates or circuit blocks are shown in Fig. 7.3b. DUT1 is a logic gate with FO = 1 and DUT2 drives an additional capacitive load C_L. The value of C_L can be obtained from circuit models or independently determined from ring oscillator measurements as described in Section 6.3.1. The logic gate widths and C_L are sized such that $\delta t\ (\approx r_{sw}C_L)$ is in the range of 20–100 ps. The load capacitor may be a MOSFET gate or a stacked metal capacitor. Here, the switching resistance of the gate r_{sw} is assumed to be the same for both DUTs.

For circuit sizing purposes, let us consider the standard inverter parameters in Appendix A, with $R_{sw} = 2{,}000\ \Omega\,\mu m$ and gate capacitance $C_{gT} = 1$ fF/μm. For the standard inverter design with $(W_p + W_n) = 2\ \mu m$, and a 2.0 μm wide gate load ($C_L = 2$ fF), $\delta t\ (= R_{sw}C_L/W)$ is only 2 ps. To obtain $\delta t = 20$ ps, a C_L value of 20 fF, corresponding to a FO = 10 for DUT2, is required. The width of the gate load is then 20 μm, corresponding to 20 PS fingers and a total area of 8 μm^2. If a four metal layer (M3–M6)-stacked capacitor with $C_w \sim 1.5$ fF/μm^2 is used, the load capacitor area including wiring is ~15 μm^2. The MOSFET gate capacitor provides a higher capacitance per unit area, but its capacitance has a weak dependence on V_t, and in turn on V_{DD}. It is also more susceptible to random variability. The metal capacitor, on the other hand, is subject to systematic process variations across wafer and from wafer to wafer.

PU and PD delays of DUTs with multiple inputs are characterized by changing the relative timing of the input signals. In logic gates with multiple inputs, measured δt may vary with the relative arrival times of the input signals. This is illustrated in Fig. 7.11a for a NAND2 with top input A and bottom input B. As input signal B is shifted with respect to signal A, and the difference in their arrival times ΔT changes from a negative to a positive value, the NAND2 transition changes from bottom input switching to top input switching. Simultaneous switching of all the inputs occurs as ΔT approaches zero. In the case of a PU transition, δt is substantially reduced with all p-FETs switching simultaneously as illustrated in Fig. 7.11b.

Measurement of PU and PD delays as a function of rise and fall times of input signals is carried out by varying the analog inputs AJN and AJP. Although the exact rise and fall times of the signals are not directly measured, their values may be obtained from circuit simulations.

The change in PU and PD delays with voltage and temperature stress provides information on bias temperature instability (BTI) and hot-e degradation in MOSFET current drive strengths. For DC voltage stress, the V_{DDE} of the EXPT is raised to a

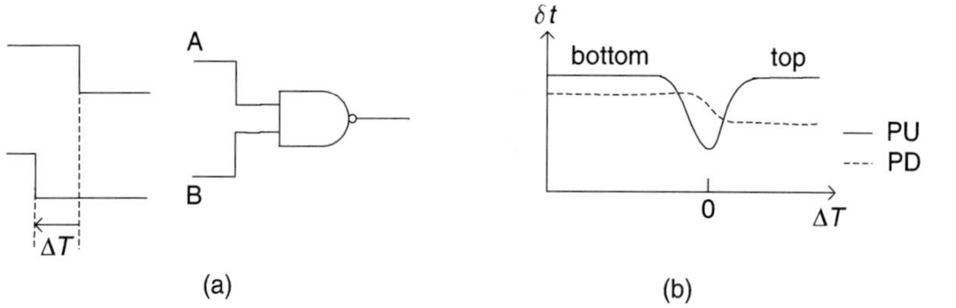

Fig. 7.11 **a** A NAND2 gate and signal waveforms for its inputs A (*top*) and B (*bottom*), shifted in time by ΔT. **b** δt of PU and PD transition as ΔT changes from a negative value to a positive value

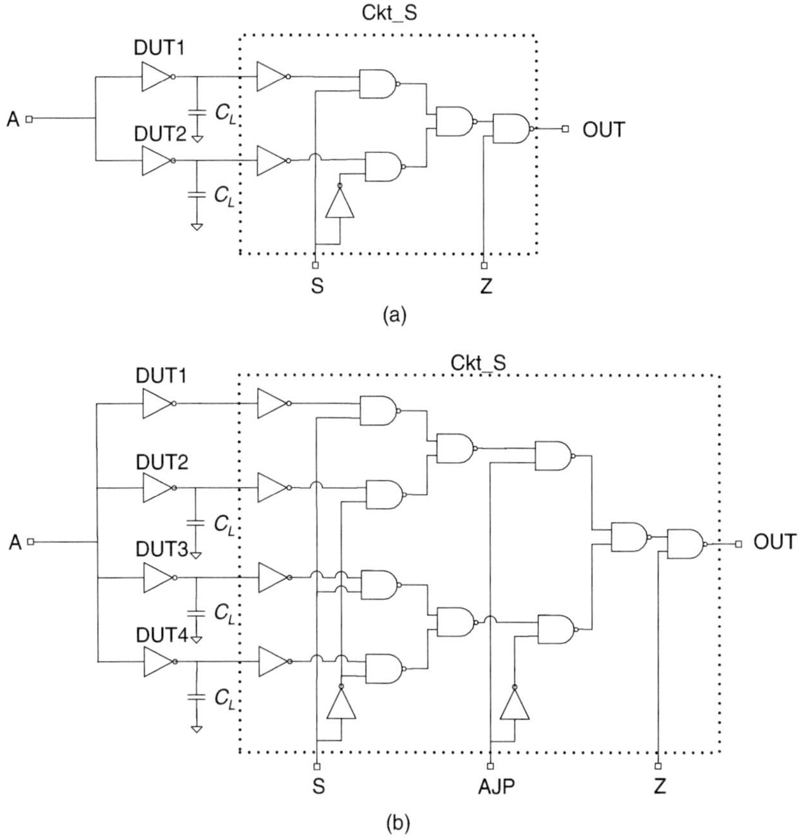

Fig. 7.12 Differential delay measurement scheme for **a** a matched pair of inverters and **b** three nominally identical DUTs using one reference DUT

higher value. Measurements of δt are carried out before and after stress to determine the delay degradation. For AC stress, V_{DDC} may also need to be raised during stress to match the input signal voltage levels.

Circuit delay variability statistics may be collected by measuring the difference in delays of pairs of nominally identical DUTs. The circuit schematic of a matched pair of gates is shown in Fig. 7.12a. Alternately, path delays through three different DUTs are compared with a common reference DUT, as shown in Fig. 7.12b. In this case, only input AJN is included in Ckt_F and DC inputs AJP and S are used for selecting the output signal paths through the four DUTs. The number of DUT pairs in a single EXPT can be increased to eight if AJN as well as AJP and S are used as DC control signals for high-speed signal path selection through different DUTs.

A second-order effect in δt is the increase in the rise and fall times of the output signal of DUT2 with a capacitor load. As illustrated in Fig. 7.13, the signal fall time at the output node of DUT2 is longer than that of DUT1, thereby increasing the r_{sw}

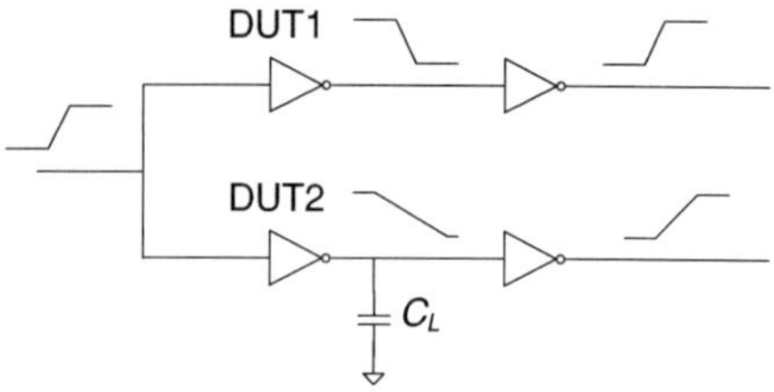

Fig. 7.13 Signal waveforms at circuit nodes preceding and following DUT1 and DUT2

of DUT2. The rise and fall times of the paths through DUT1 and DUT2 equalize after passing through two to three lightly loaded (FO $\leq$ 3) logic gates. The net result is that the delay of PD transitions has a small component of additional delay of the PU component from the following stage. These second-order effects are taken into account in model-to-hardware correlation by extracting δt from circuit simulations, using a netlist of each EXPT extracted from its physical layout.

7.4.2 Example 2: Macro for Coupling Capacitance

Modulation of effective capacitance between two long metal wires, with change in relative phases of the signals travelling through the wires, gives a measure of signal coupling. A compact test structure for model-to-hardware correlation of coupling capacitance is shown in Fig. 7.14a. DUT1 and DUT2 are nominally identical inverters with DUT2 driving one set of fingers of a metal wire comb or a serpentine wire running parallel to another as shown in Fig. 7.14b, c, respectively. The coupled wire segment is independently driven via a high-speed signal B. Metal layers above and below the signal wires form GND planes.

Measurement of δt between DUT1 and DUT2 paths is made by toggling input S between "1" and "0." With input F set at "1," the relative delay between signals A and B is varied with an external mechanically adjustable delay line or by using a programmable pulse generator. The cross talk varies from a constructive value when these two signals are perfectly in phase to a maximum negative value when they are 180° out of phase as shown in Fig. 7.15. Signal B can also be held at "1" or "0" to measure the delay with neighboring wires at V_{DDE} or GND, demonstrating the full range of cross talk between these wires.

Four such DUT pairs are placed within one EXPT. Wire widths and spaces are varied between different DUTs to map a range of wire geometries. Measured data are compared with model predictions.

7.4.3 Example 3: Macro for Latch Metastability Characterization

In this third example, the high-speed macro template design is exploited for characterization of a static CMOS latch [6]. A latch symbol along with clock (CLK) and

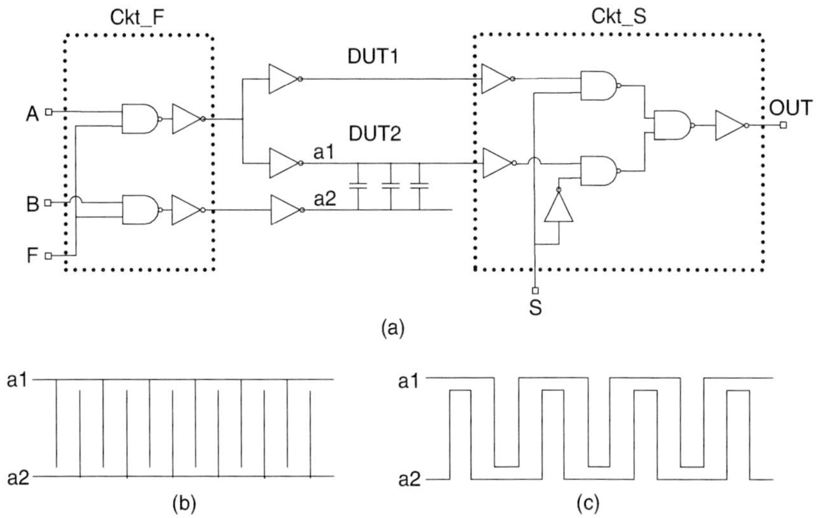

Fig. 7.14 **a** Schematic of a circuit for measuring interconnect cross talk. Coupled wire layouts for **b** an interdigitated comb and **c** a serpentine

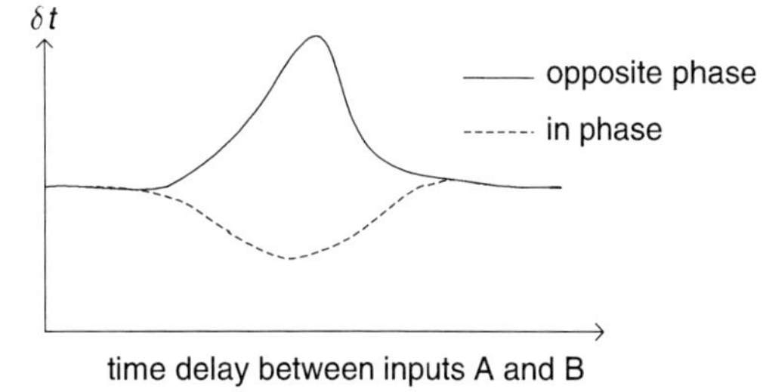

Fig. 7.15 Change in δt as a function of relative timings of input signals A and B for opposite phase and in-phase configurations

data (DAT) input signal waveforms is shown in Fig. 7.16a. The difference in time between the arrival of the rising or falling DAT signal and the falling edge of the CLK signal is indicated as ΔT_d. The propagation delay of the DAT signal through the latch to the LOUT node, τ_l, is a function of ΔT_d as shown in Fig. 7.16b. With $|\Delta T_d| > T_s$, where T_s is defined as the setup time of the latch, τ_l remains constant at τ_{lo}. As $|\Delta T_d|$ is decreased, τ_l begins to increase exponentially, theoretically reaching infinity at a singular metastable point. In practice, the metastable region has a finite width and during this time window, the latch output may be erroneous. Beyond the metastable state, the OUT signal no longer responds to changes in DAT input.

In CMOS circuit designs, it is recommended to keep $|\Delta T_d| > T_s$ in timing-sensitive paths and to ensure the latch delay is not increased beyond τ_{lo}. Characterization of the latch is carried out to measure both T_s and τ_l to compare these values with the model predictions. Measurement of the width of the metastable

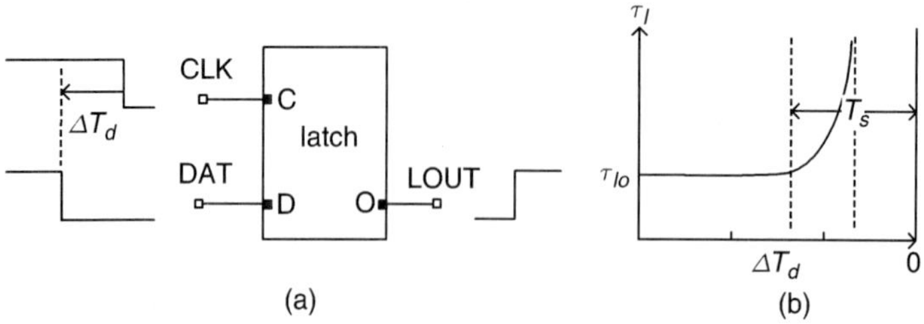

Fig. 7.16 **a** Symbol of a CMOS level-sensitive latch along with CLK and DAT inputs and output, OUT, signal waveforms. **b** Latch delay τ_l as a function of ΔT_d

region is also of interest as it gives a measure of the jitter in the system. With a jitter noise of <1 ps, such a latch can be used to measure relative time differences of its input signals with sub-ps accuracy. The test structure described here is configured to measure the characteristics of a level-sensitive latch, as well as the width of its metastability region.

In Fig. 7.17a, b, circuit schematics of an EXPT for characterization of a latch are shown. There are two high-speed pulsed inputs for CLK and DAT signals and two

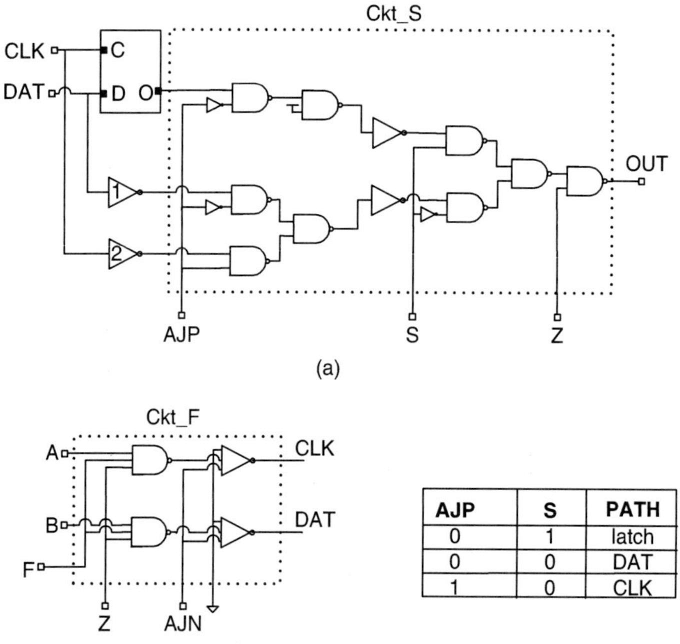

AJP	S	PATH
0	1	latch
0	0	DAT
1	0	CLK

Fig. 7.17 Circuit schematics of **a** a latch DUT and Ckt_S, and **b** Ckt_F. **c** Truth table for selecting latch, CLK, or DAT signal paths. Reproduced from [6], with permission, © 2008 IEEE

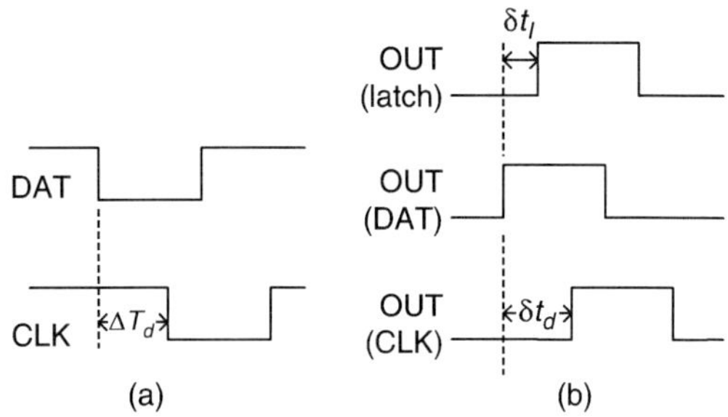

Fig. 7.18 Signal waveforms of **a** CLK and DAT inputs, and **b** OUT node for latch, DAT, and CLK paths. Reproduced from [6], with permission, © 2008 IEEE

DC inputs, AJP, and S to select a signal path to the OUT node. The CLK and DAT signals are steered through the latch, or a direct path to the OUT node is provided by setting AJP and S as shown in the table in Fig. 7.17c. The fall time of the CLK and DAT signals can be varied with analog input AJN. Input signal Z from a decoder in the macro selects one of the EXPTs and enables the OUT signal from this selected EXPT to propagate to the I/O pad, as described previously.

Input and output waveforms for the CLK, DAT, and OUT signals are shown in Fig. 7.18. The relative timing of the falling edges of CLK and DAT signals $\Delta T_{\rm d}$ is varied with the external adjustable delay lines. This time difference is determined by selecting the direct CLK and DAT paths to the sampling oscilloscope using the logic shown in Fig. 7.17c. The latch delay is measured by the time difference through the latch and DAT paths. Using the external delay line, $\Delta T_{\rm d}$ is reduced until the latch delay begins to increase. This gives the setup time $T_{\rm s}$ of the latch for the falling edge of the DAT path. DAT and CLK are derived from the same output of a pulse generator. The latch resets at the rising edge of the CLK.

The apparatus for latch characterization is shown in Fig. 7.19a. There are two independently controlled mechanically adjustable external delay lines for high-speed inputs A (CLK) and B (DAT), calibrated in ps. Time resolution of 0.1 ps is readily achievable with such delay lines. In addition to the sampling oscilloscope, a frequency counter is connected to the OUT terminal to count the number of transitions from a "0" to a "1" at the output.

In each time period of the input pulse, the OUT signal transitions from a "0" to a "1" and this is recorded by the counter. With $|\Delta T_{\rm d}| > T_{\rm s}$, the counter reading is then equal to the frequency of the input DAT and CLK pulses. As $\Delta T_{\rm d}$ is reduced, such that the latch begins to operate in its metastable region, the number of counts begins to decrease as shown in Fig. 7.19b. This indicates that for a fraction of the transitions, the latch fails to operate correctly. Finally, as $|\Delta T_{\rm d}|$ is reduced further, the latch output remains fixed at "0." If only the DAT signal is varied with respect to the CLK signal, a direct measure of the width of the metastability region is obtained.

There are several sources of experimental errors that need to be considered. There is one additional inverter in the DAT and CLK paths each, indicated as 1 and 2 in Fig. 7.17a. The signal delay through these inverters can be determined from circuit simulation and added to the measured δt for latch delay from node D through O.

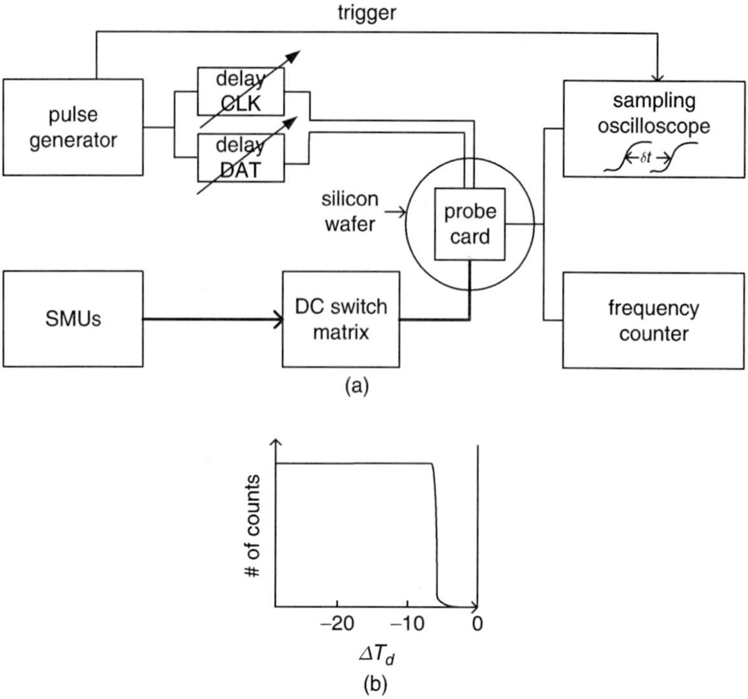

Fig. 7.19 **a** Apparatus for measurement of latch setup time and metastability. **b** Frequency counter reading for "0" to "1" transitions is plotted as a function of ΔT_d

A small error is introduced by the random delay variations in the inverter and the three NAND2 gates in the CLK and DAT paths before combining. Any backlash error in mechanical delay lines can be avoided by maintaining unidirectional travel. Overall, the time resolution with this common mode rejection scheme can be <1 ps [6].

The circuit scheme described above is for measuring the τ_1 for the falling edge of the DAT signal. With additional logic and another DC control signal, the technique can be extended to both the falling and rising edges of CLK and DAT [6].

7.4.4 Example 4: M1 Testable High-Speed Macro

The differential high-speed macro designs described in Examples 1, 2, and 3 are implemented with three or more metal levels. A simplified version of the macro template can be implemented at the M1 metal level for circuit characterization in the early phase of technology development. This macro design is useful for most of the applications described in Section 7.4.1, except that the number of high-speed inputs and DUT complexity at M1 are limited. It uses multiple dedicated power supplies to activate one EXPT at a time, similar to the M1 testable ring oscillator macro

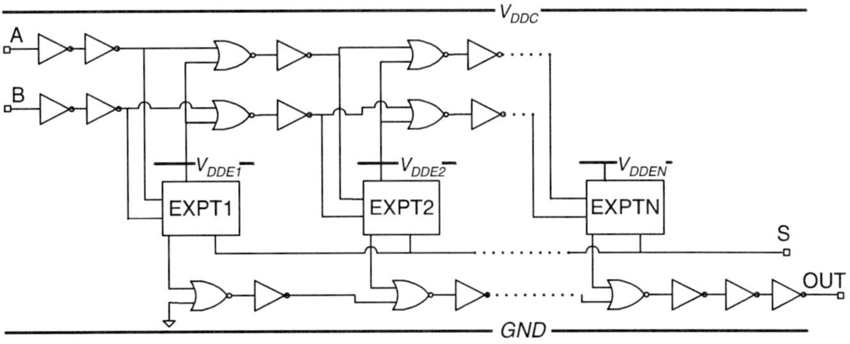

Fig. 7.20 Simplified top-level circuit schematic of the macro with *N* EXPTs sharing common input and output signals

design discussed in Section 6.2.2. The I/O circuit configuration is modified to enable delivery of multiple high-speed inputs to a selected EXPT. These modifications are necessary for wiring with a single metal level. The circuit design and the physical layout feature low resistance metal wires for high-speed signal paths (except a few crossovers in PS or DF layers) and for robust power delivery. Sections of DC signal lines can be in high resistance PS and DF levels. Adequate decoupling capacitance is provided to deliver charge to the EXPT during switching.

A simplified circuit schematic of the macro is shown in Fig. 7.20. EXPTs 1 through *N* are independently powered by V_{DDE1} through V_{DDEN}, respectively. These power supply voltages also act as logic inputs to the I/O circuitry. The I/O circuits are powered by an independent supply, V_{DDC}. All power supplies share a common GND. Only a single experiment is powered on at a given time. High-speed input signals A and B travel from left to right, only up to the selected EXPT. The outputs of all of the experiments are mixed together by a distributed OR gate whose output (OUT) emerges from the right side of the macro. Buffers are inserted in the signal wires traveling across the macro to maintain signal integrity. Each of the EXPTs may contain, for example, two pairs of DUTs, similar to those shown in Fig. 7.21.

The generic physical configuration of the test structure is shown in Fig. 7.22a and the I/O pad assignments for a 1×25 padset macro are shown in Fig. 7.22b. There are 11 GND pads and two V_{DDC} pads at the right and left ends of the macro. I/O pads VE1–VE8 are the power supply pads for the *N* (= 8) EXPTs in this macro. Pads A, B, and OUT are serviced by wide bandwidth 50 Ω I/O lines, while the power supply and control signal S are DC inputs. The DECAPs in the macro are positioned above and below the EXPT circuit and also outboard of the respective V_{DDE} pad. Common GND busses extend above and below the pads as shown. The I/O circuitry is positioned in slots between V_{DDC} and GND along the top (inputs A and B) and bottom (OUT and S) of the design. Additional decoupling capacitance to GND is incorporated in the dedicated probe card.

In the EXPT circuits shown in Fig. 7.21, only a single high-speed input is required for each pair of DUTs at a time. If the EXPT is populated with one

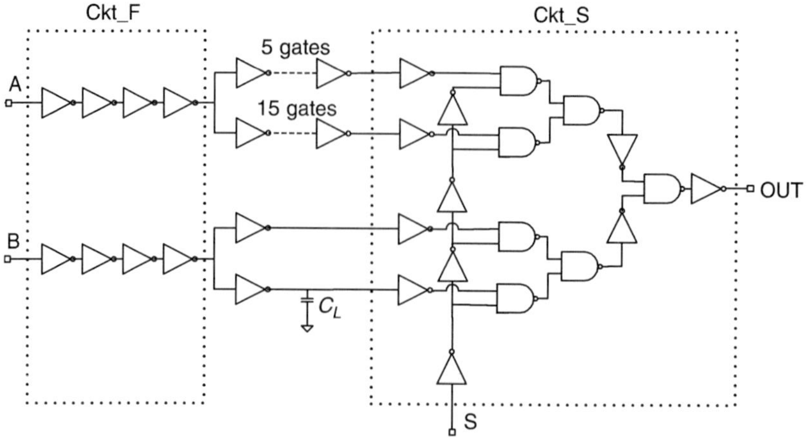

Fig. 7.21 Circuit schematic of one EXPT with two high-speed input signals, A and B, and a common output signal (OUT) for two sets of DUTs

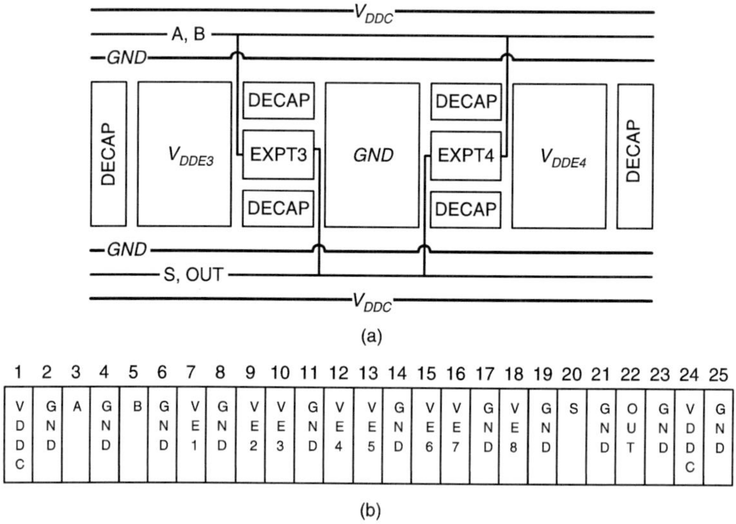

Fig. 7.22 **a** Physical layout of two EXPT blocks placed in the space between their independent V_{DDE} and common GND pads. **b** I/O pad assignments for a macro with eight EXPT blocks

DUT pair with two high-speed inputs, and precise control of their relative timing is required, there may be significant capacitive coupling between these long wire segments running parallel to each other. In this case, the circuit schematic and physical layout can be modified as shown in Fig. 7.23. Input signals A and B now travel from opposite sides of the macro, with only a very short common parallel run between them for any selected EXPT.

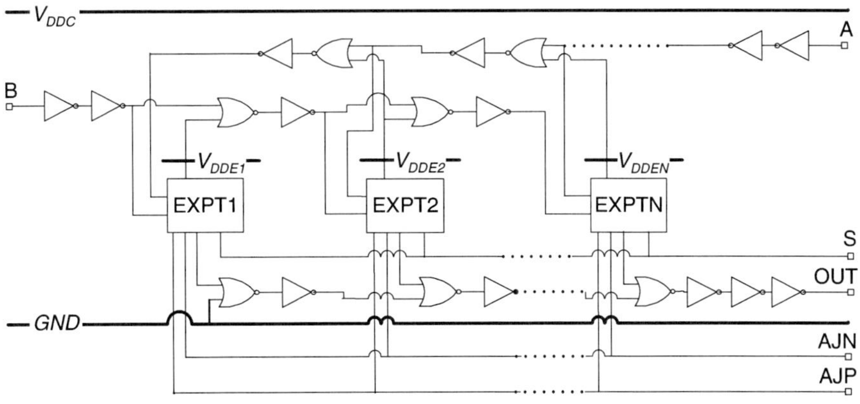

Fig. 7.23 Circuit schematic of a high-speed M1 testable macro configured for minimum cross talk. AJN and AJP are two additional DC inputs

Also shown in Fig. 7.23 are two capacitively decoupled low-speed inputs AJN and AJP. These are utilized as either analog inputs to various circuits in the EXPTs or inputs to a decoder (not shown) to pack more DUTs in the EXPTs. These additional inputs require pads that otherwise serve as V_{DDE} pads, reducing the number of EXPTs in the macro.

Finally, it should be noted that while the purpose of this template design is to enable functionality at the M1 metal level for early technology learning, it may also be used at higher metal levels with more complex DUTs. In this case, a single macro template with reduced number of EXPTs is used at any metal level. With this approach, the benefits of commonality in design, probe card, and test code may outweigh the loss in macro area efficiency of an additional more complex macro template.

7.4.5 Example 5: Macro for Pulse I–V with DC I/Os

In situations where *I–V* characterization of a MOSFET using DC sources does not capture the behavior under AC switching, because of heating, charge trapping, or floating-body effects in PD-SOI technology (Section 8.1.2), measurements can be carried out in a pulsed mode. Such measurements are typically conducted as bench tests with commercially available equipment or with a customized lab test setup. The test setup requires high-speed probes, cables, and connectors. The I/O pads are arranged in a S–G–S (signal–GND–signal) configuration as shown in Fig. 7.24 for an n-FET [7].

The principle of a high-speed pulse *I–V* test setup is as follows. A pulse generator is used to apply a low duty cycle voltage pulse, with width in the ns range, to the gate terminal of a MOSFET through a 50 Ω transmission line. The leading edge of the pulse also serves as the trigger to a sampling oscilloscope that monitors the

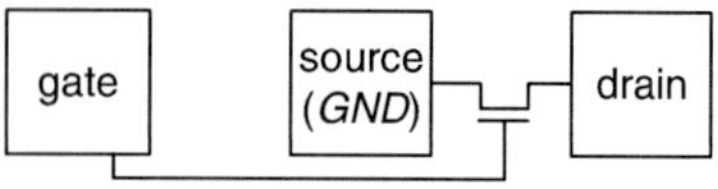

Fig. 7.24 I/O pad arrangement for high-speed pulse *I–V* measurements of an n-FET

delivered pulse via a 10× attenuated probe tap into the transmission line. At the same time, a voltage to the drain terminal of the MOSFET is applied by a DC power supply through another 50 Ω transmission line, via a bias "tee" that places a large inductor in series with the power supply. When the gate voltage pulse is applied, a voltage drop at the drain terminal is read out with another channel of the sampling oscilloscope via the same transmission line, through a large capacitor in another leg of the bias "tee." The corresponding drain current can be computed as the measured voltage drop divided by 50 Ω.

Pulse *I–V* characterization can also be carried out using only DC I/Os, thus eliminating the need of a custom high-speed probe card and the overhead of bench testing. The basic concept of such a macro design is suggested in Fig. 7.2b [8]. In this case, high-frequency signals are applied sequentially to the gates of several MOSFETs, sharing common source and drain terminals within the macro. The I_{ds} delivered by these MOSFETs appears as a DC current and may be measured using standard parametric ATE.

The schematic of a circuit for generating high-frequency signals to bias the gate terminals of 10 n-FETs is shown in Fig. 7.25. A ring oscillator (RO), with 2α stages and a NAND2 gate to enable the oscillations with an input signal EBL, serves as the

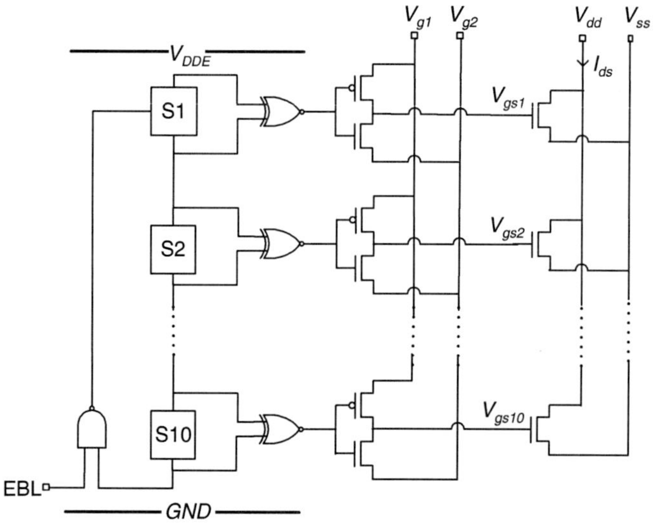

Fig. 7.25 Circuit schematic for applying high-frequency AC voltages, using an RO with multiple taps, to the gates of 10 n-FETs connected in parallel

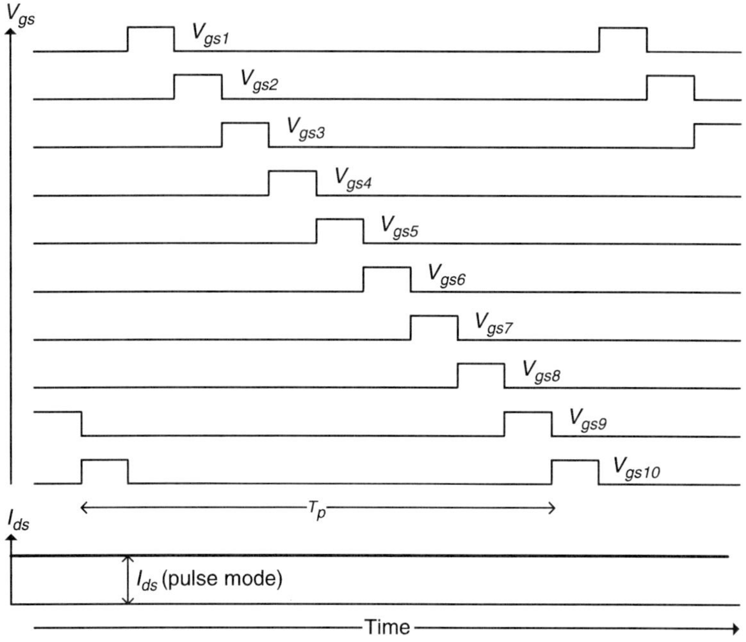

Fig. 7.26 Input signal waveforms applied to the gate terminals of 10 n-FETs connected in parallel. Reproduced from [5], with permission

source of the high-frequency signals required for AC characterization. The output voltages of 10 segments of the RO, each comprising $2n$ stages, where $n = \alpha/10$, are tapped and applied as inputs to XNOR2 gates. The outputs of the XNOR2 gates are trains of non-overlapping pulses of width $2n\tau_p$ and period $T_p/2$, where τ_p is the average delay of one stage of the RO and T_p is the period of the RO. These pulse trains, shown in Fig. 7.26, serve as inputs to inverters that in turn drive the gate terminals of the n-FETs.

The gate voltage V_{gs} of each n-FET is set by DC voltage inputs V_{g1} and V_{g2}, where $V_{gs} = \left(V_{g1} - V_{g2}\right)$. The drain and source voltages of the MOSFETs are set by two other independently controlled DC voltages, V_{dd} and V_{ss}, respectively. The current from the V_{dd} power supply has an essentially constant DC value, equivalent to the average I_{ds} of these n-FETs under pulsed conditions, with one and only one of the 10 n-FETs "on" at all times. The I_{off} of all n-FETs is measured by disabling the ring oscillator, which sets their V_{gs} values to 0.

The RO and XNOR2 gates are powered by a power supply V_{DDE}. The RO frequency may be measured with a standard frequency counter by incorporating a frequency divider and a I/O driver circuit as described in Chapter 6. This is a useful feature for validation of the macro design.

There are several advantages in measuring the average I_{ds} of MOSFETs using this scheme. The random variations are averaged over 10 (or more) MOSFETs in a

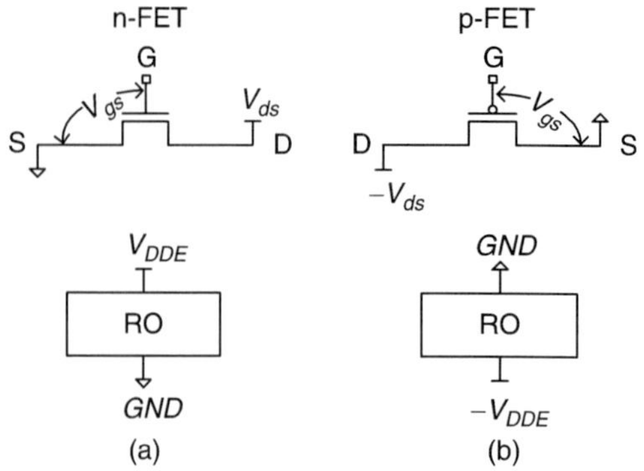

Fig. 7.27 Power supply connections, including those of the corresponding RO circuitry, for measuring pulse *I–V* characteristics of **a** an n-FET and **b** a p-FET

single measurement. At the same time, the current drawn by the V_{dd} power supply is equivalent to that of a single MOSFET, and *IR* drop in the wires connecting the MOSFETs to the I/O pads is less significant than if all MOSFETs were measured in parallel. The pulse width can be adjusted with the RO design and rise time of V_{gs} can be representative of a CMOS circuit in a product. Since the net I_{ds} drawn by the 10 MOSFETs is nearly constant, minimal decoupling capacitance is needed, even though pulse characteristics are obtained.

When the MOSFETs are biased in the saturation region, each MOSFET is on 10% of the time and off for remaining 90% of the time. This produces negligible heating in the individual MOSFETs. There is small, but sometimes significant, contribution to the measured I_{ds} from the MOSFETs in the off-state. A correction is estimated by measuring the I_{off} of all 10 MOSFETs in parallel (with the RO turned off) and subtracting $0.9\times$ of its value from the measured I_{ds}.

The DC I_{ds} of all 10 n-FETs in parallel can be measured by setting the V_{DD} of the RO to GND. By varying the gate and drain bias voltages with inputs V_{g1}, V_{g2}, V_{dd} and V_{ss}, DC *I–V* curves of the 10 n-FETs can be obtained and compared with the *I–V* curves obtained under pulsed conditions. For such DC measurements, special care must be taken to ensure power bus integrity.

The circuit schematic shown in Fig. 7.25, with appropriate polarities of voltage levels, works for the p-FET as well. The voltage levels for an n-FET and a p-FET and for the RO power supply, V_{DDE}, are shown in Fig. 7.27. In case of the p-FET, the high-voltage level is set at the common GND, and the "GND" of the RO is set at a negative V_{DDE} value.

References

1. Weste NHE, Eshraghian K, Smith MJS (2000) Principles of CMOS VLSI design, 2nd edn. Addison Wesley, Reading, MA
2. Baker RJ (2010) CMOS circuit design, layout, and simulation, 3rd edn. Wiley, IEEE Press
3. Uyemura JP (2001) CMOS logic circuit design. Kluwer Academic, Norwell, MA

4. Ketchen MB, Bhushan M, Anderson CJ (2004) Circuit and technique for characterizing switching delay history effects in silicon-on-insulator logic gates. Rev Sci Instrum 75:768–771
5. Ketchen MB, Bhushan M (2006) Product-representative "at speed" test structures for CMOS characterization. IBM J Res Dev 50(4/5):451–468
6. Bhushan M, Ketchen MB, Das KK (2008) CMOS latch metastability characterization at the 65-nm-technology node. Proceedings of the 2008 IEEE international conference on microelectronic test structures, 2008, pp 147–151
7. Jenkins KA, Sun JY-C, Gautier J (1997) Characteristics of SOI FET's under pulsed conditions. IEEE Trans Electron Dev 44:1923–1930
8. Ketchen MB, Bhushan M, Jenkins KA (2005) Circuit to measure high speed pulse *I–V* characteristics with only DC I/O's. Proceedings of the 2005 IEEE international SOI conference, 2005, pp 77–78

Chapter 8
Test Structures for SOI Technology

Contents

The MOSFETs in a silicon-on-insulator (SOI) technology are delineated in a thin silicon film, electrically isolated from the bulk silicon substrate by an oxide layer. CMOS circuits fabricated on an SOI substrate have inherent low parasitic source and drain junction capacitance and hence a potential for achieving higher switching speeds relative to circuits fabricated on bulk silicon. Because of electrical isolation and a smaller device volume, the circuits have higher immunity to latch up and to soft errors generated by incident radiation. In partially depleted silicon-on-insulator (PD-SOI) technology, electrical isolation of the MOSFET body from the underlying silicon substrate facilitates modulation of the threshold voltage, providing an opportunity for further circuit performance gain. Some high-performance microprocessor products in the marketplace are manufactured in PD-SOI technology.

Electrical and thermal isolation from the substrate introduces effects unique to PD-SOI technology. The MOSFET body potential and its threshold voltage are functions of the source, gate, and drain voltages and may change with time as a

M. Bhushan, M.B. Ketchen, *Microelectronic Test Structures for CMOS Technology*,
DOI 10.1007/978-1-4419-9377-9_8, © Springer Science+Business Media, LLC 2011

circuit undergoes multiple switching transitions. Logic gate delay and leakage power are therefore dependent on switching history. Thermal isolation of the body of a MOSFET leads to an increase in its temperature in the on-state. As a result, the *I–V* characteristics measured in the DC mode with continuous heating may be different than in an AC mode. In this chapter we describe test structures to measure the impact of changes in floating-body potential and heating on circuits fabricated in PD-SOI technology. A unique self-timed circuit for in-line characterization of switching history effect and ring oscillator-based test structures to measure active state leakage power exemplify design strategies to measure high-speed effects with parametric ATE.

A brief introduction to PD-SOI technology and circuit characterization is given in Section 8.1. Techniques for measurement of leakage power, impact of switching history on circuit delays, and the influence of self-heating effects on MOSFET characteristics are covered in Section 8.2. Examples of macro designs for measuring leakage power in the active state, delay variation with switching history and its statistics, and thermal time constants are described in Section 8.3.

Several books on SOI technology and circuit applications are now available [1, 2]. References on measurements of PD-SOI effects are cited throughout this chapter.

8.1 PD-SOI Technology

The development of SOI devices and circuits with superior radiation hardness and a higher temperature range of operation began in the 1960s for military and space applications. With a strong push toward higher microprocessor operating frequencies in the 1990s, leveraging the lower junction capacitance and floating-body effect, IBM pioneered the use of PD-SOI technology for high-performance digital CMOS applications. The first microprocessor in this technology was announced by IBM in the year 2000 [3]. Since then, the development of PD-SOI technology has followed the scaling trends in bulk silicon CMOS technology, and new products are introduced in the marketplace every year.

Cross sections of an n-FET and a p-FET on an SOI substrate are shown in Fig. 8.1a. The source, channel, and drain of the MOSFETs are delineated in a thin

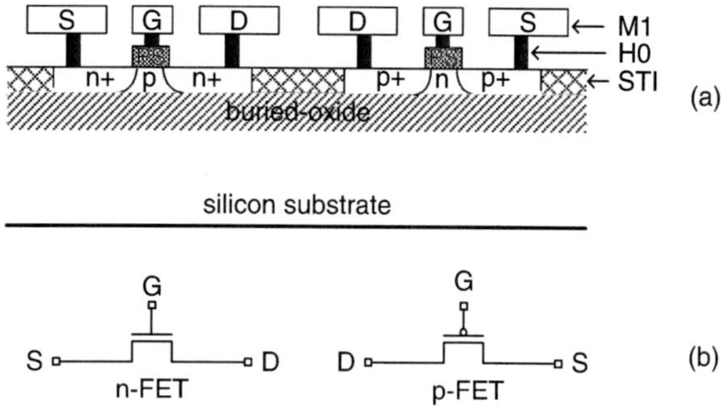

Fig. 8.1 **a** Schematic cross section of an n-FET and a p-FET on an SOI substrate. **b** Circuit symbols for an n-FET and a p-FET

silicon film separated by an oxide layer from the bulk silicon substrate. This buried-oxide (BOX) layer is typically >100 nm thick. In PD-SOI technology, the thickness of the silicon layer above the BOX is greater than the maximum depletion layer width and there is a quasi-neutral region in the body of the MOSFET. The silicon layer is fully depleted in fully depleted SOI (FD-SOI) technology. Because of an undesirable strong short-channel effect and issues with process control in manufacturing, PD-SOI technology is preferred over FD-SOI in volume manufacturing of high-performance digital CMOS products.

The floating-body MOSFET in PD-SOI technology has three terminals (source, gate, and drain) as shown in Fig. 8.1b. With the elimination of contacts to the substrate and n-well, the physical layouts of the floating-body PD-SOI circuits are more compact. MOSFET fabrication process steps on SOI substrates are similar to those for bulk silicon with some differences in process recipes. There are a few additional processing steps for making contact to the substrate and for inclusion of contacts to the body of the MOSFETs required in selected analog circuit applications. Some of the physical layout ground rules (GRs) for active circuit elements may be different than for bulk technology. If silicon wafers in both bulk silicon and PD-SOI technologies are being fabricated in the same foundry, test structure macro designs for bulk silicon described in this book can be made compatible with PD-SOI by including dummy n-well and substrate contacts in the physical layout and by following other bulk silicon-specific GRs related to metal contacts to the gates of the MOSFETs.

Electrical isolation of the MOSFET body from the silicon substrate introduces several differences in the device and circuit behavior from that of similar MOSFETs fabricated on bulk silicon substrates. The parasitic junction capacitance between the source and drain and the substrate is reduced and higher switching speeds of circuits may be obtained. The balance in charge leakage in and out of the MOSFET body affects the body potential, in turn modulating the threshold voltage V_t, leakage power, and circuit delays. Thermal isolation of the body of the MOSFET from the substrate results in a rise in body temperatures at high current densities. With thermal time constants of ~100 ns, the *I–V* characteristics of MOSFETs under DC bias may be different than under AC operation with <1 ns time periods. These effects are included in the models used in circuit simulations of CMOS products in SOI technology. Validation of circuit models requires special test structure designs to isolate and capture these effects.

For applications in which variability induced by floating-body effects is detrimental to circuit functionality, contact to the bodies of the MOSFETs can be made to provide bulk silicon like MOSFETs. Additional silicon area and process steps are required to accommodate the body contacts. Physical layouts of inverters with and without body contacts are shown in Fig. 8.2a, b. In these examples, the body of the p-FET is tied to V_{DD} and that of the n-FET to GND. The details of silicon diffusion areas for making contact to n-body for p-FET and p-body for n-FET are shown in Fig. 8.2c. The parasitic resistance and capacitance of the body contacts add to the circuit delay and generally their use in PD-SOI circuits is restricted to analog and I/O applications.

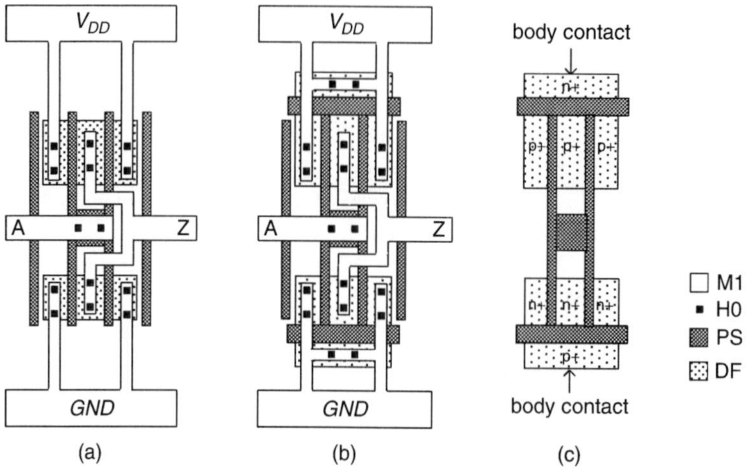

Fig. 8.2 Physical layout of an inverter in PD-SOI technology **a** with floating body and **b** with body contacts. **c** Silicon diffusion areas for source, drain, and body contacts of p-FET and n-FET

8.1.1 Junction Capacitance

Physical cross sections of an n-FET in PD-SOI and bulk silicon are shown in Fig. 8.3. The parasitic junction capacitances of n^+ silicon diffusion areas in the S and D contacts to the p-type substrate are substantially reduced by the presence of the BOX layer in PD-SOI. The reduction in circuit delay from this reduced junction capacitance in both n-FET and p-FET varies with the total switching capacitance of a logic gate and its physical layout. The benefit is larger for a logic gate with FO $= 1$ than with FO $= 3$ or 4 or for a logic gate driving a wire load. An estimated delay reduction for an FO $= 3$ inverter stage with our standard two-PS finger physical layout shown in Appendix A is $\sim$10% [4].

8.1.2 Floating-Body (FB) Effect

In a static state, the floating-body (FB) potential of a MOSFET is determined by a balance of charge generated by impact ionization and source-to-body, drain-to-body,

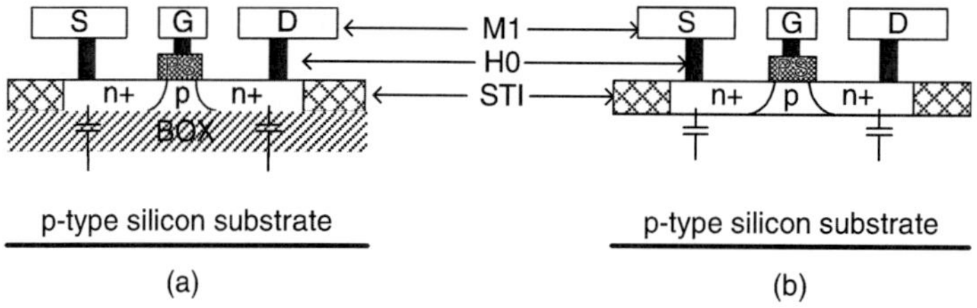

Fig. 8.3 Schematic cross section of n-FETs showing the junction capacitance **a** in PD-SOI and **b** in bulk silicon technologies

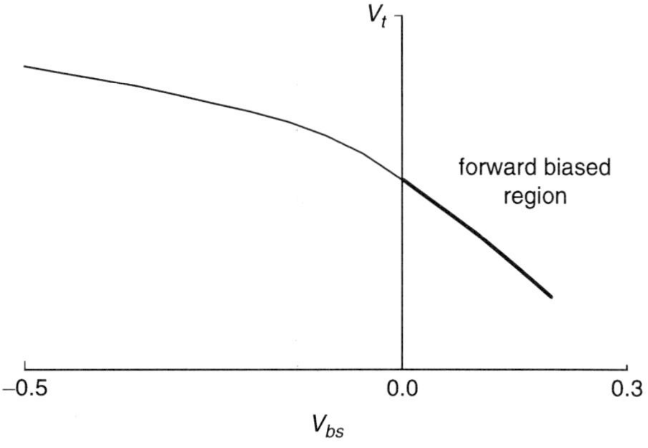

Fig. 8.4 MOSFET V_t as a function of V_{bs} for an n-FET in PD-SOI technology. The forward biased region is indicated by a *thicker line*

and gate-to-body leakage currents. In a switching transient, the body potential is further modulated by capacitive coupling to the gate and drain electrodes. When the body becomes more forward biased with respect to the source, V_t is lowered as shown in Fig. 8.4. The dynamic V_t modulation during switching transients cannot be measured directly; however, it impacts measurable quantities such as circuit delay and MOSFET leakage current, I_{off}.

Modulation of the FB potential and V_t in the p-FET and n-FET of an inverter undergoing PD and PU transitions is illustrated in Fig. 8.5, where V_{bs} and V_t values are all shown as positive. Prior to a PD transition, the voltage at the output node V_{out} is a "1" and the leakage current is determined by I_{offn} of n-FET N1. At the start of a PD transition, the n-FET body is momentarily pulled up by the gate-to-body capacitive coupling and then pulled down by the drain-to-body capacitive coupling. Immediately after the PD transition, the V_{bs} of n-FET N1* is lowered and its V_t is raised, as shown in Fig. 8.5c. The V_{bs} of p-FET P1* after the PD transition is higher and its V_t is lowered. The leakage current of the inverter, I_{offp}, is now determined by P1* in Fig. 8.5a*.

The V_{bs} and V_t values of the inverter p-FET and n-FET for a PU transition are shown in Fig. 8.5d. Prior to a PU transition, V_{out} is a "0" and the inverter leakage current is determined by I_{offp} of p-FET P2. Immediately after the PU transition, the V_{bs} of p-FET P2* is lowered, raising its V_t. The V_{bs} of n-FET N2* is raised and its V_t is lowered. The leakage current of the inverter, I_{offn}, is now determined by N2* in Fig. 8.5b*. For an interconnected chain of such inverters experiencing passage of a single isolated edge, the indicated $\delta V_t(\text{N})$ and $\delta V_t(\text{P})$ values will invoke a change in IDDQ of the chain that will persist until the MOSFET bodies settle back to their pre-switch potentials.

The V_{bs} and V_t values after a switching transition vary with time after the transition and eventually return to the pre-switch state. The time constant to reach static equilibrium state following an isolated switching transition is typically between a few μs and a few ms. The MOSFETs in an inverter switching at a constant rate as in a ring oscillator reach an equilibrium state within a few ms, and their average FB

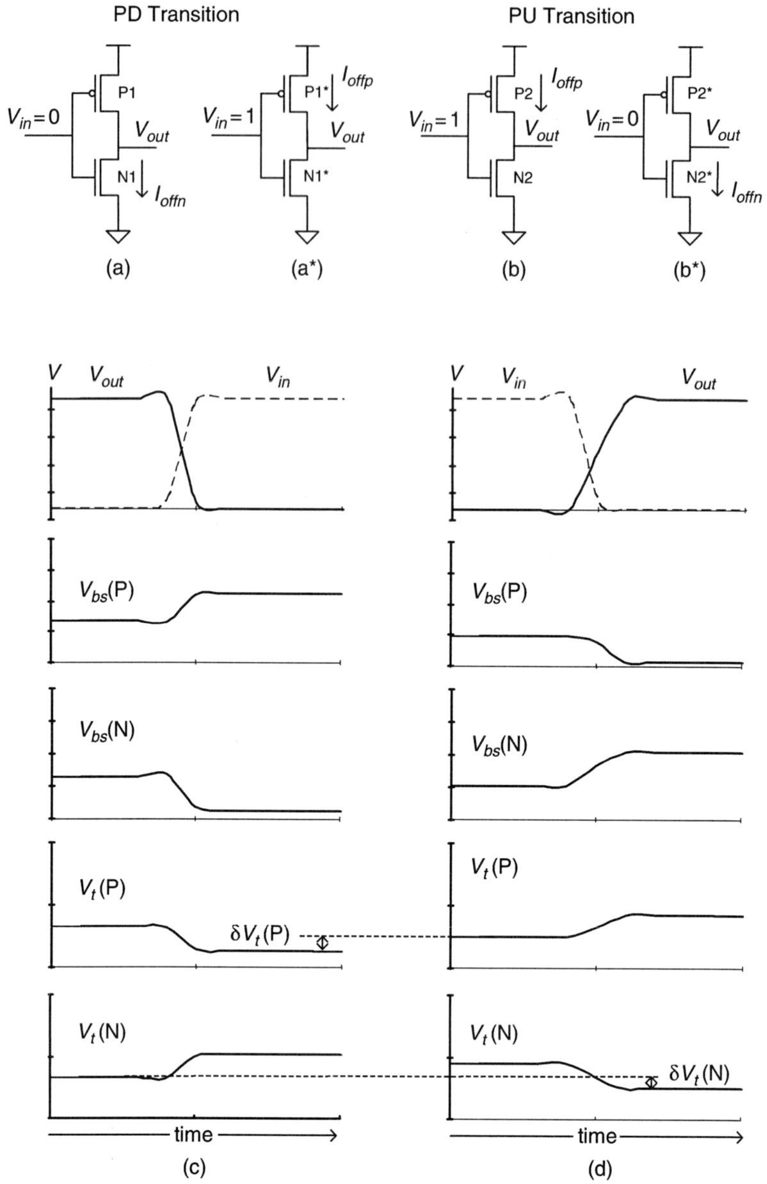

Fig. 8.5 Circuit schematic indicating I_{off} contributions for an inverter before and after **a, a*** PD and **b, b*** PU transitions. **c, d** Inverter input and output waveforms and absolute values of V_{bs} and V_{t} as functions of time for the transitions corresponding to **a, a*** and **b, b***, respectively

potential values are typically different than in a quiescent state or within a few ns of an initial switching event.

From the above discussion, it is apparent that the leakage current in the quiescent state IDDQ, and signal propagation delay τ, of a circuit in PD-SOI technology vary with switching history. When a circuit switches after being in a static idle state for a few ms or longer, the event is called a "first switch" or 1SW transition. When it switches again within a few ns of the 1SW transition, it is called "second switch" or 2SW transition. A circuit switching periodically, as in the case of a clocked circuit, is in a steady state (SS). The V_{bs} and V_t values of the MOSFETs in SS state are typically between their values in the pre-1SW and pre-2SW states.

Output PD and PU signal waveforms for 1SW, 2SW, and SS transitions of an inverting logic gate are shown in Fig. 8.6. The (1SW – 2SW) switching history effects for PD and PU transitions, H_{tpd} and H_{tpu}, can be defined as

$$H_{tpd}\,(\%) = 2\left(\frac{\tau_{1pd} - \tau_{2pd}}{\tau_{1pd} + \tau_{2pd}}\right)100 \tag{8.1}$$

and

$$H_{tpu}\,(\%) = 2\left(\frac{\tau_{1pu} - \tau_{2pu}}{\tau_{1pu} + \tau_{2pu}}\right)100, \tag{8.2}$$

where τ_{1pd} and τ_{2pd} are the 1SW and 2SW PD delays and τ_{1pu} and τ_{2pu} are the 1SW and 2SW PU delays. The MOSFETs with the dominating influence on pre-switch IDDQ and subsequent delays, in different switching transitions of an inverter, are listed in Table 8.1.

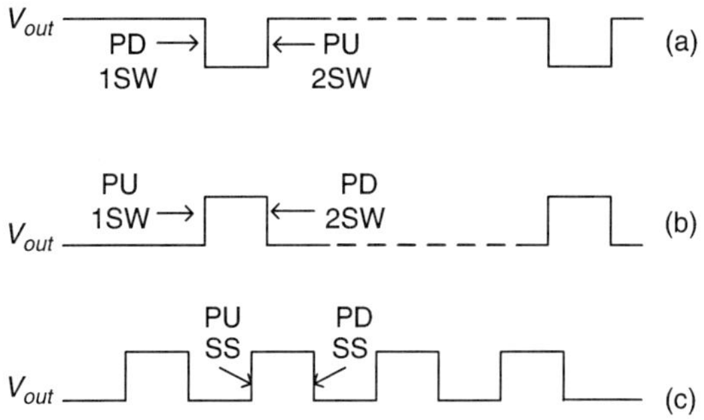

Transition	Delay
PD 1SW	τ_{1pd}
PD 2SW	τ_{2pd}
PU 1SW	τ_{1pu}
PU 2SW	τ_{2pu}
PD SS	τ_{sspd}
PU SS	τ_{sspu}

(d)

Fig. 8.6 Inverter *output* signal waveforms **a** for 1SW PD and 2SW PU, **b** for 1SW PU and 2SW PD, and **c** for SS transitions. **d** Corresponding circuit delay symbols

Table 8.1 MOSFETs dominating the pre-switch IDDQ in an inverter for different switching transitions

Transition	MOSFET
1SW PD	N1
2SW PD	N2*
1SW PU	P2
2SW PU	P1*

MOSFET symbols correspond to those in Fig. 8.5a, b

The measured (1SW − 2SW) history effect H_t in a circuit path comprising a chain of gates is the average of H_{tpd} and H_{tpu}. The values of H_t for PD or PU transitions are positive for $\tau_1 > \tau_2$ (1SW slower than 2SW) and negative for $\tau_1 < \tau_2$ (1SW faster than 2SW). The history effect varies with V_{DD} and, to a lesser degree, temperature. The MOSFETs in PD-SOI technology can be engineered to result in positive, negative, or negligibly small values of H_t in standard logic gates at a desired value of V_{DD}. Generally, H_t for standard logic gates is designed to be $\lesssim$5% at the operating V_{DD} of a product. It is, however, dependent on circuit topology and some circuits may experience H_t values of >20%.

History effect H may be more generally defined as the fractional difference in the delay values corresponding to two different pre-switch conditions, for PD or PU of a single gate or for a chain of gates. With pre-switch conditions $pc1$ and $pc2$, this can be expressed as

$$H(pc1 - pc2)(\%) = 2\left(\frac{\tau(pc1) - \tau(pc2)}{\tau(pc1) + \tau(pc2)}\right)100. \tag{8.3}$$

For example, referring to Fig. 8.6, one can define H (1SW – SS) of an inverter as the % difference in PD or PU switching delay between 1SW and SS and H (SS – 2SW) as the % difference in PD or PU switching delay between SS and 2SW. For a chain of inverters the same expressions apply where the delay values are now averaged over PD and PU. Note also that H (1SW − 2SW) $= H_t$ is approximately equivalent to the sum of H (1SW − SS) and H (SS − 2SW). In CMOS circuits in a product, the pre-switch conditions for individual gates vary widely and are typically much more complex than the situations shown in Fig. 8.6. It is usually the case that actual delays are bounded by 1SW and 2SW, and SS delays lie between 1SW and 2SW. For most situations, H_{tpu} and H_{tpd}, as defined in Eqs. (8.1) and (8.2) give upper bounds of delay variation arising from history effect.

8.1.3 Self-Heating

I–V characteristics of MOSFETs in PD-SOI technology are intrinsically different under static DC conditions than under transient switching because of floating-body and heating effects. The difference between MOSFET currents measured under DC and pulse excitation is larger at higher bias voltages and higher current densities where significant V_t modulation and self-heating occur. A comparison of an n-FET drain current I_{ds} as a function of V_{gs} with and without self-heating is shown in Fig. 8.7.

8.2 PD-SOI-Specific Measurements

Physical models for MOSFETs in PD-SOI circuits are used to predict the FB potential and body temperature. These models are validated by characterization of MOSFET and circuit parameters under DC and high-speed switching

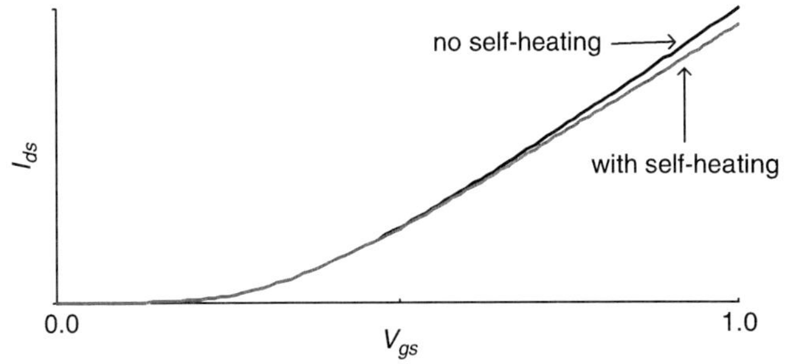

Fig. 8.7 I_{ds} as a function of V_{gs} for an n-FET, with and without self-heating, in PD-SOI technology

conditions. Test structures for high-speed PU and PD delay measurements and pulse I–V characterization described in Chapter 7 are applicable to PD-SOI technology. New design and measurement techniques specific to PD-SOI technology are described in the following sections.

8.2.1 Measurement of Active State Leakage Power

Leakage power of bulk silicon circuits is derived from a straightforward measurement of IDDQ which is the DC current drawn by a circuit in its quiescent state. When a circuit block comprising a number of MOSFETs is switching periodically, the measured current IDDA, is the time average sum of its active switching current and the leakage current when it is idle between transitions. If the switching delay is much smaller than the average time between transitions, the background leakage current is well represented by the IDDQ value.

In PD-SOI, the threshold voltages of MOSFETs and in turn their I_{off} values are dependent on the switching history. The leakage current of a circuit sitting in a quiescent state for a long time is different than immediately after undergoing a transition or when it is switching periodically (steady state), as is the case of a ring oscillator or a clock circuit. As an example, a change in V_t of 20 mV corresponds to ~60% shift in leakage current.

The leakage current of a PD-SOI circuit switching in steady state can be determined by measuring the current IDDA drawn by the circuit at different frequencies at a fixed duty cycle, d_{cl}. Here d_{cl} is defined as the ratio of pulse width to the pulse period. For a pulsed signal input, the measured IDDA of a chain on N inverters shown in Fig. 8.8a is the sum of the leakage currents of (N – 1) inverters and the switching current of one inverter. A plot of IDDA as a function of frequency of the input pulse is shown in Fig. 8.8b for $\tau_1 > \tau_2$ (positive H_t). The fitted straight line to the data points is extrapolated to $f = 0$ to give the leakage current IDDS in the active state. The difference in leakage currents in the active and quiescent states, ΔIDDQ, is then extracted as indicated in Fig. 8.8b. It is assumed that the

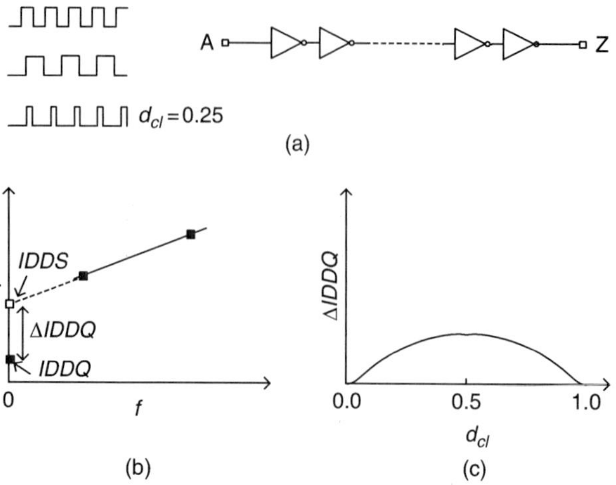

Fig. 8.8 **a** Circuit schematic and input pulse waveforms for measuring active state leakage power. **b** Measured current (*I*) as a function of frequency *f* for ΔIDDQ extraction. **c** ΔIDDQ as a function of duty cycle d_{cl} at a fixed *f*

time interval between the input signal edges is always larger than the signal propagation delay through the chain so that only one gate in the chain is switching at any instant.

Measurement of ΔIDDQ, repeated for different values of d_{cl}, is shown in Fig. 8.8c. With d_{cl} close to 0 or 1, IDDS in the active state is nearly equal to the quiescent or pre-1SW value. It increases with increasing d_{cl} ($\tau_1 > \tau_2$), reaching a maximum at $d_{\mathrm{cl}} = 0.5$, and then symmetrically decreases back to its pre-1SW value. Hence, measurement of ΔIDDQ at $d_{\mathrm{cl}} = 0.5$ gives an upper bound of IDDS in the active state. Macro designs for leakage measurements based on this circuit scheme are described in Section 8.3.1.

8.2.2 Measurement of History Effect

A unique test structure design to measure the average PU and PD history effect for (1SW – 2SW) transitions H_{t}, utilizes the feature that the width of a pulse as it propagates along a long chain of *N* logic gates is altered when $\tau_1 \neq \tau_2$. When a pulse of width T_{wi} is launched on an inverter chain in bulk silicon technology, as shown in Fig. 8.9a, the output pulse width T_{wo} is the same as T_{wi}. In PD-SOI technology, the first edge of the pulse experiences a delay of $N\tau_1$ while the second edge is delayed by $N\tau_2$ as shown in Fig. 8.9b. Hence,

$$T_{\mathrm{wo}} = T_{\mathrm{wi}} - N(\tau_1 - \tau_2)\,. \tag{8.4}$$

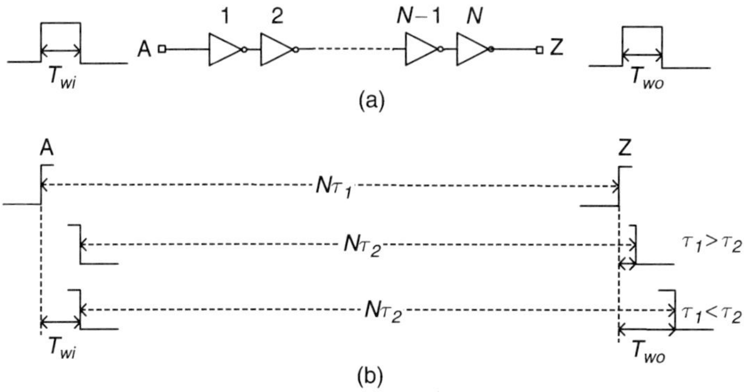

Fig. 8.9 **a** A pulse of width T_{wi} emerges with a width T_{wo} after traveling through a chain of N logic gates. **b** First and second edges of the pulse showing narrowing of T_{wi} for $\tau_1 > \tau_2$ and widening for $\tau_1 < \tau_2$

As the pulse travels down the chain, it narrows if $\tau_1 > \tau_2$, widens if $\tau_1 < \tau_2$, and is unchanged if $\tau_1 = \tau_2$. For $N(\tau_1 - \tau_2) > T_{\mathrm{wi}}$, the pulse is annihilated in the chain before it reaches the far end [5].

A circuit schematic comprising a chain of n inverters and a NAND gate followed by an inverter to generate a sharp pulse of width $n\tau_1$ from an input signal edge is shown in Fig. 8.10a. This pulse is generated within the test structure and launched on the long delay chain of N inverters in Fig. 8.9. If the inverters in the two chains in Figs. 8.9 and 8.10a are nominally identical and n is increased until the pulse is just annihilated ($T_{\mathrm{wo}} = 0$), then

$$H_{\mathrm{t}}(\%) = \frac{n}{N} \times 100 = \left(1 - \frac{\tau_2}{\tau_1}\right) \times 100, \qquad (8.5)$$

where the definition of H_{t} has been slightly modified here to correspond to the experimental output. In this situation, H_{t} is simply the ratio of the number of inverters in

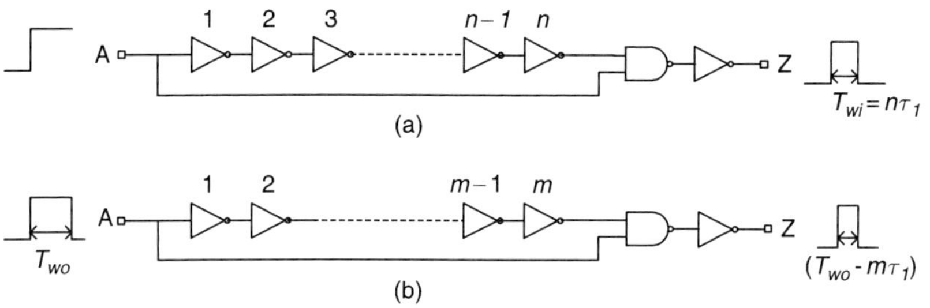

Fig. 8.10 Circuit schematic **a** to generate a pulse with a single-input signal edge and **b** to generate an offset for the output pulse width to measure negative values of H_{t}. Both n and m must be even numbers

the two chains and is therefore self-calibrating. It is also self-timed as the pulse is generated within the test structure and the measurement of H_t is independent of the timing of the pulse. With a scheme to generate a number of different initial pulse widths and a method to detect the presence or absence of the output pulse, H_t can be characterized with a precision determined by the incremental increase in input pulse widths, the highest resolution being equivalent to the delay of two logic gates. This type of test structure can be implemented with only DC I/Os. As the measurement is self-timed and self-calibrated, it requires no sophisticated time measurement instrumentation. History effect can thus be measured in the manufacturing line with a standard DC probe card.

In the case where $\tau_1 < \tau_2$, the pulse width increases as it propagates along the inverter chain and the circuit shown in Fig. 8.10b is added to the end of the primary delay chain with N inverters. Its effect is to introduce a zero-offset of $(m/N) \times 100\%$ and enable the measurement of negative values of H_t $(\tau_1 < \tau_2)$.

A macro design to measure H_t using these ideas is described in Section 8.3.2 (Example 2). The generation and detection of the input and output pulses is accomplished with level-sensitive latches. Measurements are carried out with DC I/Os and a number of experiments comprising delay chains with different logic gates are accommodated in a single macro implemented at the M1 metal level.

History effect for any arbitrary switching pattern is determined by measuring PU and PD delays under any two different pre-switch scenarios. The differential scheme for circuit delay measurement under high-speed switching conditions described in Chapter 7 can be used for measuring average of PU and PD delay values to determine H in a chain of logic gates or for individual PU and PD delays and H values of a single logic gate. Pulse width and pulse period of the waveforms shown in Fig. 8.6 can be varied to measure H as a function of the time differences between various transitions and thereby determine FB relaxation times. When the time between transitions is longer than the time for the FB potential to come to an equilibrium state (~10 μs to a few ms), all delays revert to their 1SW values. A high-speed probe card, pulse generators and/or logic pattern generators, and a sampling oscilloscope are required for such measurements. This approach is not suitable for monitoring of H_t in the manufacturing line on a routine basis, but is ideal for obtaining detailed information for model build.

A simplified version of the circuit for measuring H is shown in Fig. 8.11. Here DUT1 and DUT2 are two nominally identical logic gates, with DUT2 driving an additional capacitive load C_L. Input S is toggled between "1" and "0" to select the signal path from input A to OUT through DUT1 or DUT2, respectively. The shift in the OUT signal δt is related to the difference in delays of DUT1 and DUT2 and is to a first order given by

$$\delta t = r_{swp} C_L \text{ or } \delta t = r_{swn} C_L, \tag{8.6}$$

where r_{swp} and r_{swn} are the switching resistances of the logic gate for PU and PD transitions, respectively. The impact of the FB effect on delay is primarily through modulation of r_{swp} and r_{swn}. The value of δt is measured for each of the PU and PD

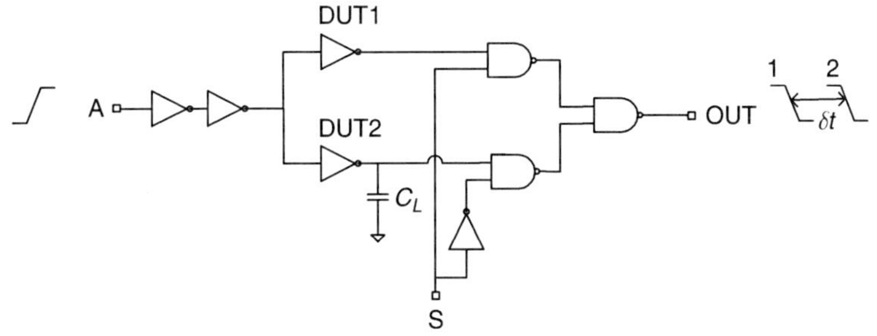

Fig. 8.11 Differential time measurement circuit for determining history effect H of a logic gate

transitions indicated in Fig. 8.6 and H values are calculated using Eq. (8.3), where the measured δt values are used in place of values of τ. It is assumed that r_{sw} values of the two DUTs are nearly identical. If C_{L} is a fixed metal capacitor, history effect is confined to the r_{sw} of the logic gate. If a MOSFET gate capacitor is used instead, a small error may be introduced because of the dependence of C_{L} on FB potential and in turn its V_{t}. Random variation of threshold voltages in the logic gates also introduces an error in this differential measurement scheme, as has been discussed in previous chapters.

The logic gates in the signal paths for DUT1 and DUT2 in Fig. 8.11 are common except for the two NAND2 gates immediately following the DUTs. The transition in these gates is a PU when the DUT undergoes a PD transition and a PD for a PU transition in the DUT. Any history effect in these gates is then mixed with that of the DUT itself, the observed value typically being about 80% or more dominated by the DUT. This source of error can be eliminated by implementing the NAND2 gates with body-contacted MOSFETs which exhibit much smaller history effect. Alternatively the measured value can be corrected based on simulation results, assuming the device models capture the FB effect to a reasonable degree.

Average of PU and PD history effect is measured with DUTs comprising chains of logic gates. If DUT1 comprises a chain of 5 logic gates and DUT2 comprises a chain of 15 logic gates, with C_{L} removed, the measured H is averaged over 10 logic gates. This arrangement reduces the effect of random variation in MOSFET properties but H values for PD and PU transitions cannot be individually determined.

In a third scheme for measurement of PU and PD history effect for (1SW – 2SW) transitions of a single logic gate, the timings of the DUT output signals for 1SW and 2SW transitions are determined with respect to a reference signal. The circuit for this differential time measurement technique includes a calibrated variable delay line and a level-sensitive latch. With the reference signal from the delay line serving as clock input to a latch, changes in the arrival time of the DUT signal at its data port can be measured with sub-ps precision. A test structure based on this technique is described in Section 8.3.4 (Example 4). The basic design can be implemented with only DC I/Os. More complex macros for digital ATE are designed to collect H_{t} variability statistics on a large number of gates.

8.2.3 Measurement of Heating Effects

The impact of self-heating in PD-SOI MOSFETs becomes evident at I_{ds} values exceeding ~800 μA/μm. Self-heating is generally stronger in n-FETs, although with recent improvements in p-FET current drive, it is observed in p-FETs also. The effect of self-heating on I–V characteristics is evaluated by comparing measurements under static (DC) and pulsed (AC) conditions.

A macro design for pulse I–V characterization of MOSFETs described in Section 7.5 can also be implemented in PD-SOI technology. This macro design with DC I/Os can accommodate only a few MOSFETs in a standard template. Hence, such macros are typically used for more detailed device characterization in early technology development and for MOSFET model build.

The increase in MOSFET temperature at a constant I_{ds} value may be extracted from the change in resistance of the PS gate measured in the direction orthogonal to the current flow in the channel. Calibration of gate resistance as a function of temperature is obtained by varying the wafer chuck temperature while the MOSFET is in the off-state. Care should be taken to ensure that the heat drawn from the metal wire connections to the gate does not introduce appreciable cooling.

Measurement of thermal time constant of the MOSFET body is more challenging as it requires sampling a temperature-sensitive parameter immediately after switching the MOSFET on or off. Using the differential time measurement scheme described earlier, the thermal time constant of a MOSFET can be determined by exploiting the temperature sensitivity of inverter delay at known time intervals after application of a signal edge. A test structure based on this principle for measurement of thermal time constants in a silicon island with multiple MOSFETs is described in Section 8.3.5 (Example 5).

8.3 Macro Designs for PD-SOI Circuit Characterization

Modulation of MOSFET temperature and V_t from self-heating and FB effects typically produces changes in circuit delays of the order of ~10% or less. Detection of these small changes requires accurate measurements of time delays. It is a good practice to include DUTs comprising body-contacted MOSFETs for validating a macro template design. Absence of any PD-SOI effect in these DUTs helps establish design and measurement integrity.

All macro designs follow the physical ground rules (GRs) for PD-SOI technology. The macros with FB MOSFETs can be made compatible with bulk silicon technology if dummy n-well and substrate contacts are included. One I/O pad in each macro is dedicated to silicon substrate contact through the BOX layer. The substrate is normally held at the GND potential during test but can be biased to create a back channel in the MOSFET, if desired.

In this section, five macro design examples for measurement of PD-SOI-specific effects are provided. Macros for measurement of dynamic leakage power and history effect in Examples 1 and 2 utilize DC I/Os. Simplified versions of these macros are implemented at the M1 metal level for early monitoring in the technology development cycle or in manufacturing. A macro design for measurement of statistical variations in history effect, leveraging the capabilities of digital ATE, is described in Example 3. High-speed macro designs suitable for detailed characterization in a laboratory bench test setup, for measuring PU and PD delays and history effects with an arbitrary switching sequence and for measuring thermal time constants, are covered in Examples 4 and 5, respectively.

8.3.1 Example 1: Macros for Dynamic Leakage Power

Measurements of dynamic leakage power in PD-SOI circuits as described in Section 8.2.1 are carried out with either a ring oscillator or a chain of logic gates. A simple modification of an RO circuit enables IDDA measurements at the fundamental and third harmonic of the RO. The change in IDDQ in the active steady state with a duty cycle d_{cl} of 0.5 is obtained by extrapolation of the IDDA values to zero frequency as shown in Fig. 8.8b. Measurements at different d_{cl} values, pulse widths, and periods experienced by circuits in a product are made with a delay chain test structure.

In standard RO measurements, spontaneous generation of higher harmonic oscillations corrupts data collection at the fundamental frequency, and these higher harmonics are undesirable. In Section 6.2.1.4, a circuit to eliminate higher frequency harmonics in an RO is described. A controlled generation of higher frequency harmonics in an RO, however, facilitates measurements of leakage power in the active state in PD-SOI circuits [6].

Shown in Fig. 8.12a is a circuit (CU3) for generating a third harmonic in an RO. The number of stages in the longest path through this circuit is 4α, whereas the number of the same stages in the RO to be initiated is $6\alpha + 1$. Three equally spaced signal edges at the output terminal EX are generated from a single-input edge at the IN terminal, with the time intervals between the edges corresponding to one-sixth of the fundamental period of the RO. The output of this CU3 circuit block serves as an enable signal for the RO to initialize a third harmonic oscillation.

The CU3 circuit is integrated in the RO test structure as shown in Fig. 8.12b. Input signal A0 is set to "1" or "0" to select either the external EBL signal or that from the CU3 unit to generate the fundamental or third harmonic in the RO. The leakage current IDDQ of the RO in the quiescent state is measured. This is followed by enabling the RO in its fundamental and third harmonic states and measuring its frequency and IDDA values, (f_1, IDDA1) and (f_3, IDDA3), respectively. The change in leakage current during switching, ΔIDDQ, is then calculated as

$$\Delta\mathrm{IDDQ} = \mathrm{IDDQ} - \mathrm{IDDS}, \tag{8.7}$$

where IDDS is obtained by extrapolating the IDDA vs. f plot back to $f = 0$ as shown in Fig. 8.8b. In one fundamental time period of the RO, effectively only one stage is switching at a time, whereas three stages are switching in the third harmonic

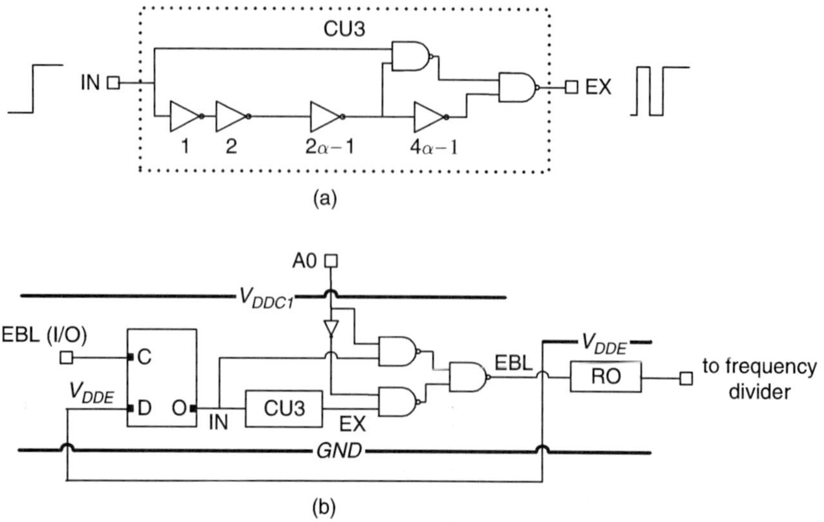

Fig. 8.12 Schematic **a** of CU3 circuit to enable a third harmonic and **b** of a test structure configured to generate either a fundamental or third harmonic oscillation in an RO. Reproduced from [6], with permission, © 2010 IEEE

state. The number of stages in the idle state is thus different in the two cases. The accuracy of the ΔIDDQ determination is improved by increasing α (>50) so that the switching time of one stage is a small fraction of the RO time period. Alternatively a correction can be incorporated to account for the different number of idle gates in the fundamental and third harmonic states.

The circuit shown in Fig. 8.12 can be implemented at the M1 metal level using an RO macro design described in Section 6.2.1, with the addition of an I/O pad for the DC input A0. The physical arrangement of the modified macro is shown in Fig. 8.13. The circuit in Fig. 8.12b (ckt), on a separate power supply V_{DDC1}, is placed between two I/O pads adjacent to the RO. The RO is powered by V_{DDE} and the frequency divider and the I/O driver are powered by V_{DDC}.

The schematic of a circuit for measuring leakage current of a delay chain of logic gates for different duty cycles is shown in Fig. 8.14a. An RO, with $(2\alpha + 1)$ stages,

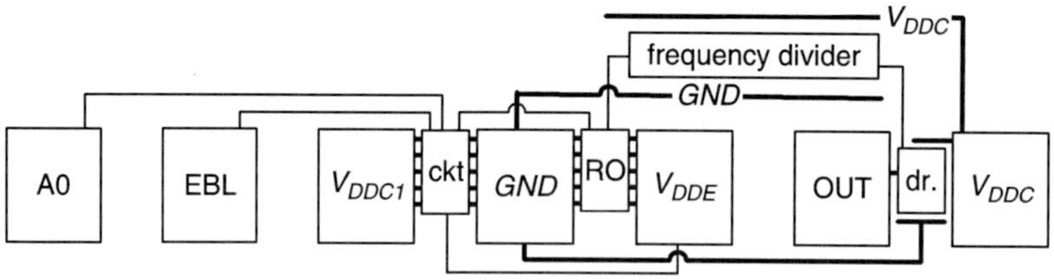

Fig. 8.13 Physical layout schematic and I/O pad assignments for PD-SOI circuit leakage measurements in the active and quiescent states

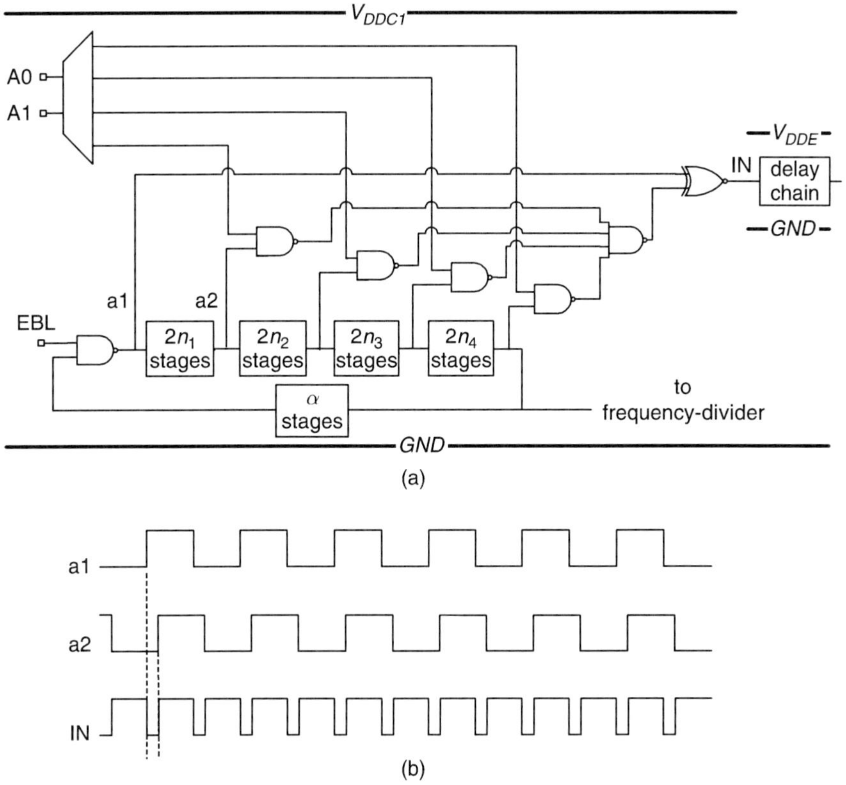

Fig. 8.14 **a** Circuit schematic to measure ΔIDDQ of a chain of logic gates with varying input pulse widths and periods. **b** Signal waveforms at nodes a1, a2, and IN for $d_{cl} = 0.75$

generates a periodic signal with a single duty cycle of 0.5. The RO is tapped at different intervals, to produce a continuous train of pulses that are launched down the delay chain. A pair of taps, at the output of the NAND2 in the RO and at $2n_1$, $2n_2$, $2n_3$, and $2n_4$ stage intervals, is selected by a decoder. The d_{cl} value for a tap across $2n_1$ stages is given by

$$d_{cl} = 1 - \frac{2n_1}{2\alpha + 1}. \tag{8.8}$$

Signal waveforms at nodes a1, a2, and IN for $d_{cl} = 0.75$ are shown in Fig. 8.14b. Note that the signal frequency at IN is twice that of the RO and that d_{cl} values of 0.75 and 0.25 are equivalent as the signal traverses a delay chain.

The frequency of the RO is varied by changing its V_{DD}. Alternatively, an RO comprising current-starved inverters is used and its frequency modulated with additional analog input signals applied to the gates of the current-starved n-FETs and p-FETs. The IDDQ and IDDA values of the delay chain are measured at two different frequencies and ΔIDDQ extracted using Eq. (8.7).

The delay chain may comprise nominally identical logic gates or a large circuit block. The number of gates in the delay chain must be sufficiently large (typically 100–200) to generate a significant change in leakage current, thereby minimizing measurement errors. However, the minimum edge spacing of the pulse train must be larger than the propagation delay through the chain to ensure that only one switching event occurs at a time. This time constraint places an upper and a lower limit on d_{cl} at any given frequency.

The circuit for variable duty cycle can be accommodated in a similar way as the RO test structure shown in Fig. 8.13, with an independent power supply V_{DDC1}. The delay chain replaces the RO and an additional I/O pad is required for the decoder input signal A1. In both schemes, the RO frequency signal is fed to a frequency divider and frequency measurements may be made with a frequency counter.

8.3.2 Example 2: Macros for H_t Measurements Using DC I/Os

In PD-SOI technology, in the presence of history effect, the width of a signal pulse may change as it travels through a chain of logic gates. The basic idea of a technique utilizing this property to measure (1SW – 2SW) history effect H_{t} is described in Section 8.2.2. The technique is self-timed and self-calibrated, with all the high-speed action taking place within the macro itself. The range of H_{t} values and the measurement resolution are set by the details of the test structure design.

The macro design described here incorporates four independent test structures for measuring H_{t} of different logic gate types. The test structures, or EXPTs, have a common template populated with different logic gate designs. The H_{t} values are measured with a resolution of 1/8th of the full range value of H_{t}. As an example, with a full H_{t} range of 20%, the resolution is set at 2.5%. The macro design requires only DC I/Os and is suitable for measurement with parametric or digital ATE in the manufacturing line. A compact version of this design described here is implemented at the M1 metal level.

A block diagram of an EXPT circuit is shown in Fig. 8.15. The essential elements of the EXPT are as follows:

- a circuit to launch a sharp signal edge (launch_ckt),
- a circuit to generate eight different pulse widths from eight segments in a reference chain (p1 to p8) of $n = 20$ logic gates each,
- a decoder and an output "OR" circuit to select one of eight pulse widths,
- a primary delay chain with N (= 800) logic gates,
- a circuit to capture the output of the delay chain (capture_ckt), and
- DECAPs to deliver charge during switching.

The launch_ckt and capture_ckt blocks shown in Fig. 8.16a, b are each implemented with a single level-sensitive latch and associated circuitry. The latches are initialized to hold a "0" with all three DC inputs AL, SD, and SR set at "0" (clock input = "1" and data input = "0"). With SR = SD = "1," when AL is raised to a "1," sharp rising

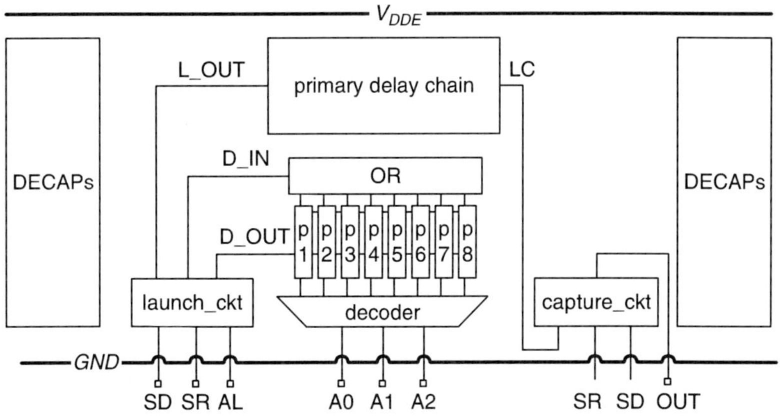

Fig. 8.15 Circuit schematic of an EXPT for measuring H_t

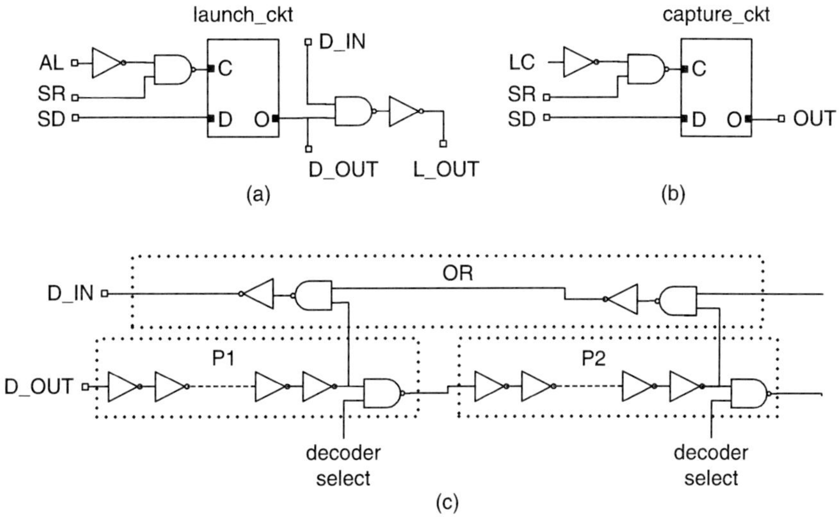

Fig. 8.16 Circuit schematics **a** of pulse launch circuit, **b** pulse capture circuit, and **c** a section of the reference chain and "OR" circuit

signal edges generated at D_OUT and L_OUT nodes are launched at the reference and primary delay chains, respectively. The output of the reference delay chain is fed back into D_IN to create the second edge of the pulse launched on the primary delay chain. The pulse travels down the primary delay chain and either emerges at the far end or is annihilated before it gets there.

The reference chain has eight segments, each segment comprising a chain of 17 logic gates of the same design as the logic gates in the primary delay chain and 3 additional logic gates with equivalent delays for a total of $n = 20$ gates. A decoder selects the number of segments in the reference chain. The pulse width is then $n\kappa\tau_1$, where κ is the number of segments and τ_1 is the average 1SW delay of the logic gates in the chain. The number of logic gates in the primary delay chain is $N = 800$. In this design, n/N is selected to be 0.025, and H_t is measured in 2.5% steps as given by Eq. (8.5).

The output of the primary delay chain is captured in the latch of the capture_ckt block. If a pulse from the primary delay chain appears at the clock input LC of the latch, with SR = SD = "1," the clock signal is turned on and the OUT node is subsequently raised to a "1." If the launched pulse is annihilated in the primary delay chain, the OUT node remains at a "0." The output voltage level is recorded and both the launch and capture latches are reset with AL = SR = SD = "0."

The DC input signals AL, SD, and SR for each of the measurement steps are listed in Table 8.2. In the first three steps, the latch is initialized and the output high- and low-voltage levels are observed to check the validity of the design and test setup and to verify hardware functionality. Steps 4 through 7 are repeated for each decoder output to select from one to all eight segments of the reference chain, thereby generating pulses of sequentially larger widths. In these four steps, first with the CLK (C) at "1" and DAT (D) at "0," the latch is initialized to a "0" state. Next, the CLK is lowered to "0," DAT is set to "1," and then the CLK is raised to a "1" again to generate a signal edge at the latch output that initiates the pulse formation process as shown in Fig. 8.10.

An output voltage level of "1" at the OUT terminal in Fig. 8.15 at step 7 indicates the detection of a pulse at the end of the primary delay chain. Conversely, the pulse is annihilated in the chain if the voltage level at the OUT terminal is a "0." Starting with the narrowest pulse width (one segment in the reference chain), the bit pattern for the OUT voltage is recorded. For $\tau_1 > \tau_2$, and $12.5\% > H_t > 10\%$, the pulse is annihilated when $n\kappa/N$ is $\lesssim 0.10$, and the OUT voltage for $\kappa = 1, 2, 3$, and 4 is a "0." For $\kappa = 5, 6, 7$, and 8, the OUT voltage is a "1." The output bit pattern, read from right to left, for the eight steps is then "11110000."

The floorplan of an EXPT is shown in Fig. 8.17. At the 45 nm technology node and beyond, it is possible to place all the circuit blocks for an EXPT between two I/O pads (V_{DDE} and GND). The GND bus travels along the width of the macro below

Table 8.2 Test voltage levels for the decoder and latch inputs and output for initialization and measurement of one H_t bit

	Decoder inputs			Control inputs				
STEP	A2	A1	A0	AL	SD	SR	OUT	Comments
1	0	0	1	0	0	0	0	Initialize
2	0	0	1	0	1	0	1	Test "1"
3	0	0	1	0	0	0	0	Test "0"
4	0	0	1	0	0	0	0	CLK = "1"
5	0	0	1	0	0	1	0	CLK = "0"
6	0	0	1	0	1	1	0	DAT = "1"
7	0	0	1	1	1	1	0 or 1	Record H_t bit

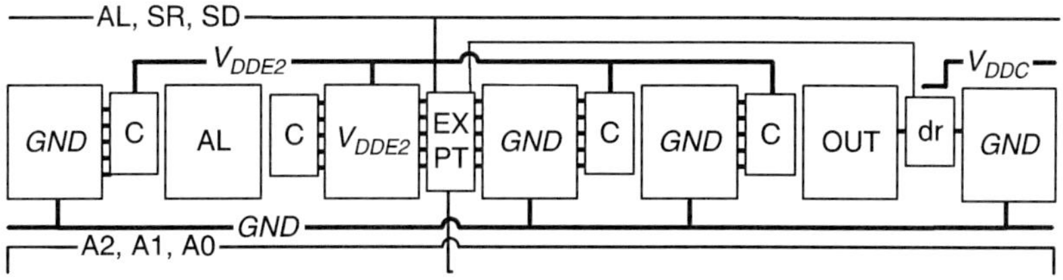

(a)

1	2	3	4	5	6	7	8	9	10	11	12	13	14	15	16	17	18	19	20	21	22	23	24	25
A0	A1	GND	VE1	GND	SD	GND	AL1	VE2	GND	GND	OUT	GND	VDDC	GND	GND	VE3	AL2	GND	SR	GND	VE4	GND	A2	Sub

(b)

Fig. 8.17 **a** Floorplan of an EXPT with four banks of DECAPs (C). **b** I/O pad assignment of a 1×25 padset macro with four history EXPTs, each with an independent V_{DDE} power supply (VE1–VE4)

the I/O pads. Short sections of V_{DDE} bus, one for each EXPT, are located above the I/O pads. The decoder inputs A0, A1, and A2 and latch inputs SD, SR, and AL carry DC signals and travel across the width of the macro. These signals are shared among two or more EXPTs.

There are four banks of DECAPs for each EXPT to supply charge during the transient switching activity. This is consistent with the detailed discussion of DECAP placement included in Section 2.4.6.

The I/O pad assignments for a standard macro template, with a 1×25 linear pad array, is shown in Fig. 8.17b. There are four EXPTs with independent V_{DDE} power supplies and a common V_{DDC} power supply for the output driver and buffers. All the circuit blocks within each EXPT shown in Fig. 8.15 are powered by a dedicated V_{DDE} power supply. There are two AL signal pads (AL1 and AL2), one on each side of the macro, and pad 25 is used for substrate contact. With a common OUT I/O pad, only one EXPT is measured at a time, while the power supplies of the other three EXPTs are turned off.

The physical layout schematic of an EXPT placed between two I/O pads is shown in Fig. 8.18. The power grid consists of interdigitated power and ground busses, drawn in wide M1 metal layer and emanating from the two I/O pads. The wiring from the reference delay chain to the OR circuit and other horizontal signal wires is drawn in the M1 metal level. Wires carrying DC signals including decoder input and output wires, traveling in the vertical direction, are drawn in the PS level. Vertical signal wires are drawn in M1 metal, with short PS or DF segments for crossing horizontal power busses and other horizontal wires.

In this compact EXPT design, occupying a 40 μm × 60 μm area between the I/O pads in the 45 nm technology node, the primary and reference chains comprise inverters with FO = 1. In total, there are nearly 1200 logic gates, wired with PS,

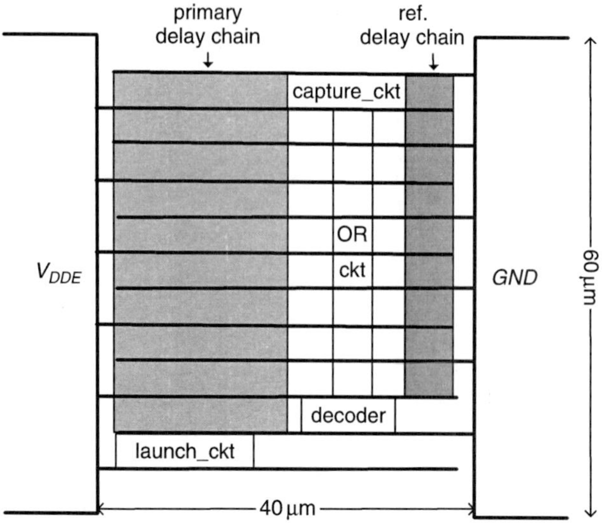

Fig. 8.18 Physical layout schematic of a history EXPT situated between its V_{DDE} and GND I/O pads and implemented at the M1 metal level

DF, and M1 layers. There is adequate space to accommodate a zero-offset circuit described in Section 8.2.2 for enabling negative H_t ($\tau_1 < \tau_2$) measurements. With further technology scaling, more complex logic gates (NAND and NOR) can also be accommodated.

8.3.3 Example 3: Macros for PU and PD History Effect

Macro designs for high-speed differential time measurements described in Section 7.4 are easily adopted for measuring history effect H, in PD-SOI. The differential time measurement scheme allows independent measurements of PU and PD delays and FB effect attributed to p-FETs and n-FETs, respectively. A straightforward determination of H is made from the signal waveforms shown in Fig. 8.6 with measurements of the delay differences of DUT1 and DUT2 in Fig. 8.11. Measurements are made for a fixed pulse width and varying time periods to get 1SW, 2SW, and SS delay differences from which H_{tpu}, H_{tpd}, and other H values are calculated. Alternatively PU and PD delay differences can be measured at a fixed period, long compared with the body relaxation time (~10 ms), over a wide range of pulse widths (500 ps to a few ms), as well as for SS delays at 1 GHz. From these measurements, the H values are determined using Eq. (8.3), and the time dependence of the FB relaxation can be extracted.

Waveforms for any arbitrary switching history are generated with a programmable digital pattern generator. In logic gates with two or more inputs, H may

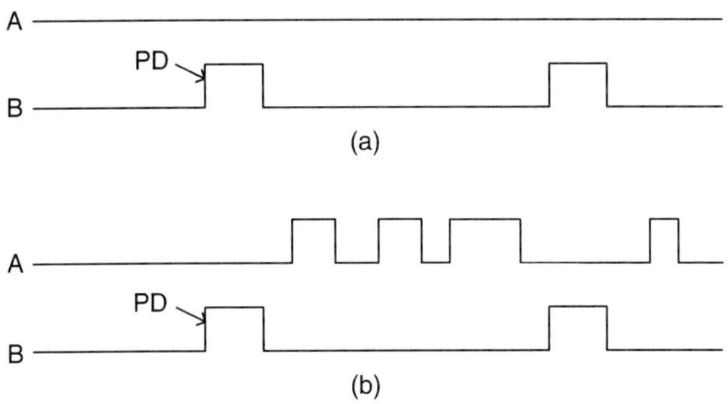

Fig. 8.19 Input signal waveforms for a NOR2 gate **a** with input B switching and **b** with both input A and B switching. Delay measurements are carried out for a PD transition initiated by input B in both cases

be measured with any one of the inputs switching at a time or all inputs switching simultaneously. Other combinations exemplified in Fig. 8.19 for a NOR2 DUT are also possible. In Fig. 8.19a, input A is fixed at "0," while input B is switching periodically. In Fig. 8.19b, input A may have multiple transitions in any arbitrary sequence in between the switching transitions initiated by input B. Such signals may be provided by a digital pattern generator or by a pulse generator operating in burst mode.

In the case of an inverter driving a passgate, because of the large sensitivity of delay through a passgate to its V_t, the history effect may be stronger than for an inverter. The schematic of a circuit for measuring n-passgate PU and PD delays and H values is shown in Fig. 8.20. With independent controls of the arrival times of A and B input signals, the gate of the n-passgate may turn on before or after the inverter switching transition. Hence, the S, G, and D voltages of the n-passgate and in turn its FB potential are dependent on the timing sequence of A and B signals. There are 8 possible combinations of S, G, and D pre-switch potentials each for

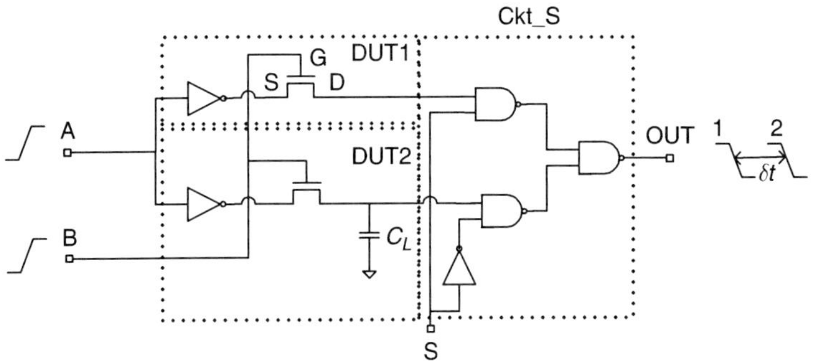

Fig. 8.20 Circuit schematic for measuring history effect H for an inverter driving an n-passgate with adjustable relative timing of high-speed inputs A and B

PU and PD transitions, out of which a total of 14 configurations are stable [7]. The propagation delays and history effect can be different in each of these 14 cases.

Care must be taken to ensure that the sampling oscilloscope is triggered synchronously with the input signal initiating the transition. As an example, with both A and B at "0" for several ms, if signal B transitions from "0" to "1" before signal A, the G terminal of the n-passgate will be at "1" prior to 1SW transition. The sampling oscilloscope should be triggered by the signal edge A. If, on the other hand, again with both A and B at "0" for several ms, signal A transitions from "0" to "1" before signal B, the G terminal of the n-passgate will be at "0" prior to the 1SW transition and the sampling oscilloscope should be triggered by input signal B.

Measurements of *H* values for PU and PD transitions of the inverters and n-passgates in an SRAM cell are carried out by reconfiguring the SRAM cell to wire each component as described in Section 6.4.3. Unloaded and loaded pairs of inverters or inverters driving n-passgates are placed in the EXPT template as DUT1 and DUT2 for measuring *H* for the PU and PD transitions. DUT1 and DUT2 may also be configured as delay chains for measuring the average *H*.

An SRAM cell configured to measure "write" delay and *H* is shown in Fig. 8.21a. In Fig. 8.21b, this SRAM cell block is inserted in the differential high-speed template. The number of SRAM cell blocks in DUT1 and DUT2 in Fig. 8.21b is three and five, respectively. The number of SRAM cell blocks may be increased to eliminate random variations in the cells. Alternatively, a number of DUTs with a small number of SRAM cell blocks may be measured to obtain random variations in delay and *H*.

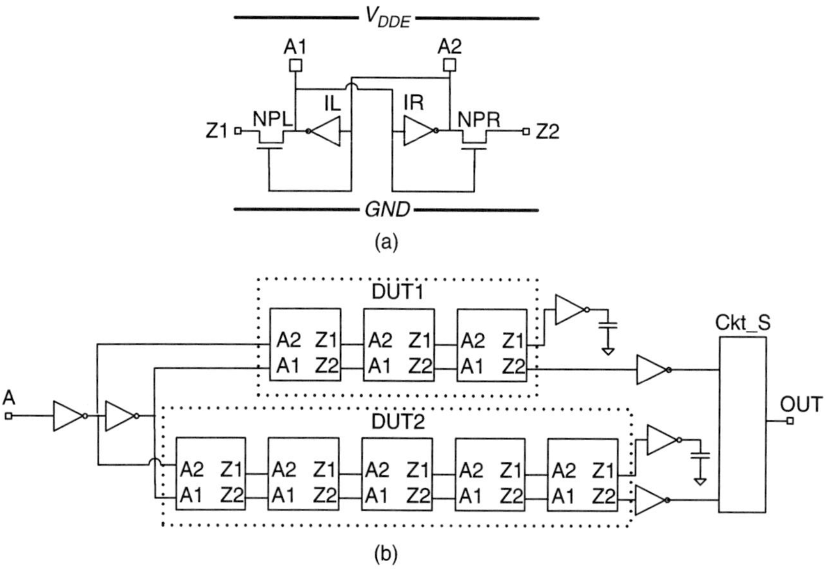

Fig. 8.21 Circuit schematics **a** of an SRAM cell configured to measure "write" delay and **b** differential high-speed measurement of SRAM delays and *H*

8.3.4 Example 4: Macro for H_t Statistics

Measurement of variability in the history effect of nominally identical logic gates and collecting the related statistics is more challenging. The difference between 1SW and 2SW delays of logic gates, configured as shown in Fig. 8.11, is of the order of a few ps. Hence, detection of any variation in H_t requires a measurement accuracy of <1 ps. Such accuracy can be obtained in an in-line or bench test environment, with no sophisticated test apparatus, through the use of a level-sensitive latch. As described in Section 7.4.3, the sensitivity of the output voltage of a level-sensitive CMOS latch to its clock-to-data path delay, specifically the width of its metastable region, can be in the sub-ps range. In this example a test structure design utilizing this property of a latch to collect H_t statistics is described [8].

The simplified schematic of a circuit to illustrate the technique of measuring changes in arrival times of data input to a latch is shown in Fig. 8.22a. Input A drives a variable logic delay chain (LDC) circuit block and a DUT comprising a capacitively loaded logic gate, such that the delays of the two paths are comparable. The LDC and DUT outputs provide the clock (CLK) and data (DAT) inputs, respectively, to a level-sensitive latch. The clock-to-data delay ΔT_d is adjusted by varying the delay of the LDC circuit with an analog signal V_{AJ}. At some value of ΔT_d, the latch output voltage V_{out} transitions from a "1" to a "0." This reference value of ΔT_d can be set with sub-ps accuracy at the center of the transition shown in Fig. 8.22b. When a change in the delay of the DUT occurs, for example, from a 1SW transition to a 2SW transition, as shown in Fig. 8.22c, the delay of the LDC circuit is re-adjusted using V_{AJ}, by a time δt_a, to again position the latch in the metastable region. The actual value of δt_a, determined from the calibration curve of

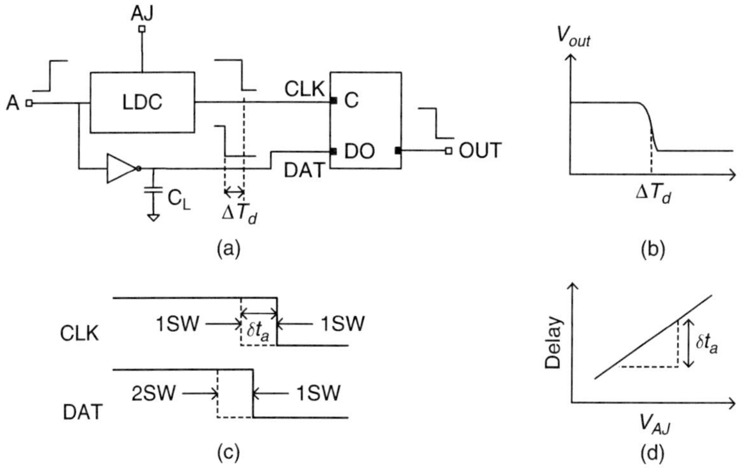

Fig. 8.22 **a** A circuit schematic to measure time delay ΔT_d with sub-ps precision. **b** V_{out} as a function of ΔT_d. **c** CLK and DAT waveforms for measuring 1SW – 2SW delay (=δt_a). **d** Delay calibration plot for the LDC circuit

LDC delay as a function of V_{AJ}, gives the difference between 1SW and 2SW delays of the DUT. It is essential that the LDC be comprised of body-contacted gates or that it be preceded by a control circuit to ensure that it always experiences a 1SW delay when measurements are being made.

The LDC circuit may be a chain of inverters on an independent power supply, with the value of V_{DD} used to vary its delay. Alternatively, a chain of current-starved inverters with one or two analog signal inputs for varying the delay can be used. It is convenient to bias the LDC circuit such that its delay varies linearly with V_{DD} or other analog control inputs. A calibration curve shown in Fig. 8.22d can be obtained from a companion voltage-controlled ring oscillator (VCO) comprising the same logic gates as in the LDC circuit block and measuring its delay/stage as a function of V_{AJ}. A small correction may be necessary to account for the difference of the SS delay/stage of the RO compared with the delay/stage of the LDC in a 1SW configuration.

Using this approach, test structures can be designed to measure PU and PD history effects in single gates or average history effect in a chain of gates. The schematic of a circuit for measuring PD history effect of one DUT, using the above scheme, is shown in Fig. 8.23. A waveform generator (WFG) circuit is coupled with a history element (HE) circuit block and provides the input signals that exercise 1SW and 2SW transitions in the DUT.

A sharp signal edge is generated at the output node B1 of LatchA in the WFG, with DC inputs C1 and D1, using the scheme described in Section 8.3.2

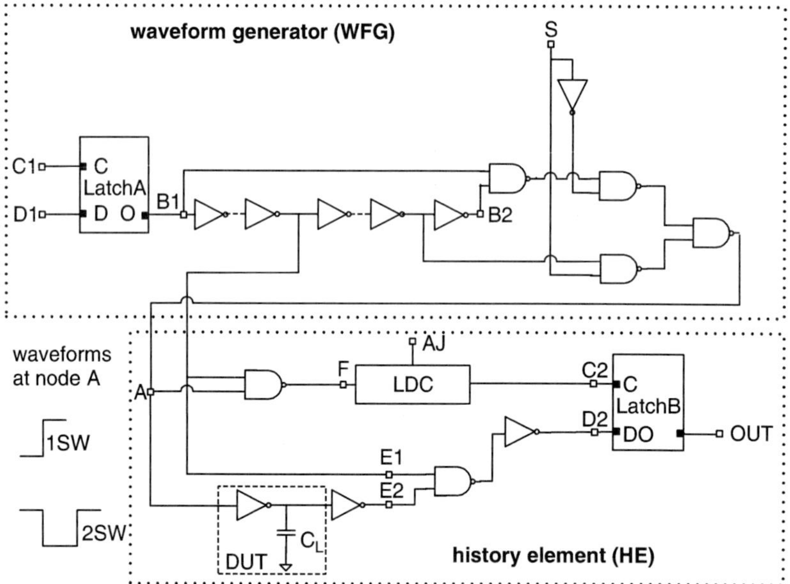

Fig. 8.23 Circuit schematic of a waveform generator (WFG) coupled with a history element (HE). Waveform shapes at node A for 1SW and 2SW transitions are included

(Example 2). With input S = "0," the signal from B1 arrives at A in the HE circuit block, traveling along the upper path in the WFG circuit. The LDC is tuned to match the 1SW delay of the DUT and establish a reference time for determining δt_a by monitoring the output of LatchB. When S is set to a "1," a pulse of width equal to the delay from B1 to B2 arrives at input A to induce a 1SW PU followed by a 2SW PD transition in the DUT. The LDC is now tuned to the 2SW transition of the DUT and the two settings provide the difference in the 1SW and 2SW delays through the DUT. The signal waveform shapes at node A for the two cases are included in Fig. 8.23.

With appropriate timing delays, the signal at node E1 from the WFG allows the rising edges to propagate through both logic paths to LatchB. However, it prevents the initial falling edge from passing the subsequent NAND2 gates in the S = "1" case. Thus in the S = "1" case, just prior to the 2SW transition, the DUT experiences a 1SW transition while the LDC does not. For both the S = "0" and S = "1" cases, the LDC delay is a 1SW event.

The signal waveforms at node B1 and at nodes A and F for 1SW and 2SW PD transitions of the DUT are shown in Fig. 8.24. The pulse width for measuring the 2SW transition of the DUT, equal to the delay from B1 to B2, is designed to be in the ns range. The inputs C1 and D1 to LatchA are cycled such that the repetition rate of the experiment is a few ms and the reset of LatchA, corresponding to the return of node B1 to "0," occurs a ms or more after its initial "0" to "1" transition.

For an estimate of the measurement integrity, let us assume the DUT delay is designed to be 100 ps, and that it dominates the delay from node A to D2. The delay from node A to C2 is dominated by the 1SW delay of the LDC block. The offset between the CLK and DAT signals necessary to be at the latch transition region is ~15 ps in the 65 nm technology node [9]. For a H_t value of 10%, the LDC delay has to be adjusted by approximately 10 ps. The precision with which the delay can be determined by positioning the latch in its metastable region has been demonstrated to be <1 ps [9]. This provides a robust signal against which to evaluate the variability.

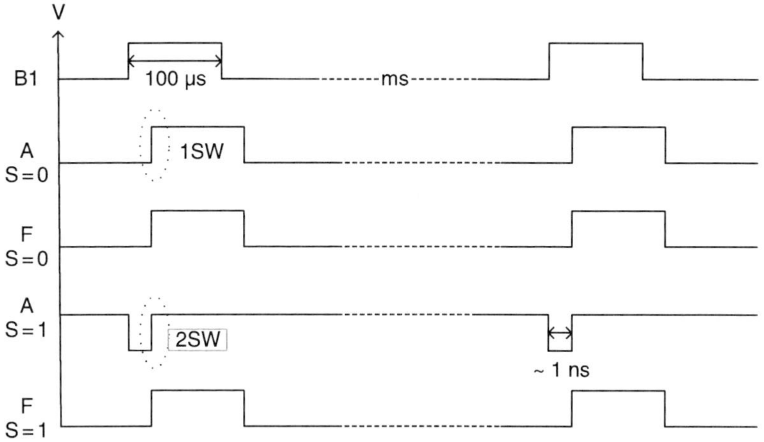

Fig. 8.24 Signal waveforms at nodes B1, A, and F in the circuit shown in Fig. 8.23

The measurements of H_t in the test structure design shown in Fig. 8.23 can be conducted with DC I/Os. It is more expedient to group a number of HE circuits together with a common WFG circuit to enable acquisition of H_t statistics, with test time efficiency achieved by parallelism. The test structure design can then be configured to leverage the capabilities of digital ATE for rapid measurements. A history block HB with n HEs is shown in Fig. 8.25a. The *f*_out signal from the companion VCO is used for direct calibration of LDC delay differences. All HEs are addressed simultaneously. Their output signals are directed into a $k \times n$ shift register bank as shown schematically in Fig. 8.25b, where practical values for k and n are 64 and 16, respectively. Note that with a modest I/O count, several such circuits can be accommodated in a single macro, further exploiting parallelism.

The circuit is initialized with S = "0," and all of the HEs are exercised in parallel through a set of $k/2$ 1SW sequences, each with an incrementally different LDC setting. The resultant "0" or "1" from each of these experiments is fed into the shift register bank, filling half of that bank after the first $k/2$ sequences. Input S is then switched from "0" to "1" and $k/2$ more sequences are run, rendering a full $k \times n$ shift

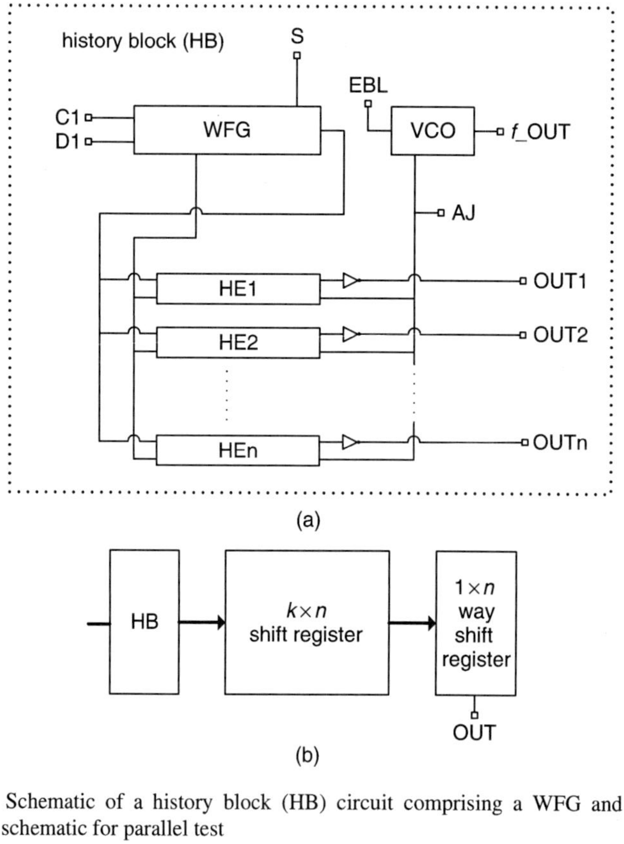

Fig. 8.25 **a** Schematic of a history block (HB) circuit comprising a WFG and n HE units. **b** High-level schematic for parallel test

register bank, containing an entire set of 1SW/2SW data for the n HEs from which δt_a and resultant H_t values can be obtained. The contents of the shift register bank are then read out sequentially, one bit at a time, through a single output node via the two-way $1 \times n$ shift register. First a column of data is moved left to right from the register bank into the two-way register. Then the n bits are shifted vertically to the OUT port of the two-way shift register. After n such operations, another column is shifted in from the register bank. This process continues until the register bank is empty and at the same time completely reset to zero values. With digital ATE the time for sequential readout of the data is negligible compared with the measurement time.

8.3.5 Example 5: Macro for Measuring Thermal Effects

The high-speed differential time measurement macro template described in Section 7.2 can be modified to measure thermal time constants of MOSFET structures [10]. The technique involves parallel MOSFETs (multiple PS fingers), delineated in the same silicon diffusion area, and isolated from the substrate by STI and BOX layers. One of these MOSFETs forms the n-FET or p-FET of an inverter. The silicon island is heated by turning on any one of the other MOSFETs with a sharp signal edge. The inverter delay is measured as a function of the elapsed time after the application or removal of the heater signal. Since the inverter delay is a function of temperature, the thermal time constants for heating and cooling are determined by measuring the inverter delay as a function of elapsed time.

An example physical layout of an n-FET configuration for measuring the thermal time constant of elements situated on the same DF island is shown in Fig. 8.26. Input A is connected to the gates of the p-FET and n-FET of the inverter. Input signals B1, B2, and B3 are applied to the gates of the other three n-FETs, serving as

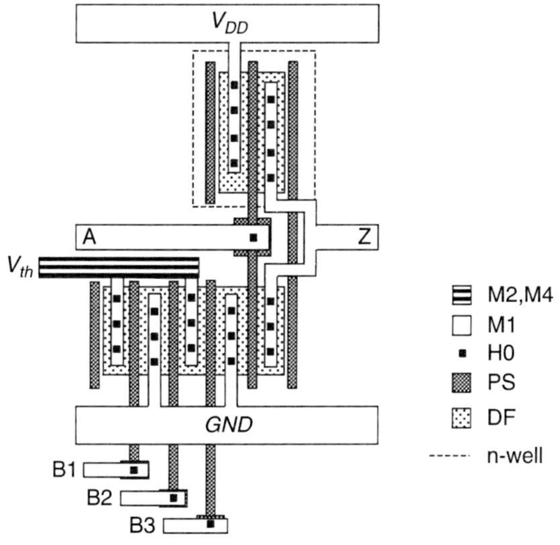

Fig. 8.26 Physical layout of an inverter with three n-FET heater elements on the same DF island as the n-FET of the inverter

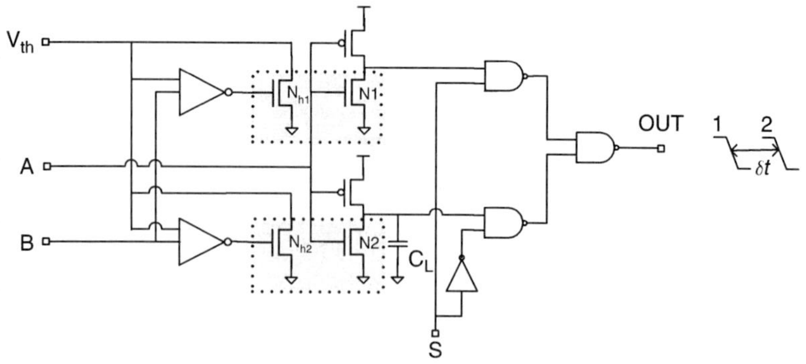

Fig. 8.27 Schematic of a circuit for measuring inverter delay as a function of time before and after applying a heating pulse

independently controlled heater elements within the DF island. This configuration is suitable for measuring thermal time constants as well as for measuring the effect of spatial separation between the heater element and the n-FET of the inverter.

The circuit schematic of an experiment for measuring the thermal time constant with a single heater element is shown in Fig. 8.27. The difference in delays of an unloaded inverter and an inverter with a capacitor load C_L is measured as described in Example 3. The gates of both the inverters are driven by a high-speed signal A. Each of the n-FETs in the pair of inverters has one n-FET heater element on the same DF island, indicated by the shaded area in the figure. The drain of the n-FET heater is connected to a separate power supply V_{th} and its gate is driven by high-speed input signal B through an inverter, also powered by V_{th}.

The timing diagrams of inputs A and B are shown in Fig. 8.28. Input signal B is raised to a "1" for heating just before the inverter PD transition occurs. A complementary B signal is used to turn the heater off prior to the PD transition. The heating or cooling time prior to the measurement is varied externally by adjusting the time delays between signal edges B and A.

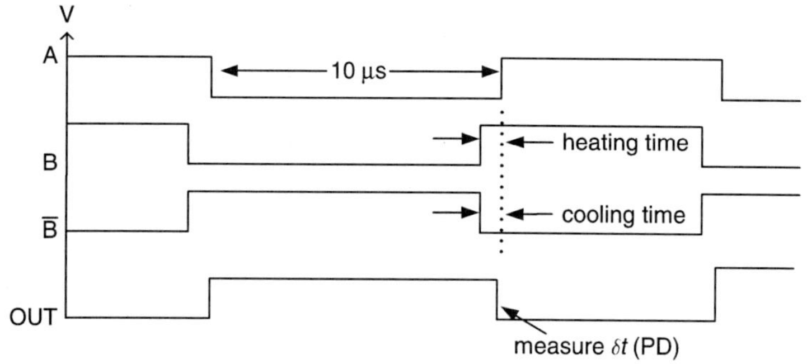

Fig. 8.28 Input signal waveforms for the heater pulse (B) and inverter input and output (A, OUT). Reproduced from [10], with permission, © 2007 IEEE

With the availability of more than two high-speed signals, other heater elements as shown in Fig. 8.26 can be independently controlled. The size of the DF island can be varied to study the spatial variation of temperature across a large island. Similar experiments may be carried out for the p-FET.

The V_{th} power supply draws a DC current of $\sim$1 mA, when the input signal B is set at "1." Robust metal wiring with four or more metal layers is provided to minimize the voltage drop in this power supply bus. It is a good practice to add a control experiment with the heater and the inverter on isolated DF islands to validate the design. In this case, voltage level at input B should have negligible impact on the inverter delay.

8.4 Model-to-Hardware Correlation

Modeling of PD-SOI devices and circuits is more complex than for bulk silicon because of the presence of floating-body effects. The body potential prior to 1SW is determined by a DC solution balancing the effects of impact ionization, gate oxide tunneling, and drain-to-body and source-to-body diode leakage currents with time constant of the order of 10 μs to a few ms. The body potential prior to a 2SW transition, within a few ns of the 1SW transition, is determined by capacitive coupling between the gate, source, and drain. Circuit behavior for these two types of transitions can be simulated with compact PD-SOI models in a straightforward manner [11].

If, on the other hand, the second transition during measurement in a test circuit occurs within a few ns to a few ms of the 1SW transition, the body potential may be somewhere between the pre-1SW and pre-2SW states. If a circuit is switching periodically in steady state (SS), as in the case of a ring oscillator, the time to reach equilibrium may be in the ms range. The circuit simulation time for modeling the behavior of such events can become very long. Special techniques are applied to reduce the simulation time to achieve SS in circuits [12].

Model-to-hardware correlation of 1SW, 2SW, and SS delays and corresponding history effect values at an appropriate set of voltages and temperatures covers the range of delay variation for most circuits. The macro design in Example 3 can be used for measuring PU and PD delays for any arbitrary switching sequence. If the delays for a switching sequence fall outside the range of 1SW, 2SW, and SS cases, circuit simulations reproducing the switching sequence may be carried out to ensure that the FB effect in the model correctly captures the circuit behavior.

References

1. Bernstein K, Rohrer NJ (2007) SOI circuit design concepts. Springer, Berlin
2. Marshall A, Natarajan S (2002) SOI design: analog, memory and digital techniques. Kluwer Academic, Boston, MA
3. Shahidi GG (2002) SOI technology for the GHz era. IBM J Res Dev 46:121–131

4. Taur Y, Ning TH (2009) Fundamentals of modern VLSI devices, 2nd edn. Cambridge University Press, New York, NY
5. Ketchen M, Bhushan M, Pearson DJ (2005) High speed test structures for in-line process monitoring and model calibration. Proceedings of the 2005 IEEE international conference on microelectronic test structures, 2005, pp 33–38
6. Bhushan M, Ketchen MB (2010) Generation, elimination and utilization of harmonics in ring oscillators. Proceedings of the 2010 IEEE international conference on microelectronic test structures, 2010, pp 108–113
7. Ketchen M, Bhushan M, Bermon S (2005) Switching delay variability in NMOS and PMOS PDSOI passgate circuits. IEEE VLSI-TSA-Tech international symposium on VLSI technology proceedings, 2005, pp 68–69
8. Bhushan M, Ketchen MB (2009) Acquisition of silicon-on-insulator history effects statistics. US patent pending
9. Bhushan M, Ketchen MB, Das KK (2008), CMOS latch metastability characterization at the 65-nm technology node. Proceedings of the 2008 IEEE international conference on microelectronic test structures, 2008, pp 147–151
10. Ketchen MB, Xiu K, Bhushan M (2007) Measurement of thermal time constant in 65-nm PD-SOI technology with sub-ns resolution. Proceedings of the 2007 IEEE international SOI conference, 2007, pp 53–54
11. Goo J-S, William RQ, Workman GO, Chen Q, Lee S, Nowak EJ (2008) Compact modeling and simulation of PD-SOI MOSFETs: current status and challenges. IEEE 2008 custom integrated circuits conference, CCIC 2008, pp 265–272
12. Joshi RV, Kroell K, Chuang CT (2004) A novel technique for steady state analysis for VLSI circuits in partially depleted SOI. Proceedings of the 17th international conference on VLSI design, 2004, pp 832–836

Chapter 9
Test Equipment and Measurements

Contents

The test structures described in Chapters 3, 4, 5, 6, 7, and 8 are designed for DC, low-frequency ($\lesssim$10 MHz), or high-frequency ($\sim$GHz) measurements. DC characterization of MOSFETs and resistance measurements can be carried out with constant voltage or constant current power supplies, voltmeters, and ammeters. Capacitance measurements, in the absence of any on-chip active circuitry, require dedicated impedance meters operating over a range of frequencies. Frequency counters, spectrum analyzers, oscilloscopes, and pulse generators are used for AC measurements in the frequency and time domains. Silicon fabrication facilities are equipped with parametric and digital automated test equipment (ATE) coupled to computer-controlled wafer handling and data storage systems. Customized test equipment and fixtures may be designed for special applications.

There is a large variation in the complexity of test equipment, from that used in an elementary laboratory to the ATE used in silicon manufacturing and product test facilities. However, the basic principles of measurement remain the same. From

M. Bhushan, M.B. Ketchen, *Microelectronic Test Structures for CMOS Technology*,
DOI 10.1007/978-1-4419-9377-9_9, © Springer Science+Business Media, LLC 2011

test structure design and data analysis perspectives, it is important to examine and understand the I/O requirements, measurement range and accuracy, and sources of errors.

In this chapter, a brief description of standard parametric and high-speed test equipment is given with emphasis on measurement speed and accuracy requirements. In Section 9.1, electrical test requirements and measurement terms are introduced. Basic test equipment specifications and configurations for DC and AC parameter measurements are described in Section 9.2. ATE for in-line parametric, digital, and memory tests are covered in Section 9.3. Test equipment for laboratory bench tests is described in Section 9.4. Methodologies for test automation and test-code generation, including output parameter formats to facilitate data analysis, are discussed in Section 9.5.

Principles of test equipment and procedures used in semiconductor technology are covered in several books [1, 2]. For the equipment installed in a test facility, specifications and application notes provided by the equipment manufacturers are the most relevant source of information for setting up electrical tests.

9.1 Electrical Tests and Measurement Terms

The invention of the voltaic cell or battery by Alessandro Volta in the year 1800 led to a widespread use of electricity in the later part of the 19th century. The concept of a moving coil galvanometer to measure current was introduced by Hans Oersted in 1820 and an instrument of this type was used commercially by the telegraph companies in the late 19th century. Measurement of voltage and current on a regular basis began with the Weston's portable instrument patented in 1888. Electrical test instrumentation has since followed the trend in electronics, from vacuum tubes to transistors and integrated circuits, filling the growing need of electrical measurement capability for a wide range of applications.

DC characterization of CMOS circuits can, in principle, still be done with a battery and an ammeter constructed from a moving coil galvanometer with a low resistance shunt and a few discrete resistors to extend the measurement range. The added complexity in today's commercial test equipment provides higher precision over a wide range, fast data acquisition, digitized data, and computer interface for instrument operation and data transfer. The test equipment is an integral part of or placed in close proximity to a wafer probe station that is computer controlled to land probes on silicon with a spatial precision of a few μm.

Macros for DC and low-frequency CMOS characterization are typically tested with a DC probe card and commercially available test equipment. DC current measurements cover a range from less than a 1 pA to several hundred mA, while voltage measurements range from 1 μV to a few tens of volts. Precision resistance measurement range is typically from a few mΩ to a few kΩ for circuit elements and spans a wider range for electrical shorts and opens in yield test structures. External capacitance meter range, limited at the low end by the measurement capability of the instrument to a fraction of a pF, extends to a few μF in the upper end. Capacitances in the 1 fF range are measured with charge-based capacitance measurement (CBCM) techniques requiring DC current and voltage sources and meters, and an external or internal clock generator. Ring oscillator frequency measurements, with a divided frequency of $\lesssim$10 MHz, are made with an off-the-shelf external frequency counter or an oscilloscope coupled to parametric test equipment.

Digital testers with the capability of supplying external clocks on multiple channels can be used to generate programmable bit-stream digital patterns of the type "01100011000" and record the corresponding output pattern. The minimum current measurement capability of commercial digital testers is typically in the μA to mA range. Recently this is being extended at the low end to the pA range to leverage the parallel testing capabilities of digital testers for parametric tests.

Test capabilities are further expanded for laboratory bench testing. The test frequency range is extended beyond a few MHz with special probe cards, shielded signal cables, and decoupling capacitors. External high-frequency programmable pulse generators, adjustable delay lines with sub-ps resolution, and sampling oscilloscopes are used for test structures described in Chapters 7 and 8.

The range, accuracy, and resolution for a test parameter such as voltage or current are important considerations in selecting the right equipment for an application and in setting up a test procedure. The full functional range of a test parameter is typically split into smaller sub-ranges, which are set by a range selector switch in the test equipment. The resolution and the accuracy of the test equipment are specified within each sub-range in a data sheet provided by the test equipment manufacturer. These specifications are valid under specified temperature, humidity, and external noise level limits on the surrounding environment and after allowing for sufficient equipment warm-up time.

Resolution is the minimum increment displayed by the equipment. This may be specified in absolute value or as the LSB (lowest significant bit) of an analog-to-digital converter (ADC). For an *N*-bit ADC, the resolution is one part in 2^N. As an example, in a 12-bit ADC, the resolution is 1 part in 4,096 or 0.0244% of the full range value.

Repeatability or inherent measurement precision is the variability in repeated measurements using the same equipment under nominally identical test conditions. In many cases, repeatability error may be dominated by fluctuations external to the test equipment such as changes in probe contact resistance or changes induced by aging of a DUT.

Reproducibility is the variability in measurements on the same DUT made on different test equipment, test setup, and by different operators.

Accuracy of a measurement is specified as a magnitude of *error* between the displayed and actual values. The accuracy in any force and measure value is typically different for each sub-range and also varies within a sub-range. The accuracy is typically specified as

$$\text{Accuracy} = a \times \ (\text{output value}) + b \times \ (\text{output range}) + c,$$

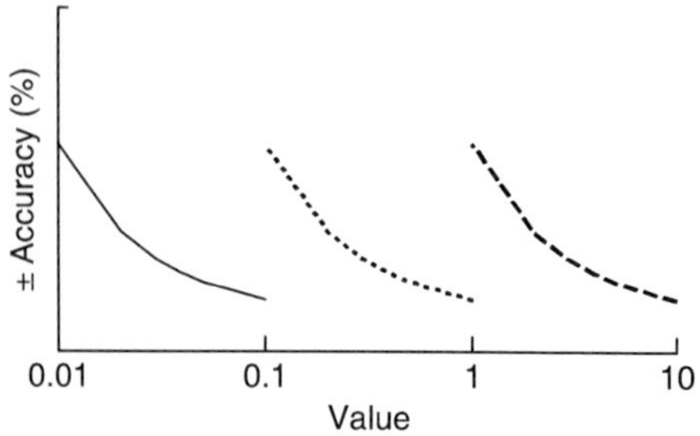

Fig. 9.1 Accuracy (measurement error) of an instrument as a function of force or measured value in three sub-ranges

where a, b, and c are constants listed in the equipment data sheet. This variation in accuracy for three different sub-ranges, each covering a decade, is illustrated in Fig. 9.1. For non-zero values of b and c, the accuracy as a % of the output value is best (most accurate) at the upper end of the range and poor (least accurate) at the lower end of the range. The worst case measurement accuracy within a sub-range is specified as that at the lower end of the sub-range.

Test equipment accuracy is worse than resolution and includes errors in repeatability and reproducibility of the equipment itself. Equipment manufacturers may specify accuracy as anywhere from $\pm 3\sigma$ to $\pm 6\sigma$ of the total error distribution. Overall accuracy is further affected by test setup, cables and contacts, and other external sources of variations.

Calibration of the test equipment is performed periodically to ensure its operation within specifications. The test equipment is factory calibrated with NIST (National Institute of Standards and Technology) standards and self-calibrated on the test floor. The calibration data obtained on the test floor may be stored and applied to the tester to automatically correct for any drift from factory calibration in the tester with aging or changes in environmental conditions.

9.2 Standard Test Equipment

In this section, a description of test equipment for standard in-line measurements is given with emphasis on measurement accuracy, range, and test time.

9.2.1 Source Measure Unit (SMU)

All electrical test structures require a source of power and instrumentation to measure voltage and current levels. Either constant voltage or constant current power supplies may be used, although for the majority of the test structures described in the previous chapters, constant voltage sources are preferred. The power supply and measurement functions are combined in a single compact unit, commonly referred to as a source measure unit (SMU). These programmable SMUs are essential for high throughput in a semiconductor test facility. ATE may have multiple SMUs which are connected to I/O pads on a macro via a programmable switch matrix described in Section 9.2.2. Stand-alone SMUs are also available for laboratory bench tests.

The function of an SMU is illustrated in Fig. 9.2. The SMU can be configured to output a constant voltage (and measure current) or a constant current (and measure voltage) or a GND potential. The SMU also serves to measure or sense a voltage or a current at an I/O pad. For sensing a voltage, the SMU is configured as a constant current source with zero current output. Conversely, to measure current, the SMU is configured as a constant voltage source with zero voltage output. The combined functions of source and measure are either voltage force current measure (VFIM) or current force voltage measure (IFVM). These modes are also sometimes referred to as force voltage measure current and force current measure voltage.

An SMU has multiple voltage and current sub-ranges with each range covering a decade or less. If the limits of expected measurement values are known and fall well within a sub-range, the sub-range may be preset in the SMU. If the expected range is not known or falls within several sub-ranges, it may be automatically set by the SMU. Auto-setting of the measurement range gives more flexibility but may result in an increase in test time as the instrumentation has to search for the correct sub-range. The test time may be relieved somewhat if the minimum sub-range is specified, and the auto-ranging is activated only if the measurement value exceeds the upper limit of that sub-range.

The maximum current and voltage delivered by the SMU are set by the compliance limits. It is important to set the current compliance limit to a value consistent with the current handling capability of the probes to prevent probe damage. This limit may be further lowered for experiments where excessive currents may damage the DUT itself. Voltage compliance limits are set to prevent dielectric breakdown or overheating the DUT.

A resistor measurement can be performed using the VFIM feature in an SMU connected to one terminal with the second terminal of the resistor connected to GND. MOSFET measurements, for example, an n-FET with the source terminal connected to GND, require a minimum of three SMUs, one each for gate, source,

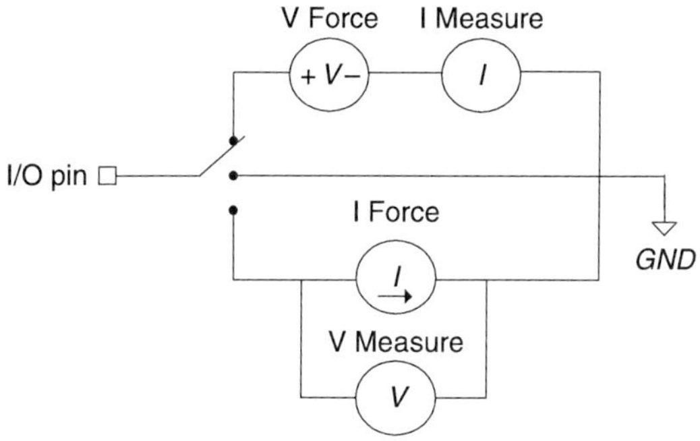

Fig. 9.2 Source measure unit (SMU) configurations to force a constant voltage and measure current (VFIM), or force a constant current and measure voltage (IFVM), or to connect the I/O pin to GND

and drain terminals. Parametric testers with 1–8 SMUs and 12–48 I/O pins are offered in the market place for greater flexibility in measurement configurations. In these testers, a low-noise switch matrix is programmed to connect any SMU to any one of the I/O pads on the test structure. The throughput is increased substantially with a per-pin SMU architecture and a parallel test approach [3].

The accuracy of force voltage and current settings is affected by noise. Noise sources are characterized as normal mode and common mode. Normal mode noise is present in the SMU output and modeled as a voltage source in series with the DUT for VFIM mode and as a current source in parallel with the DUT for IFVM mode. Common mode noise is present between the external GND and the low voltage terminal of the SMU or its internal GND. Any undesired voltage drop between the external and internal GND levels acts as a source of current noise.

The effect of random noise is reduced by taking an average of a number of measurements or readings. The effect of a known noise source such as AC power line cycle (PLC) is reduced by taking readings over a multiple of one full cycle (16.67 ms for 60 Hz line cycle frequency). To reduce the test time, some test equipment may have a built-in feature of initializing the measurements at a fixed phase in the AC power cycle. This ensures that PLC noise cancellation is identical in all measurements.

Another source of error is user-defined wait time between forcing a voltage or a current and recording the measured (or sense) value. There is a rise time associated with a programmed force voltage or force current value because of the *RC* time constant of the SMU and the wires. Any data taken in the settling phase will be erroneous because of incorrect value of force voltage or current being applied. If the *RC* time constant of the system is known, the settling time following a step application of force voltage is determined to the desired accuracy as

$$\text{Settling time} = 2.3RC\left\{\log_{10}(100/\%\ \text{error})\right\}.$$

For a 1% error, the settling time is 4.6*RC*, increasing to 9.2*RC* for a 0.1% error. To reduce this time, an initial higher force voltage is applied to charge the lines and then allowed to settle to the set point V_o as shown in Fig. 9.3.

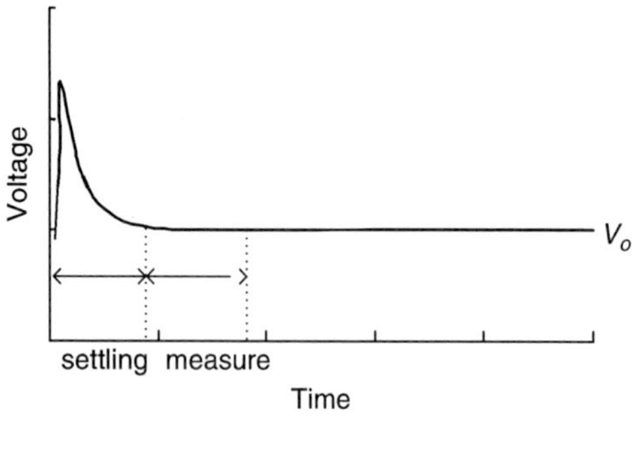

Fig. 9.3 Force voltage as a function of time, showing the time for settling of the voltage signal and optimum time window for measurement

The measurement sequence follows after the SMU output has stabilized to a desired accuracy limit. Too long a wait time increases the total test time and decreases the throughput. The optimum time for the measurement sequence to begin is empirically determined by changing the wait time and observing the measurement accuracy with a calibrated DUT. The total measurement time for N readings is given by

$$\text{Measurement time} = \text{wait time} + N \times \ (\text{acquisition time}) + \text{data transfer time},$$

where the acquisition time is the time to make a single measurement at a fixed source voltage or current. The measurement value is averaged over N readings. Typically, the settling time is of the order of a few ms and the acquisition time is in the μs range. The measurement accuracy is dependent on the speed at which tests are performed and hence there is a trade-off between accuracy and measurement time.

9.2.2 DC Switch Matrices

Generally, the number of pads on a test structure exceeds the number of available SMUs. Test efficiency and flexibility are achieved with a switch matrix interface between the SMUs and the test structure. The switch matrix concept is shown in Fig. 9.4 with solid circles representing electrical connections. Any SMU may be connected to any pad with either mechanical switches or programmable solid-state relays. The cable connections between the switch matrix, the SMUs, and the probe card connected to the test structure remain fixed, and the same test setup can accommodate many different macros on silicon. The switch matrix can be extended in both X- and Y-directions to include more SMUs or I/O pads. A programmable switch matrix with solid-state relays can make and break connections in a few ms. In order to maintain a high switching speed, the number of I/O channels is limited (< 50) to keep the stray capacitance between channels under a few pF. Good electrical isolation between the channels is important, particularly when making current measurements in the fA to pA range.

Although programmable switch matrices are very useful for high throughput, there is a burden associated with writing the test program before any measurements can be taken. In test structure design validation and debug on a laboratory bench, it is sometimes convenient to use manually operated mechanical switches as this eliminates programming time and any programming error interference with design validation.

A rotary switch configuration to construct a switch matrix is shown in Fig. 9.5a. Here, each switch is connected to one I/O pad on the test structure, connecting to any one of four SMUs or to GND. The GND connections are located between the SMU connections so that an I/O pad connection is switched to an SMU by always making a connection to GND first.

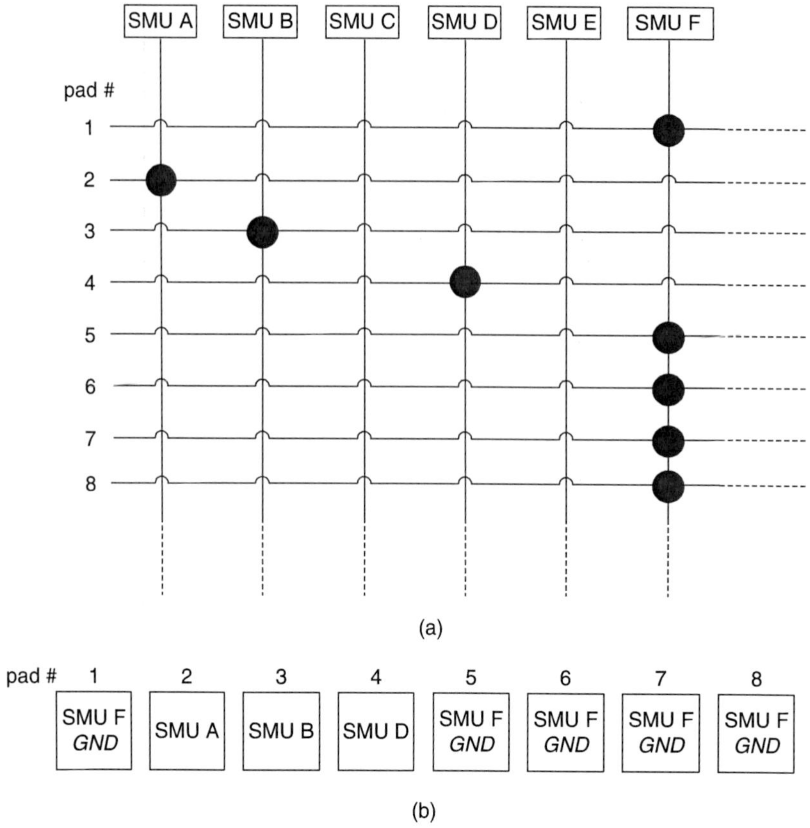

Fig. 9.4 **a** DC switch matrix with six SMUs and eight outputs. **b** Eight I/O pads with SMU assignments. *Solid circles* denote electrical connections

The face panel of a manual switch matrix for a standard 1×25 padset macro (Appendix A) is shown in Fig. 9.5b. It comprises a bank of rotary switches of the type shown in Fig. 9.5a and cable connections to SMUs and to I/O pads. All internal connections are hard wired on a printed circuit board. The user has complete freedom to connect any one of four SMUs or power supplies to any I/O pad location. This type of manual switch matrix design may be extended to include additional SMUs and I/O pad connections. The setup is useful for initial debug of macro designs over the full range of operating conditions with little or no investment in software development.

9.2.3 Impedance Meters

Precision impedance meters employ an AC source and measure vector impedance at a fixed frequency to determine capacitance, inductance, and parasitic resistance

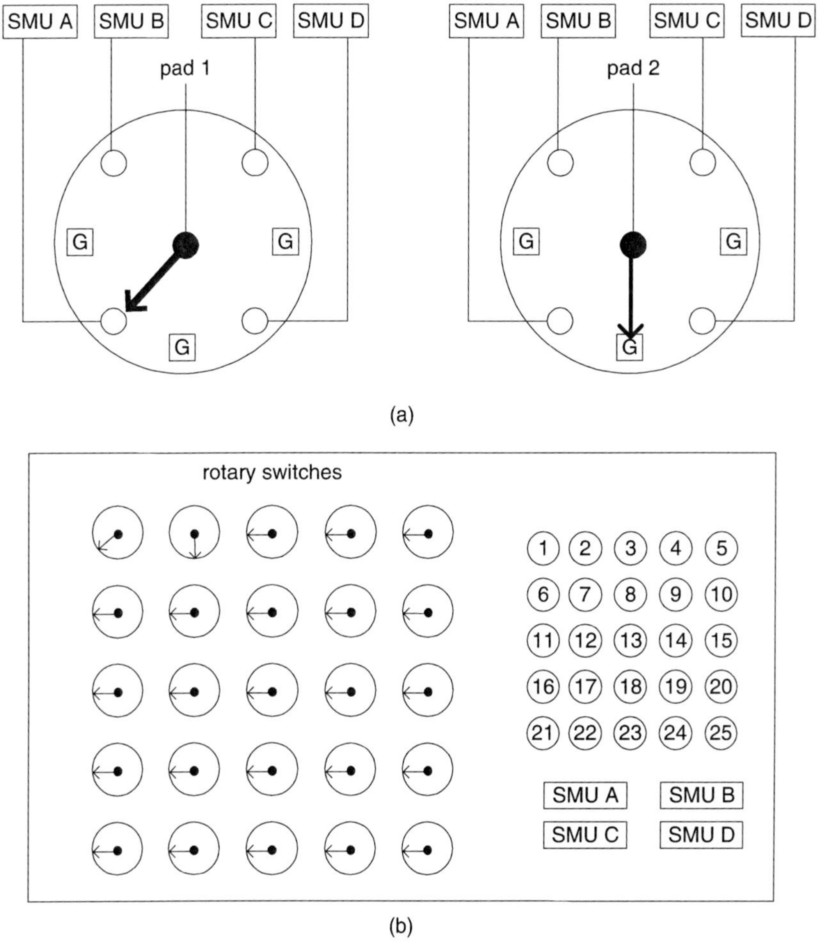

Fig. 9.5 **a** Rotary switch configuration with each I/O pad connected to any one of four SMUs or to GND (G). **b** Front face panel of a manual switch matrix for testing a macro with 25 I/O pads

components [4]. In CMOS manufacturing, impedance meters are used primarily to determine capacitance values. An impedance meter is selected on the basis of the

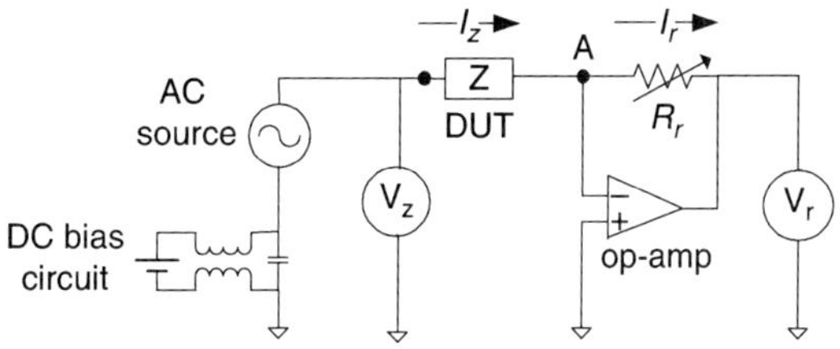

Fig. 9.6 Simplified circuit of auto-balancing bridge utilized in an impedance (LCR) meter

range of capacitance and associated parasitic element values of the DUTs to be measured, desired measurement accuracy, and measurement frequency range. The measurement frequency for in-line tests with DC (low frequency) probes is limited to ~10 MHz.

A simplified circuit for an impedance (LCR) meter is shown in Fig. 9.6. It utilizes an auto-balancing bridge method. The range resistor R_r is adjusted to maintain the potential of node A at GND, balancing the currents I_z and I_r. The magnitude of impedance Z is calculated from the measured amplitudes of AC voltages V_z and V_r:

$$Z = R_r \frac{V_z}{V_r}. \tag{9.1}$$

A DC bias circuit is included to set the voltage bias of the DUT. The time-dependent AC signal and the DC bias voltage applied to the DUT are shown in Fig. 9.7. This is a useful feature for applications where the DUT capacitance is voltage dependent, such as MOSFET capacitances. Care should be taken to ensure that any DC current flow through the DUT at voltage bias points is minimal.

The voltage amplitude of the AC source is application dependent. A high voltage amplitude improves the signal-to-noise (S/N) ratio. However, in applications where the capacitance is sensitive to the voltage across the DUT, as in the case of MOSFET capacitances, the voltage amplitude is kept <50 mV.

Typically, there are parasitic resistances associated with capacitor DUTs. Metal electrodes, interconnects and cables and, in addition, silicon diffusion and contacts for MOSFET capacitors add resistances in series with the capacitor being measured. Any leakage current through the dielectric acts as a resistance in parallel with the capacitor. This parallel resistance may be significant in thin gate oxide capacitors and in the presence of defects in ILD layers in metal capacitors.

At a given angular frequency ω, the measured impedance (magnitude and phase) of a circuit comprising capacitors and resistors can be modeled as either a two-element series or parallel circuit shown in Fig. 9.8a, b. Impedance meters provide series and parallel measurement modes. Appropriate selection of the measurement mode and frequency is therefore important for obtaining accurate measurements [4]. The relationships between the measured effective capacitance value C_{sm} in series

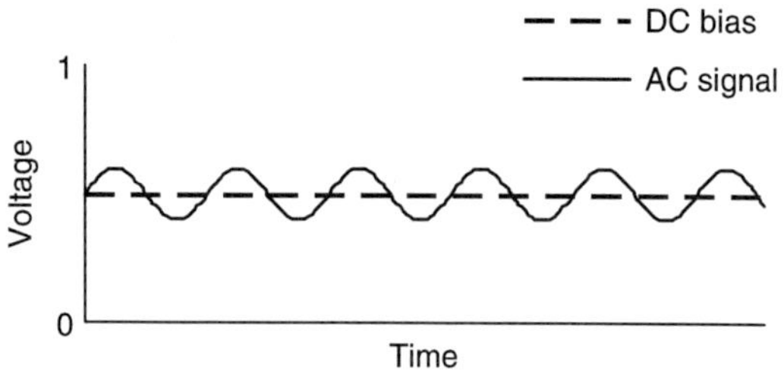

Fig. 9.7 DC bias and AC signal voltage amplitudes applied to the DUT as a function of time

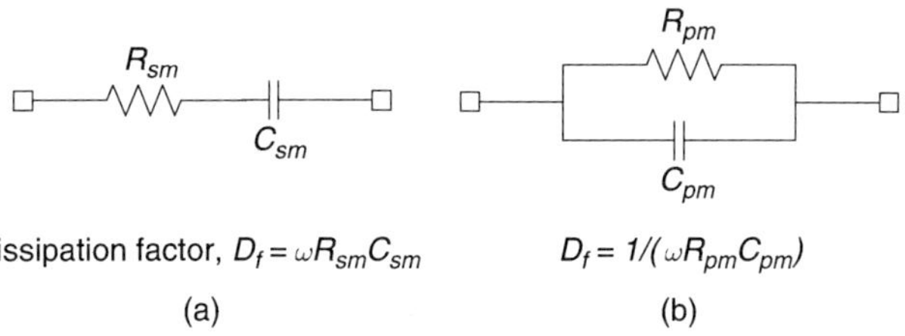

Fig. 9.8 Capacitance measurement modes: **a** series and **b** parallel

with resistance R_{sm} in series mode, and measured effective capacitance C_{pm} in parallel with R_{pm} in parallel mode are expressed in terms of the dissipation factor D_{f}:

$$C_{\mathrm{pm}} = \frac{C_{\mathrm{sm}}}{\left(1 + D_{\mathrm{f}}^2\right)}, \tag{9.2}$$

$$R_{\mathrm{pm}} = R_{\mathrm{sm}}\left(1 + \frac{1}{D_{\mathrm{f}}^2}\right), \tag{9.3}$$

where

$$D_{\mathrm{f}} = \omega C_{\mathrm{sm}} R_{\mathrm{sm}} \quad \text{(series mode)}, \quad D_{\mathrm{f}} = \frac{1}{\omega C_{\mathrm{pm}} R_{\mathrm{pm}}} \quad \text{(parallel mode)}.$$

The dissipation factor is a measure of the purity of the capacitive reactance, with $D_{\mathrm{f}} = 0$ corresponding to a pure capacitance $\left(R_{\mathrm{sm}} = 0,\ R_{\mathrm{pm}} => \propto\right)$. The ratio $R_{\mathrm{sm}}/R_{\mathrm{pm}}$ increases with increase in D_{f} and for $D_{\mathrm{f}} > 0.1$, R_{sm} begins to approach R_{pm}.

The two measurement modes give equivalent capacitance values when $D_{\mathrm{f}} << 1$, corresponding to $1/(\omega C_{\mathrm{sm}}) >> R_{\mathrm{sm}}$ in the low-impedance region and $1/\left(\omega C_{\mathrm{pm}}\right) << R_{\mathrm{pm}}$ in the high-impedance region. These regions are shown graphically by shaded areas in the impedance vs. frequency plot in Fig. 9.9. As a general guideline, at a frequency of 1 MHz, a 10 pF capacitor can be measured in series or parallel modes, whereas for a 10 fF capacitor, parallel mode is preferred.

If the circuit being measured is a pure capacitor, C, in series with a pure resistor R_{s}, then their corresponding measured values in the series mode, C_{sm} and R_{sm}, will be equal to the actual values C and R. If the same capacitor is measured in a parallel mode, the measured value, C_{pm}, will be different than the actual value C.

As a specific example, consider the case of a DUT comprising a pure capacitor of 1 pF in series with a 50 kΩ resistor. The measured values at 1 MHz in series mode are $C_{\mathrm{sm}} = 1$ pF, and $R_{\mathrm{sm}} = 50$ kΩ and $D_{\mathrm{f}} = 0.314$. The measured values in the parallel mode from Eqs. (9.2) and (9.3) are $C_{\mathrm{pm}} = 0.91$ pF, $R_{\mathrm{pm}} = 557$ kΩ, and $D_{\mathrm{f}} = 0.314$. In this case, there is an error of 9% in capacitance value in the parallel

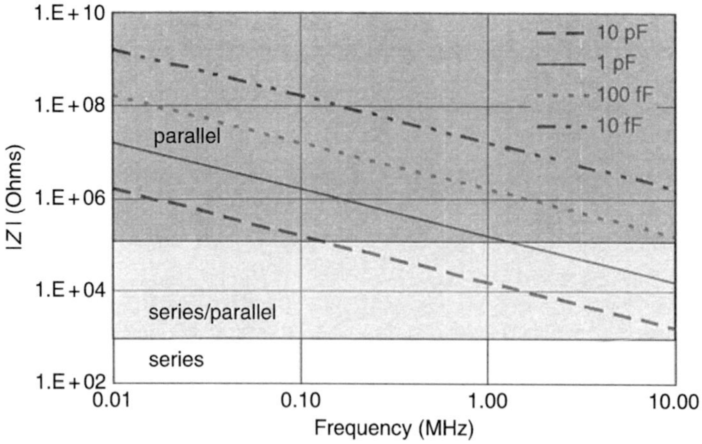

Fig. 9.9 Impedance $|Z|$ and frequency regions for measurement of different capacitor values in the series or parallel modes

mode. If the measurement frequency is lowered to 0.10 MHz, with $D_{\mathrm{f}} = 0.031$ and $C_{\mathrm{pm}} = 0.9991$ pF, the error in capacitance value in the parallel mode is reduced to ~0.01% and either series or parallel modes may be used.

The impedance measurement accuracy range for an LCR meter is illustrated in Fig. 9.10. The gray rectangles are specified accuracy regions for the LCR meter. The accuracy deteriorates as the capacitance, frequency points move away from the center of the rectangle. For example, in the chart shown in Fig. 9.10, at a frequency of 0.10 MHz, the measurement accuracy of a 10 pF capacitor is ±0.1% and that of

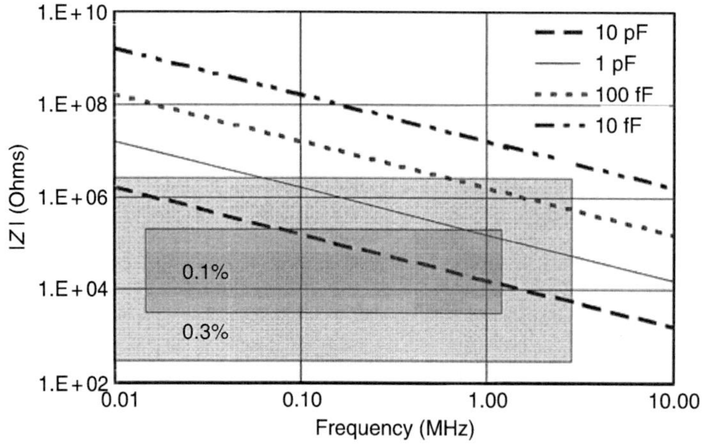

Fig. 9.10 Impedance $|Z|$ as a function of frequency for a pure capacitor. For illustration purposes, 0.1 and 0.3% measurement accuracy regions are shown as gray rectangles

a 1 pF capacitor is ± 0.3%. Such charts provided by the equipment manufacturer should be consulted for selecting the appropriate measurement mode and frequency range.

Let us consider two typical examples in the CMOS technology to make rough estimates of the desired measurement frequency range. From Fig. 9.10, it is apparent that to achieve a measurement accuracy of 0.1% at 0.1 MHz with this LCR meter, the impedance should be in the range of $\sim 10^4$–$10^5\ \Omega$.

Inter-level dielectric material in metal wire stack has inherently low leakage and metal interconnect capacitance can be measured in the series mode. For a 1 pF capacitor, and a desired capacitance measurement accuracy of 0.1%, the impedance at 0.1 MHz is $10^5\ \Omega$. The series resistance can be as high as $10^4\ \Omega$ for $D_f \leqslant 0.01$.

In a MOS capacitor with silicon oxide dielectric, the gate-tunneling current becomes significant as the gate oxide thickness is decreased below 3 nm. If at 1.0 V, the gate-tunneling current is 1 nA/μm and the gate capacitance is 1.0 fF/μm, for a total gate width of 1,000 μm, $C_{pm} = 1$ pF, and $R_{pm} \approx 1\ M\Omega$. The measurement frequency is increased to 10 MHz to get $D_f \approx 0.016$. At 10 MHz, the impedance $1/(\omega C_{pm}) = 1.6 \times 10^4\ \Omega$ and the measurement accuracy would be in the range of > 0.3% from Fig. 9.10. Measurements at frequencies $\gtrsim$ 10 MHz require high-frequency probes and a GND–signal–GND (G–S–G) I/O pad arrangement.

When both series and parallel resistances of comparable magnitude are present, a three-element model shown in Fig. 9.11 is used. Measurements are made at two different frequencies in both the series and the parallel modes to extract the correct capacitance value. Impedance meters may also provide equivalent circuit analysis from the measurements obtained by sweeping the frequency over a suitable range.

The relationships between measured resistances and capacitances in the two modes and the true capacitance C for a three-element equivalent circuit are shown in the following equations:

$$C_{sm} = C\left[1 + \frac{1}{(\omega C R_{pm})^2}\right], \tag{9.4}$$

$$C_{pm} = \frac{C}{\left(1 + \frac{R_{sm}}{R_{pm}}\right)^2 + (\omega C R_{sm})^2}, \tag{9.5}$$

$$D_f = \omega C R_{sm} + \frac{1}{\omega C R_{pm}}\left(1 + \frac{R_{sm}}{R_{pm}}\right). \tag{9.6}$$

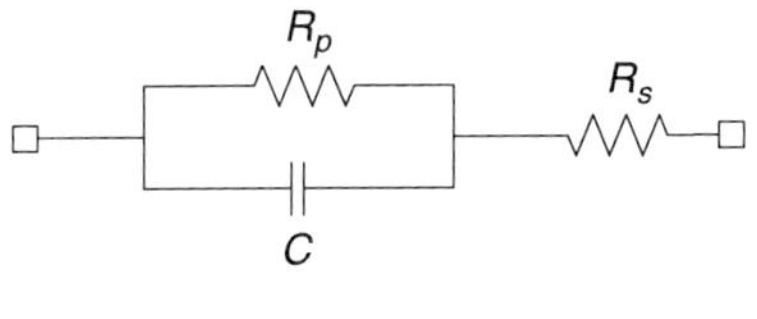

Fig. 9.11 Three-element equivalent circuit of a capacitor DUT

The range function in an impedance meter is set corresponding to the DUT impedance parameters. As illustrated in Fig. 9.1, measurement error is lower at the upper end of a sub-range. Also, there may be a discontinuity in the measured values at the sub-range boundaries. It may, therefore, be prudent to avoid making measurements at the lower end of a sub-range and skip to the next sub-range.

In addition to the frequency, the DUT dissipation factor, and the measurement mode (series or parallel), the measurement accuracy also depends on the allocated times for measurement and data acquisition. These times comprise the following:

- settling time of the AC source frequency and voltage level
- settling time of DC bias voltage
- integration time for averaging over a number of samples

For accurate capacitance measurements, a calibration is performed to eliminate stray capacitance of cables, connectors, and the wafer chuck.

9.2.4 Frequency Counters

Frequency counters are commonly employed in CMOS characterization for measuring ring oscillator frequencies. These instruments typically have a frequency measurement range of a few Hz to a few GHz. Frequency counters may also be used for measuring the outcome of any repetitive event, such as the number of output pulses generated in a fixed time period. One example of such an application is with the latch metastability experiment described in Section 7.4.3.

A frequency counter operates in the time domain. It counts the number of transitions between two adjustable voltage levels within a fixed time interval. The internal clock of the frequency counter is set by a calibrated quartz crystal oscillator. The measured frequency is equal to the number of transition events from the low to the high voltage levels divided by the time interval, as illustrated in Fig. 9.12a, b. The resolution of the frequency counter is determined in part by the time period of the internal oscillator. The accuracy of the frequency counter is dependent on the calibration accuracy and the stability of the internal crystal oscillator. For accurate measurements, the crystal oscillator must be held at a constant temperature by allowing sufficient warm-up time for the instrument.

The voltage trigger level for counting the number of transitions can be set automatically either by taking the average of the minimum and maximum voltage levels of the input signal or by using the point of maximum slew rate on the waveform. For ring oscillator measurements using DC probes, it is preferable to preset the trigger within a specified voltage window to filter any spikes in the voltage waveform, as shown in Fig. 9.12c. The voltage amplitude of the input signal should be sufficiently high to get a clean measurement point on the waveform. When RO frequency measurements are made at low V_{DDE} values (<700 mV), the I/O driver to the frequency counter is operated at a higher voltage (>0.8 V) to ensure signal integrity

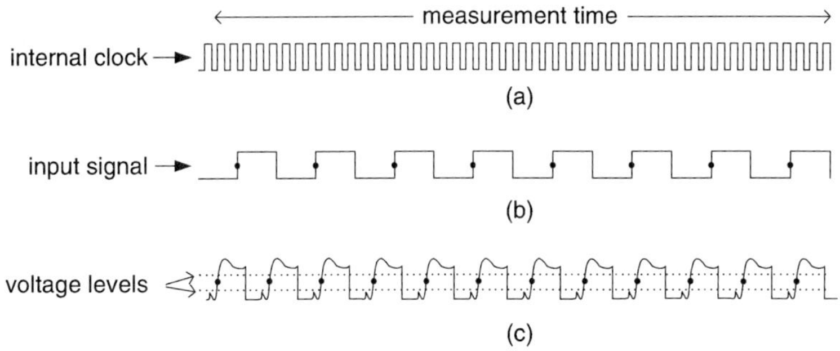

Fig. 9.12 **a** Internal time base of the frequency counter. **b** External input signal waveform. **c** Reference voltage levels for setting the trigger. *Dark circles* indicate measurement trigger points

(Section 6.2.1.3). The output waveform is observed on an oscilloscope to validate the test setup and to specify the voltage levels for the trigger point.

There are limits on input signal frequency measurement using a frequency counter. An upper limit is set by the internal oscillator frequency and a lower limit is set by the maximum allowed time interval. The time interval is generally long enough to cover many cycles to get an average frequency, and errors arising from spurious noise or end effects are minimized.

Frequency counters are generally equipped with an external trigger function. When this function is used, the frequency measurement begins each time an external trigger signal is received and terminated when either the measurement time interval is reached or a second external trigger signal is received. This external trigger function in a frequency counter is very useful when the signal frequency is changing with time as in the RO macro design for variability described in Section 6.2.4.

Frequency counters may also be used for measuring the number of pulses in a user-specified time interval or for detecting noise. The input signal waveforms for periodic transitions, pulsed transitions, and the potential at the GND of the circuit being measured are shown in Fig. 9.13. The counter reports the number of voltage crossings at the set point within the measurement time window.

Precision frequency counters may have built-in statistical functions which are useful for test structure designs for variability and for monitoring signal stability.

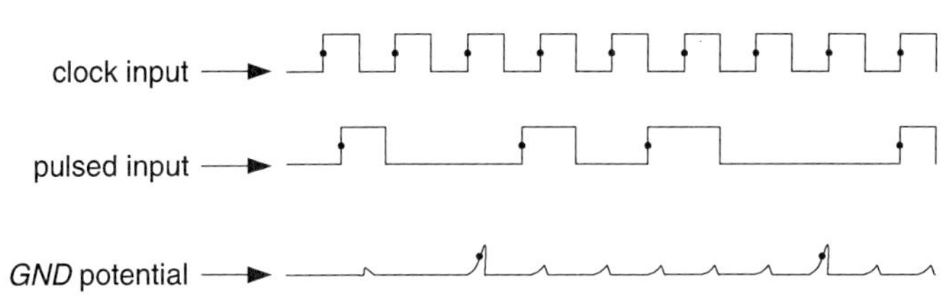

Fig. 9.13 Input signal waveforms for a clock (periodic) signal, pulsed signal, and GND potential. *Dark circles* indicate measurement set point

The frequency measurements are carried out repeatedly within an integration time specified by the user. The maximum, minimum, mean, and standard deviation σ of the frequency measured within this time are reported. Measurement accuracy is derived from σ/mean (%) which includes error introduced by any instability in the signal from the test structure. The statistical functions are also useful when the signal frequency is varied by design to give a measure of variability in a circuit element (Section 6.2.4).

9.2.5 Pulse and Clock Generators

Pulse generators as stand-alone units or coupled with parametric or digital ATE perform the function of a pulse or a clock generator with programmable pulse width and time period. These units operate over a large frequency range, from as low as 100 Hz or less to as high as several GHz. A pulse generator may have a single output channel or multiple channels with programmable time offsets between the channels. A number of options to set the pulse properties are available. Enabling a square wave option turns the pulse generator into a clock generator. When a pulse generator option is available in parametric ATE, a single sharp external signal edge to enable a ring oscillator circuit can be used, as discussed in Section 6.2.1.4. In bench tests, pulse generators are used for high-speed measurements described in Chapters 7 and 8.

The output of a single-channel pulse generator is shown in Fig. 9.14a. The pulse width, repetition rate (frequency), and delay with respect to an internal timing device

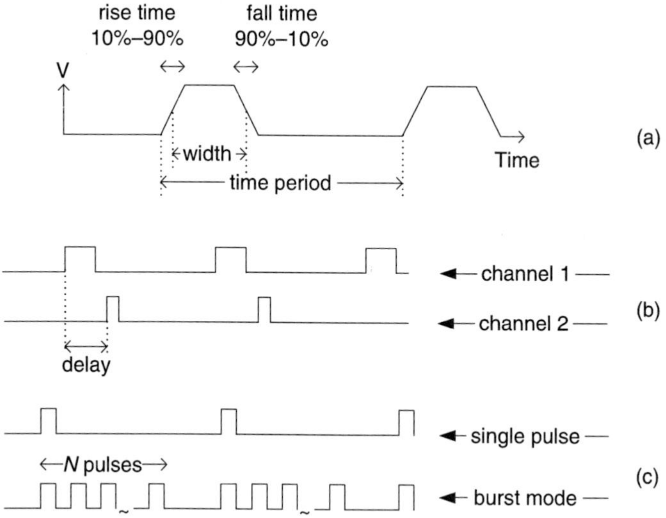

Fig. 9.14 **a** Properties of pulse waveforms. Two-channel pulse generator waveforms **b** for different pulse widths and **c** for single pulse and burst mode with *N* pulses

are set by the user. The outputs of two channels with the same time periods but different pulse widths and relative timings are shown in Fig. 9.14b. In some configurations the jitter between two corresponding edges of these channels can be < 1 ps. A burst option may be available to generate either a single pulse or a number of pulses within a cycle as shown in Fig. 9.14c.

The output from a channel may be used as a trigger signal which, when coupled to an oscilloscope or a sampling scope, serves as a time reference for an event initiated by the same or another synchronized pulse. Examples of such applications for high-speed differential delay measurements are described in Section 7.3.

9.3 Automated Test Equipment (ATE)

ATE for parametric tests, providing high measurement resolution and accuracy, evolved in parallel with digital and memory ATE for functional tests. Digital, memory, and analog tests are now being combined to a large extent in system-on-a-chip (SoC) ATE. The parametric capability of SoC ATE is being expanded to cover lower current and voltage ranges. The parallel test capability of the ATE provides high test throughput. Tester clock and digital test pattern generation functions allow more flexibility in test structure design, and 2D array test structures become attractive from area and test time efficiency points of view. However, independent parametric ATE and digital ATE are still present on the test floor of most semiconductor facilities.

The cost of ATE is much higher than laboratory bench test equipment. Once an investment in equipment and infrastructure in a silicon test facility has been made, only minor upgrades to existing equipment may be possible. Transitioning or mixing the use of parametric ATE with SoC ATE on a test floor would require test structure designs to be compatible with both types of ATE. One possible scenario is shown in Fig. 9.15, where with appropriately designed test structures, parametric ATE is used at the M1 test stop and SoC ATE for all test stops beyond M1, meeting the requirements of both high accuracy at the M1 test stop and throughput at later test stops.

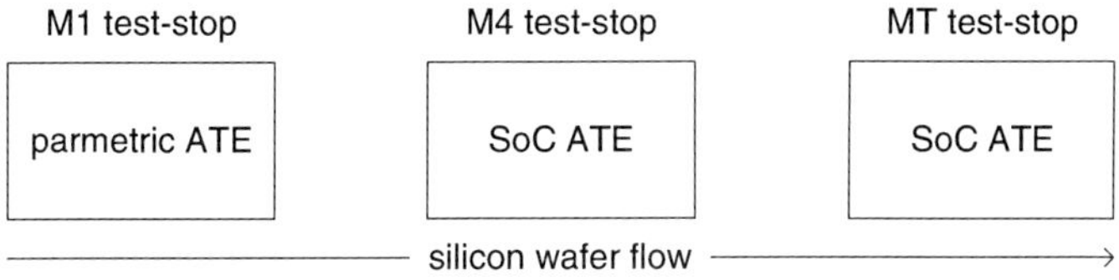

Fig. 9.15 A possible scenario showing the type of ATE used at different test stops for optimum throughput

9.3.1 Parametric ATE

Parametric ATE is designed to meet the test requirements for CMOS technology characterization. In some applications, measurements of MOSFET parameters and other discrete resistance and capacitance parameters are carried out at very low currents and voltages. As an example, in low-power CMOS technology for battery-operated consumer products, MOSFET leakage currents are in the fA/μm to pA/μm range. Metal via resistances may be < 1 Ω, giving an output voltage of a few mV for a force current of ~10 mA. If the current and voltage levels are increased by measuring a large number of devices in parallel or in series, only average values of the parameters are obtained. Increase in variability with CMOS scaling has put more emphasis on collecting statistics for all the circuit components, and individual elements need to be measured. Capacitance measurement in the presence of high gate oxide leakage currents is another challenging area where, unlike with DC parametric tests, measurement frequencies of the order of ~100 MHz are required.

Parametric ATE installed in a manufacturing line must also facilitate rapid changing of probe cards and provide high throughput, high data manipulation and transfer rates, and minimum resource requirement for test program generation. Parametric ATE attempts to cover these requirements while providing precision measurements over a wider range. ATE is customized to some degree for use in a specific CMOS manufacturing line to match the test structure design style, standard I/O pad count, and required number of SMUs.

Conventional parametric ATE employs several SMUs coupled to a switch matrix connected to a low leakage probe card, all of which are placed in a test head in close proximity to a wafer probe station. In earlier models shown in Fig. 9.16a, a capacitance meter and a pulse generator are mounted on a stand-alone rack. These instruments are connected to the probe card through additional auxiliary ports in the switch matrix.

With thinner gate oxides and the need to measure capacitances at higher frequencies, the switch matrix has been modified and the test head includes a capacitance meter (Fig. 9.16b). This approach has extended the capacitance measurement frequency range to several MHz and decreased the measurement time substantially. Introduction of CBCM techniques (Section 4.2.2) to measure capacitance using an off-chip clock and test requirements for electronic fuses blown with an electrical pulse input have led to further modification of the switch matrix with pulsed relays for shaping the pulse.

Other options with an increased number of SMUs and additional high-frequency capabilities are also available. Because of the increased parasitic capacitance of cables, connectors, and the switch matrix, the maximum number of SMUs with force and sense functions is typically limited to about eight.

Parametric testers have traditionally been used for making one measurement at a time with analog force and measure capabilities. For each value of force voltage or current level, the SMU output is first reset to GND and then ramped up to the set point. Measurements of gradients in *I*–*V* or *C*–*V* characteristics and determination of threshold voltage from extrapolation of I_{ds}–V_{gs} plots are time consuming.

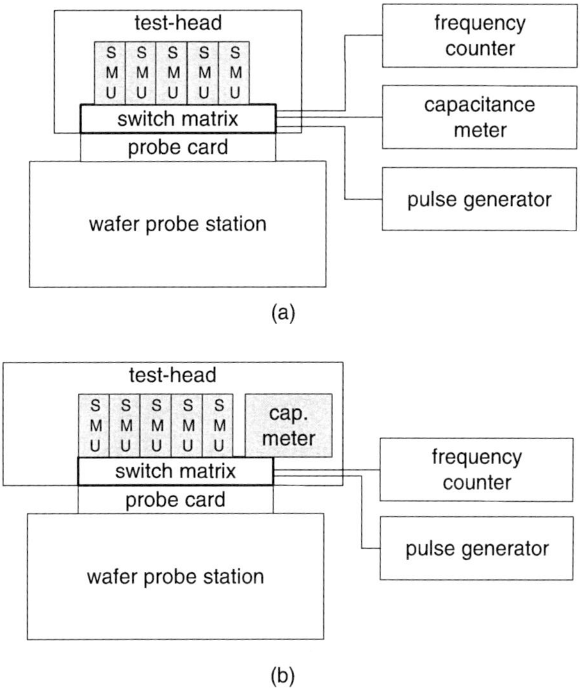

Fig. 9.16 Parametric ATE configurations: **a** with rack-mounted capacitance meter, pulse generator, and frequency counter, and **b** with capacitance meter integrated in the test head

Another drawback is in setting of decoder bits in addressable arrays. Because of a limited number of available SMUs, only two SMUs are used for any number of addressable bits and all high and all low bits are connected together. Each bit is reset to GND prior to setting each address code. This process adds to the test time. Industry demand to reduce test time has led to the development of enhanced parallel test capability in parametric ATE [3].

9.3.2 Digital and Memory ATE

Digital ATE is primarily used for CMOS product functional and fault testing [5–6]. There are a large number of books and other technical literatures on topics covering various aspects of digital testing of CMOS and silicon bipolar chips. A comprehensive list of publications on these topics is included in the work of Bushnell and Agrawal [5].

ATE for digital logic built-in self-test (BIST), memory test, and IDDQ test with varying degrees of DC parametric capabilities are commercially available. Special features of such ATE are (a) a few hundred to a few thousand independently controlled channels with digital input (test vectors), (b) DC current and voltage measurements on a per channel basis, and (c) storage capability for several Mb

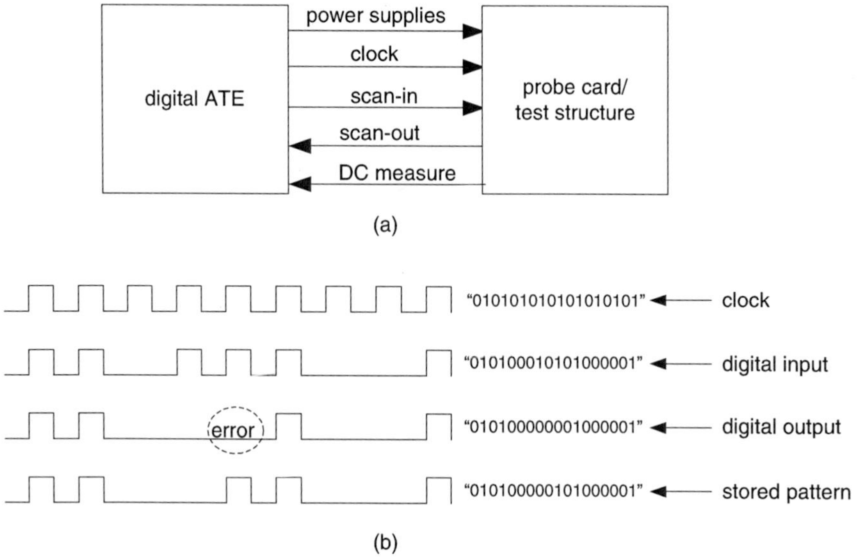

Fig. 9.17 **a** Digital ATE functions. **b** Waveforms for clock, user-specified test vector inputs, and a comparator for matching the output vector with a stored pattern

per channel to facilitate fast data acquisition. Digital testers can provide clock rates >1 GHz and can handle chip power levels of several hundred watts.

The test functions of a digital ATE are shown in Fig. 9.17. The tester supplies power, clock, and test vectors to the test structure or the product chip. The output vector is compared to a stored pattern and a fault or a fail is declared if a pattern mismatch occurs. The tests are typically not very sensitive to small changes in voltage levels of the input and output vectors.

Digital ATE is essential for testing yield test structures with a large number of DUTs or functional circuit blocks. These test structures generally do not require high measurement accuracy. Test structures for performance characterization such as ring oscillators are also well suited for digital testers.

High-density memory circuits, used as yield monitors for random defects, may require the use of memory ATE. Memory testers are similar to digital testers with the added capability of fault detection and repair. This type of ATE can handle long test vectors and special memory tests such as access time, noise immunity, and refresh.

9.3.3 System-on-Chip (SoC) ATE

Recognizing the industry need for ATE to meet a wide range of applications, the architecture of ATE for SoC applications is designed to be modular and reconfigurable, offering a high degree of parallelism in test to improve throughput. This type of ATE can be used for digital, memory, or mixed signal analog testing.

SoC ATE is now also available with parametric test capability for semiconductor characterization. The per pin measure unit (PPMU) offering has independent SMUs for each I/O pin and may have 100 or more pins [3]. Their current range has been extended down to the pA range which is sufficient for most parametric applications.

The parallel and fast measurement capabilities of such ATE may require redesigning test structures, originally configured for testing only on a parametric ATE.

9.4 Laboratory Bench Test Equipment

Test instruments designed for use in a laboratory are best suited for macro verification and test development on new or exploratory designs and for customized tests beyond the scope of ATE. Parametric tests are conducted with power supplies (SMUs) along with voltage, current, resistance, and capacitance meters having appropriate range and accuracy. A programmable switch matrix allows rapid testing of many DUTs, whereas a manual switch matrix gives more flexibility in test configurations. Some of the additional test equipment dedicated to semiconductor characterization and used for high-speed testing of CMOS test structures described in this book are mentioned below.

Semiconductor parameter analyzers (SPAs) are equipped with high-resolution SMUs for *I–V* curve tracing and device parameter extraction in the laboratory. The number of I/O is limited and generally only one device can be contacted at a time. The capability of quasi-static *C–V* characterization is also available. User-friendly, menu-driven programmable interfaces further enhance the usefulness of SPA as a laboratory benchtop equipment for semiconductor device characterization.

Oscilloscopes are used for observing and recording time-varying voltage signals. The operating frequency range varies from DC up to >10 GHz. In CMOS testing, oscilloscopes are used for making frequency measurements on ring oscillators and for detecting waveform distortion or sources of noise in the test apparatus. Oscilloscopes are especially useful for setting up the measurement apparatus and for debugging of new macro designs. Two or more channels may be available for comparing voltage waveforms of signals. Programmable digital oscilloscopes can store digitized data and provide summary statistics of input voltage variations over a period of time.

Sampling oscilloscopes are used for observing the detailed input voltage waveforms with a few ps resolution. The bandwidth is higher than that in a standard oscilloscope but dynamic input voltage range is limited to $\sim$1 V compared with $\sim$100 V for a digital oscilloscope. The jitter is typically well under a ps and measurement of time differences with an accuracy of a fraction of a ps can be made. Sampling oscilloscopes are used for differential time measurements of circuit delays for the type of high-speed test structures described in Chapters 7 and 8.

Delay lines are used for adding a known time delay to signal path. Mechanically adjustable delay lines, for both manual and automated operation, are available with resolution of 0.1 ps. These delay lines are used, for example, to change the relative arrival times of clock and data paths in the latch metastability experiment (Section 7.4.3).

RF/microwave switches replace a DC switch matrix for routing signals in high-speed tests. In a typical microwave switch unit, a single high-speed I/O can be directed to any of its six or more other high-speed terminals by action of a mechanical relay/actuator mechanism. These switch units introduce negligible signal distortion up to frequencies of 20 GHz or more.

9.5 Test Equipment Calibration

In technology development, meeting the requirements for both measurement reproducibility and measurement accuracy is of importance. Reproducibility criteria assure that measurements on a test structure are identical (within specification) on all testers in a test facility at all times and that observed variations in parameter values are not corrupted by variations in the test equipment or procedures. Absolute accuracy, which includes measurement reproducibility, is essential for technology performance benchmarking. Benchmarks based on integrated performance of a suite of devices or circuits (e.g., inverter FO = 4 delay, memory access time) provide a way to assess the technology contribution to product performance enhancements.

In order to assure that absolute measurement accuracy is within specifications, test equipment must be calibrated to NIST standards of measurement. Calibration of the test equipment itself is carried out by the equipment supplier. Full calibration on the test floor may require three or more steps as depicted in Fig. 9.18. In ATE, calibration of the I/O levels at the input to the probe card, which includes cables and connectors to the tester instrumentation, is carried out by the ATE manufacturer. The test specifications in the equipment data sheet are met under stated environmental conditions. Equipment installation and environment must therefore meet the external mechanical and electrical noise limits. Calibration can be verified on the test floor by replacing the probe card with a test board equipped with NIST-calibrated resistors, capacitors, and diodes.

Contact resistance between probes and the I/O pads, which affects measurement repeatability and accuracy, may vary with the mechanical integrity of the probe

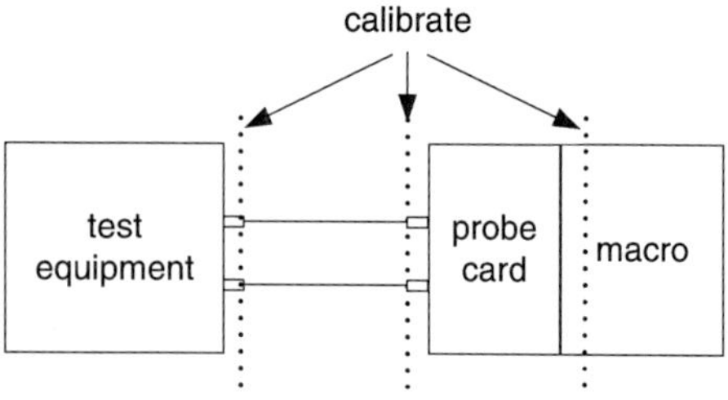

Fig. 9.18 Test setup with boundaries (*dotted lines*) indicating the calibration sequence

tip, contact pressure, wafer planarity, and metal coverage of the I/O pads. Probe contact resistance is measured on each site on a wafer using the probe check macro described in Section 3.2.4. Probe cards are replaced after a specified number of touchdowns as a preventative measure. A limited set of data measured on ATE are compared with the data obtained from a higher resolution and more accurate instrument such as a parameter analyzer. This provides calibration of the complete test setup.

Some of the parasitic circuit elements in the connectors between the test equipment and the DUT can be eliminated with four-terminal measurements and other common mode rejection techniques described throughout this book.

9.6 Test Automation

Test automation is essential for proper management of engineering test resources. Complete automation includes the following steps:

- landing probes on the macro to be tested
- measurements of all parameters with specified input settings
- data manipulation
- data transfer to a data storage unit

As shown in Fig. 9.19, the inputs to the test program are derived from design data, test algorithms or modules from a test library, and test settings to be applied to the DUT. The measured or calculated output parameters are transferred to a data storage unit and imported into a database for analysis.

The design data contains the relative (x, y) location on a reticle field of a reference point in each macro with respect to its I/O pads. The dimensions of the reticle field (X, Y) determine the stepping distance in the horizontal and vertical directions to this reference point in macros in the neighboring reticle fields. A test plan is generated to specify reticle field locations on each of the selected wafers in a cassette or a foup to be tested. The stepping sequence is optimized to minimize stepping

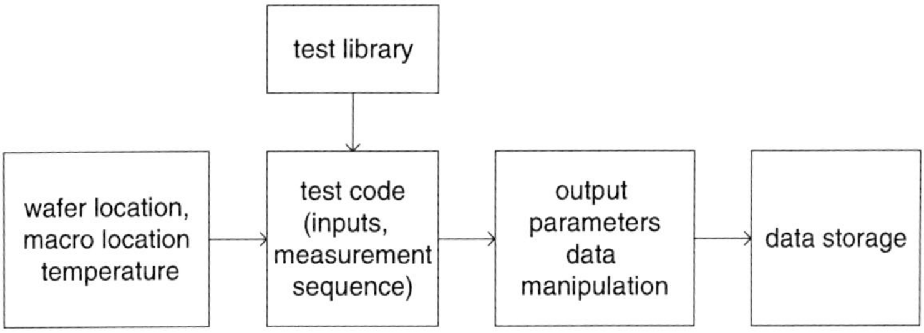

Fig. 9.19 Test program contents to locate, make contact, measure, manipulate data, and transfer output parameters to a data storage unit

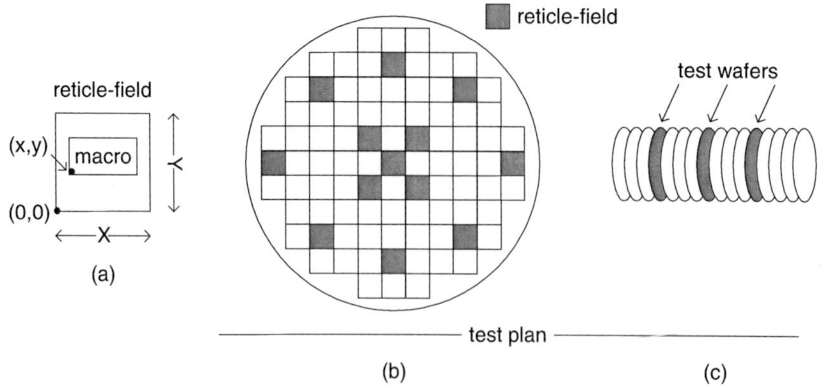

Fig. 9.20 **a** (x, y) location of a reference point in the macro. **b** Reticle field locations on a wafer to be tested. **c** Wafer locations in a cassette or a foup to be tested

or indexing time. This information, shown graphically in Fig. 9.20, is included in the test program which directs the wafer-handling system to land the probes on the macro to be tested.

Wafer temperature is set by the wafer-handling system. The wafer is held by vacuum suction on a metal chuck held at a constant temperature. Typically, the wafer temperature can be varied between –40°C and 140°C. Special probe cards and equipment are needed to set the wafer temperature outside this range. The time to equilibrate wafer temperature can be several minutes. Hence, all measurements to be made at one specific temperature on the wafers under test are grouped together.

Standard tests such as VFIM, IFVM, RO frequency, and *I–V* sweep are repeated on a large number of DUTs. A test library of algorithm modules or sub-routines is generated or provided by the manufacturer to ease the burden of writing test programs. The I/O pad assignment and input levels are defined in the test program when calling a sub-routine. A unique parameter name is assigned to each measured parameter which reflects the macro, DUT, and test conditions. ATE has some data processing capability, and calculations may be performed on the measured parameters by the tester itself. All or selected output parameters are transferred to a data storage unit. The data transfer time is minimized by storing data in the tester memory and transferring it as a bundle when the memory capacity is reached or after a specified number of tests. The tester identification and date and time of the day at which the test program was run are also preserved along with the measured data.

Test programs for macros sharing a common template for the same input voltage and current levels are identical except for the location of the macro and the output parameter names. The test program architecture should include an easy or a fully automated method of creating test programs for all such macros, including changing the input voltages, currents, and temperature directly from a document or a spreadsheet.

A test may fail when the power supply on the DUT is shorted and the tester current reaches a current compliance limit, when no contact is made to the DUT and the current is below a measurement limit, or when other tester malfunctions occur. Such measurement fails generate erroneous data which is detected and identified by error codes generated by the tester. A unique tester error code value is assigned to each type of fail to indicate shorts, opens, and other failure modes. The measured parameter for a failed test then displays the error code value. The error codes are very useful for test debug and for monitoring test yield. For easy identification, it is convenient to assign the error code values to fall outside the range of measured parameters. These values can be removed from the valid data prior to data analysis by applying data filters (Section 10.4.1).

Data integrity relies on correct functioning of the macro, the test program, and the test equipment. In the macro design verification and data validation stage, it is important to keep all possible sources of error in mind. Although test automation is extremely useful, it is somewhat cumbersome for test debug when a new macro design is implemented. It is preferable to test new macros on a laboratory bench with partial or full manual control over the instrumentation having similar or higher accuracy than provided by the ATE.

References

1. Horowitz P, Hill W (1989) The art of electronics. Cambridge University Press, New York, NY
2. Schroder DK (2006) Semiconductor material and device characterization, 3rd edn. Wiley, Hoboken, NJ
3. Chao MW-P, Liao W, Chiang J, Kuo E. SMU-per-pin system architecture supports fast, cost-effective variation characterization. http://www.keithley.com/data?asset=52571. Accessed 15 Mar 2011
4. Agilent impedance measurement handbook: A guide to measurement technology and techniques. http://www.home.agilent.com/agilent/facet.jspx?t=79831.g.1&cc=US&lc=eng&sm=g&pageMode=TM&pageMode=TM. Accessed 15 Mar 2011
5. Bushnell ML, Agrawal VD (2000) Essentials of electronic testing for digital, memory and mixed-signal VLSI circuits. Springer, Boston, MA
6. Abramovici M, Breuer MA, Friedman AD (1994) Digital systems testing & testable design. Wiley, New York, NY

Chapter 10
Data Analysis

Contents

In CMOS technology, electrical measurements are carried out on a variety of test structure macros at each test stop as outlined in Fig. 1.3. The data collected from these measurements are analyzed and the information used for technology development, routine process monitoring, and product debug. The number of measured parameters from a single macro on one reticle field repeated across a wafer may be anywhere from less than ten to over a few hundred. With hundreds of wafer starts per day, a large amount of data are collected even if measurements are made only on selected sites on a subset of wafers. Data are stored in a centralized database and

M. Bhushan, M.B. Ketchen, *Microelectronic Test Structures for CMOS Technology*, DOI 10.1007/978-1-4419-9377-9_10, © Springer Science+Business Media, LLC 2011 software tools are made available for data manipulation and graphics. Electrical test structure data are correlated with data collected from silicon process and metrology sectors as well as with product test data. Statistical data analysis and visualization techniques, customized for CMOS technology, are critical in assimilation of the information gathered in a clear and concise manner, thereby extending the benefits derived from electrical test structures.

In this chapter, data analysis techniques with special emphasis on CMOS technology are described. In Section 10.1, techniques for displaying data and important steps in data collection and consolidation are introduced. Basic statistics for process monitoring, quality control, and variability are covered in Section 10.2. Development of test plans for process tuning and quality control based on DUT designs, macro locations, and variability statistics is discussed in Section 10.3. In Section 10.4, data filters and data reduction techniques are described. In Section 10.5, five examples of graphical summaries of data collected from test structures described in this book are given.

There are a number of textbooks on statistics, statistical data analysis, and design of experiments that provide in-depth coverage of these topics [1–3]. Statistical methods are applied for manufacturing quality control using statistical process control (SPC) and Six Sigma analysis [4–6]. A historical perspective on data display and many excellent examples of data visualization techniques are described in the books authored by Edward Tufte [7–9].

10.1 Introduction

The development of present day graphical design techniques for displaying information was pioneered in Europe in the 18th century. William Playfair's graphical display of England's trade balance and the fluctuation in the price of wheat and labor cost over 256 years (1565–1821) and Minard's graphical depiction of the fate of Napolean's army in Russia still serve as excellent examples and teaching aids in the 21st century [7]. In these charts, with a combination of words, numbers, graphical lines, and shading, the clarity and relevance of large amounts of data in a multi-dimensional space has been preserved to this day. The presentation and distribution of such information in that era was done with considerable effort and expense as engraved wooden blocks were required for printing the charts. Computer technology has made it easy to generate charts; however, careful thought and planning is still needed to convey correct and relevant information in a form that can be easily assimilated.

Scientific pursuit has led to the formulation of physical laws which describe the relationships among measured electrical parameters in a well-defined mathematical form. Data collected from experiments based on well-established scientific phenomenon can be fitted to equations describing the inter-relationships of parameters or displayed in a graphical format. In a manufacturing line, time variation of properties of devices, circuits, and other process parameters contains essential information for maintaining quality control. This type of data, along the lines of economic or sports statistics, is not necessarily predictive, yet certain trends may become apparent and corrective actions taken to avoid large deviations of the parameters from their targeted values.

Let us consider data collected on metal resistor DUTs. Current through a resistor I_m is measured at three different values of force voltage V_f at 25°C and the measurements repeated at 85°C. The measured values are displayed in the first three columns of the table shown in Fig. 10.1a. The calculated value of resistance for each (V_f, I_m) pair is entered in the last two columns. A cursory look at this small table, combined with the knowledge that resistance is independent of V_f, suggests a source of noise in the measurements. The table can be expanded to include the calculated measurement error at each temperature, the temperature coefficient of resistance of the DUT between 25 and 85°C, and other such parameters of interest. When data are collected on many DUTs and the data volume becomes large, it is convenient to store data tables in a spreadsheet or a relational database for data manipulation. However, information in a table format occupying more than a typical printed page is challenging to assimilate, especially if there are large random variations in values with no obvious correlation among data points.

An *XY* scatter plot of the measured values listed in the table in Fig. 10.1a is shown in Fig. 10.1b. Here, the dependent variable I_m is plotted against the independent variable V_f and data points at each temperature are fitted to a straight line, indicating a linear relationship between I_m and V_f following Ohm's law. The inverse of the slope of the best fit line is the resistance R and the graphics indicates a change in R with temperature. Bivariate data displayed in this type of graphical format are compact, drawing the attention of the viewer toward the message to be conveyed more forcefully and directly than a data table.

A pictorial representation of computed R for each chip on a wafer is shown in Fig. 10.1c. The resistance values are placed in three bins, each bin covering a range of values, and a different shade or a color is assigned to each bin. It becomes immediately obvious that the spatial variation in R across a wafer is exhibiting a strong radial dependence overlaid with a weaker top-to-bottom dependence. This representation is particularly useful in silicon technology where process variations are related to wafer geometry. The electrical parameters of a DUT, measured at different locations on a wafer, may reveal information on spatial variations across the wafer.

A trend chart for the average value of R on all chips on the wafers measured each day is shown in Fig. 10.1d. Here, a large amount of data collected each day are reduced to a single data point. The solid line depicts the targeted value of R and the dashed lines bound the maximum allowed excursions from the target. It can be inferred in a glance that on day 13, the mean value of R fell outside the lower limit and a change in the process recipe returned the average R within the allowed limits after 4 days.

In the above examples, data reduction is carried out by combining the use of known physical laws with statistical methods. Right selection of tabular or graphical formats used in charts generated for displaying the data is important for drawing

	25°C	85°C	25°C	85°C
V_f(V)	I_m (mA)	I_m (mA)	R(kΩ)	R(kΩ)
0.10	0.500	0.420	0.200	0.238
0.20	0.995	0.824	0.201	0.243
0.40	2.015	1.669	0.199	0.240

(a)

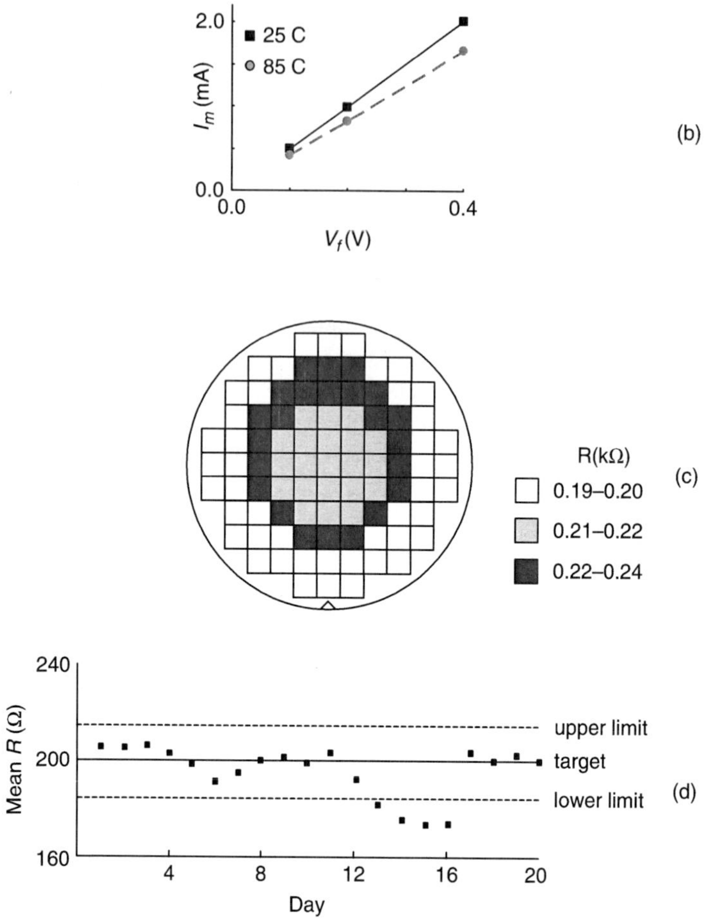

Fig. 10.1 **a** Table of measured V_f and I_m values and calculated resistance R. **b** *XY* scatter plot of I_m vs. V_f for the data shown in (**a**). **c** Wafer map with color representing R within a range or bin on each chip. **d** Trend chart of daily average R in the manufacturing line

correct inferences and for communicating information. The charts are complete only if relevant information on DUT design, placement of the macro on silicon, process recipe, and identification of wafers is presented. This information is generally included in the title or in an inset, to be used as reference, while the focus remains on the data being shown. Too long a title or large insets clutter the chart, but exclusion of this essential information reduces the value of the charts or may even render them totally useless. The graphical contents, while maintaining accuracy, must be balanced with clarity and aesthetics [7].

Our physiology is tuned toward pattern recognition, and graphical displays make good use of this human attribute. Data normalization techniques and scaling of parameters on DUTs of different designs are helpful in drawing attention toward data outside an acceptable range. A small but distinct difference in two uncluttered graphical displays that are identical in all other respects, when placed side by side, can be more easily detected than if displayed on subsequent pages. This feature is easily extended to multiple graphical displays on one page, as exemplified in Section 10.5.

The use of color in graphical displays is also helpful in focusing attention on different patterns and in distinguishing data types or classifications. Because of variations in sensitivities to different colors in different people and color blindness, extensive use of colors, specifically red and green hues, should be avoided. Gray scale, fill patterns, and shading work well if color printing is not available.

Adoption of a few carefully designed graphical templates is especially useful when data are shared among different process and design teams. The template concept introduced in Section 2.5 for macro designs and carried over to test program generation is applicable to data analysis and graphics as well. Enabling the use of graphical templates begins with macro design, documentation of DUT properties, careful planning of relative placement of macros on a reticle field, and generation of a targeted range of parameter values. A parameter naming convention, indicating useful information on DUT design, DUT location on silicon, and test conditions is important in tracking and presentation of such information.

10.2 Basic Statistics

There is inherent variability in the electrical data obtained from test structures. A small spread in the data may be present in repeated measurements on the same DUT. This spread arises from noise in the test equipment, change in the environmental conditions, and variation in probe contact resistances in repeated touchdowns due to mechanical vibrations and aging. Measurements made on multiple copies of the same DUT design in a macro may vary over a wider range because of local random variations in the DUTs and spatial variations within the macro. The data are distributed over an even wider range for measurements of such DUTs on multiple chip locations on a wafer, on multiple wafers from a lot, and on multiple lots. Statistical methods provide a way to summarize such distributions and condense them into a few variables.

Knowledge of the statistical nature of circuit elements in CMOS technology is essential for designing test structures, for developing a test plan, and for analyzing the data. Here, we begin with a brief introduction to basic concepts in statistical data analysis and then apply them to the data collected from electrical test structures.

10.2.1 Central Tendency or Mean Value

The statistical distribution of a parameter can be succinctly summarized by two parameters, one defining the center of the distribution known as central tendency and the other defining the range over which the data are distributed. The central tendency is described by the mean, median, or mode of the distribution of a set of observations or data sample. The *mean* of a distribution is the arithmetic average of all the observations. For n observations in a sample, each with a value denoted by x_i $(i = 1, 2, \ldots, n)$, the mean is given by

$$\bar{x} = \frac{\Sigma x_i}{n}. \tag{10.1}$$

The *median* is the value in the distribution that divides the observations into two halves such that half the values fall below the median and the other half above it. The *mode* is the value that occurs most often in the data sample. The *range* is the difference between the maximum and the minimum values in the data sample.

Let us examine the significance of these definitions using a random data sample of observed values. In Fig. 10.2a, the data are plotted in the sequential order of measurement. Adjacent data points are connected with lines to assist in visualization. The values with a measurement resolution of 5 are expected to fall between 30 and 70. The mean, median, and mode of this data sample, shown in Table 10.1, are 51.3, 50.0, and 55.0, respectively. The range of 250 is clearly larger than the expected value of ~40. Only one data point occurs with a value >250, contributing to the large observed range. Such anomalous observations, or fliers, may occur in data collected from test structures because of misprocessing of silicon wafers, poor probe contacts, or malfunction of test equipment. If data are collected manually, this may even be a typographical error in recording the data. Here, a statistical summary indicates the presence of fliers and a graphical plot immediately points to the single flier and any related information such as the place in the measurement sequence where the flier occurred.

In Fig. 10.2b, the high flier is filtered from the data sample and now the mean value is lowered to 49.0 but the median and mode remain the same. The median and mode values of a distribution are not corrupted by the presence of a few occurrences of erroneous data. A histogram of the data is shown in Fig. 10.2c. The value observed most often is 55.0, which is the mode of the distribution. This graphical format, a bar chart, is especially useful when there is no causal relationship in the observations.

10.2.2 Statistical Distributions and Variability

The range of a data sample gives no indication of how the data are distributed within the range. Information on the distribution of data within its range is given by the

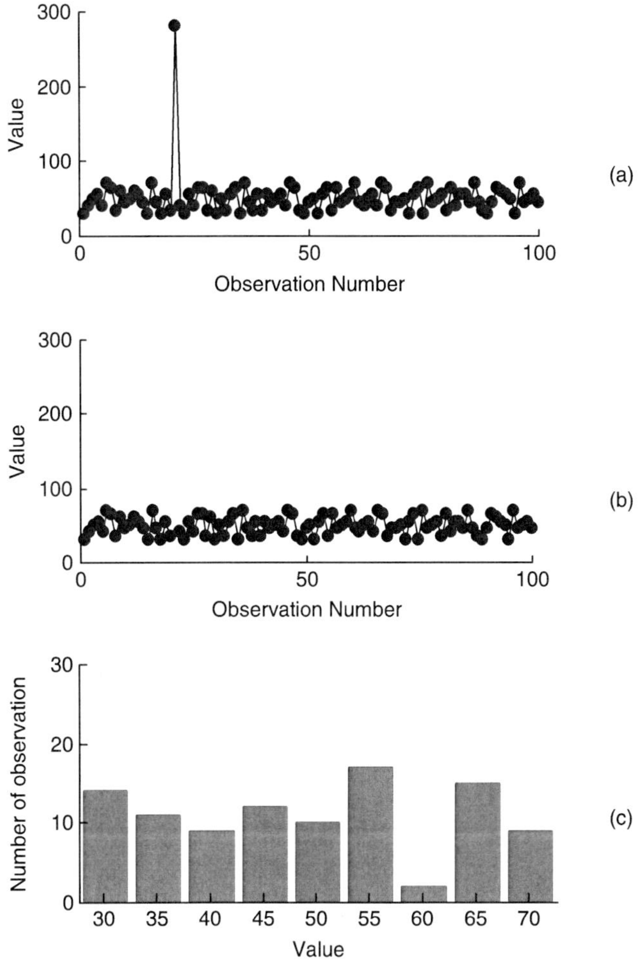

Fig. 10.2 Data sample consisting of numerical values of 100 observations **a** with one high flier and **b** after filtering the high flier. **c** Histogram showing the number of observations (frequency) for each value in the filtered data

Table 10.1 Sample properties corresponding to data shown in Fig. 10.2

Statistical parameter	Raw data	Filtered data
Mean	51.3	49.0
Median	50.0	50.0
Mode	55.0	55.0
Range	250.0	40.0

variance, which is a measure of how far each data point deviates from the mean value of the distribution. The variance s^2 is expressed in terms of the sum of squares of the difference of data values from the mean and is always a positive number:

$$s^2 = \frac{\Sigma(x_i - \bar{x})^2}{(n-1)}. \tag{10.2}$$

The square root of the variance s is known as the standard deviation. It is a very useful term, having the same units as the variable itself. When n is very large, and the sample represents the entire data population, the mean and variance of the data sample are denoted by μ and σ, respectively. The values of $\bar{x}$ and s may deviate from μ and σ, respectively, when n is small, as discussed in Section 10.3.4.

A histogram, introduced in Fig. 10.2c, is one type of graphical illustration of variability. The width of each rectangle, which is called a cell or a bin, denotes the range of values contained in it as indicated on the horizontal axis. The height of the rectangle is proportional to the number of observations or frequency of occurrence of the values within the bin. For a meaningful histogram, it is recommended to keep $n > 50$. The number of bins κ is selected such that $2^\kappa = n$, as shown in Fig. 10.3.

A vast majority of data samples follow a normal distribution as shown in Fig. 10.4, giving the probability of occurrence of an observed value with its peak centered at the mean of the distribution. The number of observations is lowered in a symmetrical fashion as the observed values deviate further from the mean value. This type of distribution occurs when there are a number of different sources of random variations in the data sample. The central limit theorem in statistics states that for such cases, the probability density function is a bell-shaped curve. The probability density of observed value x, $p(x)$, is given by a Gaussian function,

$$p(x) = \frac{1}{\sqrt{2\pi\sigma^2}} e^{-\left[\frac{(x-\mu)^2}{2\sigma^2}\right]}, \tag{10.3}$$

where $\int p(x)\mathrm{d}x = 1$. The Gaussian distribution is also referred to as normal distribution.

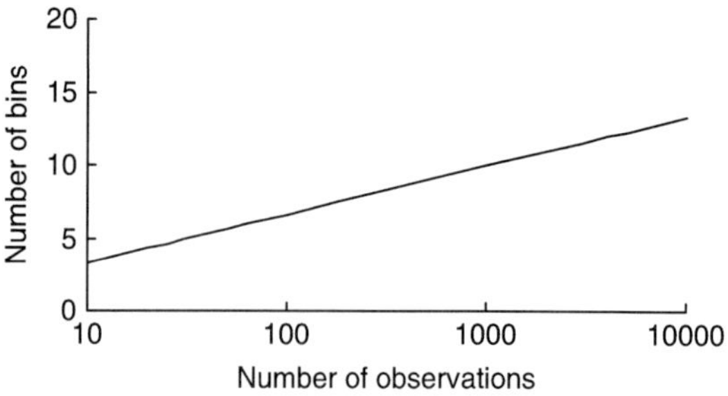

Fig. 10.3 Number of recommended bins (cells) in a histogram as a function of number of observations

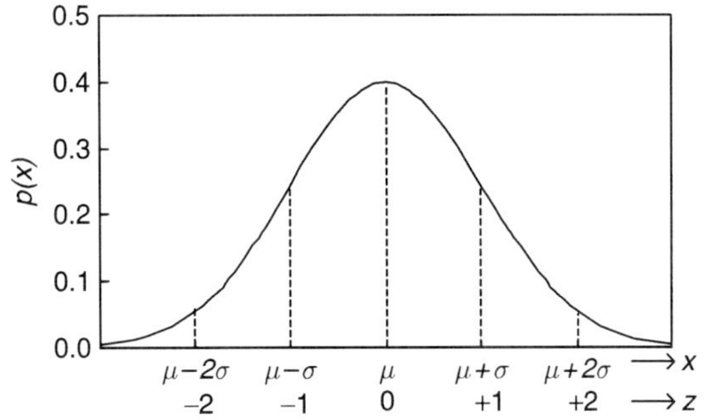

Fig. 10.4 Normal distribution showing probability density $p(x)$ as a function of x, and z, centered about its mean ($x = \mu, z = 0$)

Data collected from electrical test structures do not necessarily follow a normal distribution. Such non-normal behavior occurs, for example, when the dominant source of variation is a non-linear physical effect. Deviations from a normal distribution are discussed in Section 10.2.3.

The total area under the curve in Fig. 10.4 is 1.0, equivalent to 100% of the number of observed values in the sample. It is convenient to transform the x-axis in this plot to a variable z, where

$$z = \frac{x - \mu}{\sigma}. \tag{10.4}$$

The cumulative distribution function or C.D.F., denoted by $F(x)$, is shown in Fig. 10.5. It indicates the probability that a variable has a value $<x$. The slope of the line at any value of x is a measure of the probability density $p(x)$.

Statistical tables provide the tail area in a normal distribution in normalized units of z [1]. In Table 10.2, the % of sample population within the absolute value of z, $|z|$, and the probability of occurrence outside this range are listed. The table indicates that 50% of the sample population is within $\pm 0.67\sigma$ of the mean value μ and 95.46% of the population is within $\pm 2\sigma$. The corresponding probability that

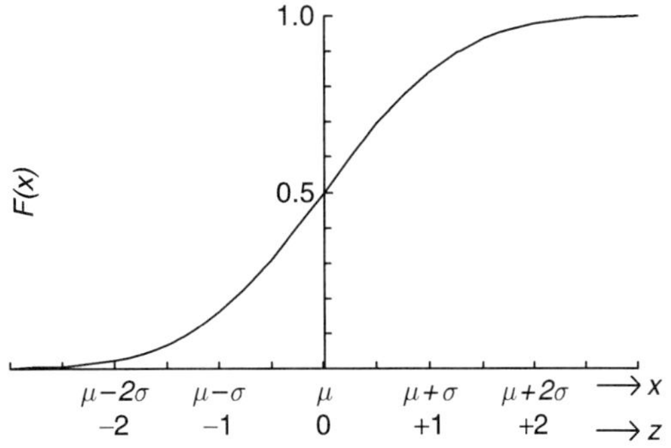

Fig. 10.5 Cumulative distribution function (C.D.F.), showing the probability $F(x)$ of a parameter having a value $<x$

Table 10.2 Probability of observations within and outside $|z|$

$\|z\|$	Percent of samples within $<\|z\|$	Probability of occurrence $>\|z\|$, 1 in
0.50	38.30	1.6
0.67	50.00	2.0
1.00	68.26	3.1
1.50	86.64	7.5
2.00	95.46	22
2.50	98.76	80
3.00	99.73	370
3.50	99.96	2,500
4.00	99.997	16,000
4.50	99.9997	160,000
5.00	99.99997	1,600,000
6.00	99.9999998	500,000,000

an observed value would be outside these limits is 1 in 2 and 1 in 22, respectively. Generally, in manufacturing, limits are set at $\pm 3\sigma$, leading to the "Six Sigma" goal, with 3 out of 1000 observations falling outside specified limits. The limits are set higher when better product yield, quality, or safety are desired.

10.2.3 Non-normal Distributions

When the sample size is small $(n \lesssim 50)$, the probability distribution may deviate from normality with $s > \sigma$. It follows a t-distribution, which is similar to a normal distribution with thicker tails. Statistical tables are available to estimate σ from a t-distribution, with $(n - 1)$ degrees of freedom, for any value of n [1].

In some cases, parameters may have inherent non-normal distributions. The deviation from normality may occur in several different ways depending on the characteristics of the dominant source of variation. It is therefore important to carry out a test for normality prior to applying the probability analysis described in the previous section. This can be done by numerically evaluating the difference in the shape of the distribution from a normal distribution. Alternatively, parameters such as skewness and kurtosis are computed to provide information on the type and the extent of deviation from normality, where

$$\text{skewness} = \frac{\Sigma (x_i - \bar{x})^3}{ns^3}, \tag{10.5}$$

$$\text{kurtosis} = \frac{\Sigma (x_i - \bar{x})^4}{ns^4} - 3. \tag{10.6}$$

Skewness is zero for a perfectly normal distribution, positive for a distribution with a long tail for $z > 0$, and negative for a distribution with a long tail for $z < 0$.

Kurtosis is positive if the distribution is more strongly peaked than normal and negative for a flatter distribution. Any asymmetry in the distribution can also be detected by calculating and comparing σ values, σ_+ and σ_-, for the positive and negative half of the distribution, respectively. In a positively skewed distribution, $\sigma_+ > \sigma_-$.

Some MOSFET parameter distributions are intrinsically non-normal. Others may have a non-normal distribution because of silicon process variations. One example of intrinsic non-normality is a positively skewed distribution of I_{ds} in the subthreshold region. An example distribution of I_{off} for MOSFET DUTs is shown in Fig. 10.6a. The source of a large increase in I_{off} on the positive side is random variation in V_t, which manifests itself as a logarithmic variation in I_{off}. This relationship, expressed in Eq. (5.1), is restated here in terms of V_t and I_{off}:

$$\frac{V_t}{SS} = \log_{10}\left(\frac{I_{dsvt}}{I_{off}}\right). \tag{10.7}$$

The subthreshold slope SS and the current at which V_t is measured, I_{dsvt}, are constants for the DUTs. The V_t distribution itself may be negatively skewed if V_t rolls off with L_p over the range of data collected. In this case, a larger increase in I_{off} spread is observed than predicted by random variability in V_t alone.

Standard definitions of μ and σ for a normal distribution, which assume all sources of variations to be random, when applied to non-normal distribution, may lead to erroneous conclusions. As an example, for the I_{off} distribution shown in

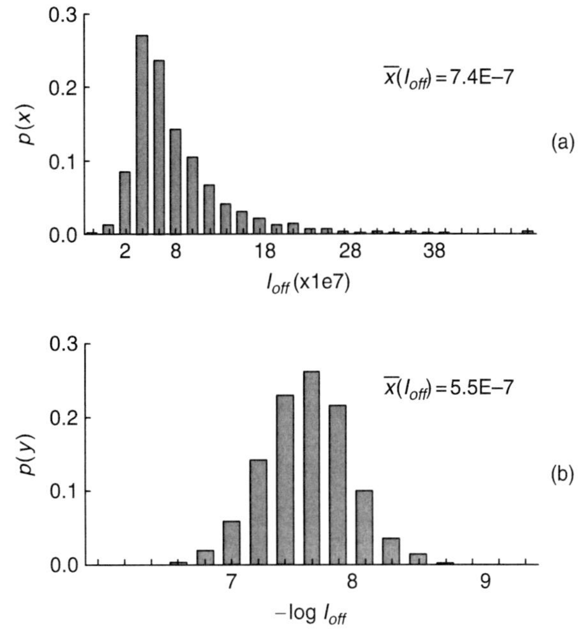

Fig. 10.6 **a** I_{off} distribution with a long tail at the positive side (positive skew). **b** $\log(I_{off})$ distribution

Fig. 10.6a, the arithmetic average of I_{off} or its $\bar{x}$ value is higher than the value at which the peak of the I_{off} distribution occurs.

A skewed distribution may be converted to a normal distribution by applying a transformation to the variable x [5]. An appropriate transformation is selected on the basis of the relationship between $\bar{x}$ and s. In a normal distribution, which is symmetrical with respect to $\bar{x}$, the value of $\bar{x}$ is independent of s. In a non-normal distribution, if there is a systematic dependence of $\bar{x}$ on s, then a transformation in x may be applied to convert it to a normal distribution. It is observed that in case of I_{off}, $\bar{x}$ varies linearly with s. A transformation of the type $y = \log(I_{off})$ then exhibits a normal distribution in y as illustrated in Fig. 10.6b. The variance of y is then transformed back to get the variance in I_{off}. This empirical method of determining the I_{off} transformation is consistent with the known logarithmic dependence of I_{off} on V_t, as indicated in Eq. (10.7).

A second example of a skewed distribution of a variable is the resistance of interconnect vias. The resistance of a single via varies inversely with its area. Typical inter-level metal via resistances are on the order of an ohm or less and small variations have little impact on circuit functionality. In the case of H0 vias in the 65 nm technology node and beyond, via resistances may fall in the 5 to >100 Ω range. Increased via resistances may occur because of reduced via dimensions and the extent of metal fill, or even an absence altogether of metal in vias. The resultant H0 via resistance distribution may be positively skewed and an appropriate transformation may need to be applied to obtain a normal distribution.

10.3 Data Collection

Careful planning in determining the number of instances and physical location of different test structure macros in the available space on a reticle field ensures that data can be collected to meet all anticipated process and product debug needs. Defining macro and parameter naming conventions in the design and test program generation phases simplifies data analysis and visualization tasks. Establishing an infrastructure for storing and analyzing data and selection of appropriate software packages for statistical data analysis are critical factors in data mining efficiency. Establishing a hierarchy in data analysis helps balance test time budget and minimizes the amount of time and resources required to extract correct information from the data collected from test structures.

10.3.1 Macro Placement

Throughout this book, we have discussed the importance of creating macro templates to improve test structure design and test efficiency. With a small number of macro templates, a large number of macros may be generated in a straightforward manner and the physical layout process automated through the use of P cells (parameterized cells). Placement of macros on silicon is based on macro contents and available area on silicon.

Proper documentation is an important step toward full utilization of a macro. Detailed documentation including circuit diagrams, physical layout description of the DUTs, I/O pad assignments, and test requirements may be prepared in a standard format for ease of viewing and importing this information into a database. Any error in documentation may take up precious resources in the macro debug phase. It is even worse if errors go undetected and the conclusions derived from the test structure data are misleading. Simple macro designs sharing a common design template, differing only in DUT descriptions, can be automatically generated from a common documentation template. Errors are avoided if the documentation is created directly from the P cell code generation defining the design of experiments (DOE) as illustrated in Fig. 10.7. In this case, the DOE is created by the end user of the macro and continuity in the entire flow from the concept phase to design and test is maintained without human intervention. Documentation for more complex macros is usually generated manually, but maintaining a standard manuscript format for all the macros is helpful in ensuring that the documentation is complete. Standardization also makes it easier to locate the desired information by end users who may not be familiar with the detailed macro design.

All macros should be classified to indicate their likely use. The test stops at which a macro is functional should be specified in the document as some macros can be tested at all levels and some only at a specific metal level. There are several categories of macro classification based on contents, where and how frequently the macros are tested, and how the data are utilized. These can be summarized as follows:

- application area: technology development, process debug, routine manufacturing, DFM, product performance, and product debug
- purpose: process monitor, product yield monitor, and GR validation
- cross-correlation: scribe line to on-product and electrical to metrology
- frequency of test: routine and infrequent (debug and GR check)

Based on this information, the number of instances of a macro on a test chip or scribe line and the physical location of each instance in a reticle field are established. As an example, for correlating ring oscillator frequency to MOSFET characteristics,

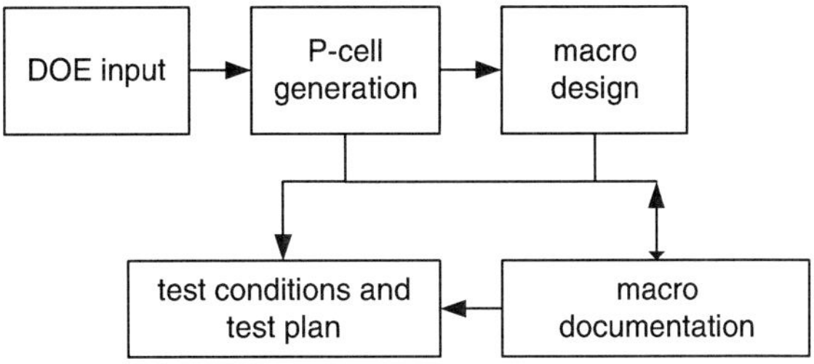

Fig. 10.7 Graphical illustration of macro design, test plan, and documentation flow from DOE input

the RO and MOSFET macros should be placed in close physical proximity to minimize the impact of spatial variations across the reticle field. Other examples of macros to be grouped together are those to evaluate the impact of layout and GR variations and macros to measure circuit delays using ROs and high-speed differential measurements. Several instances of a few key macros may be placed on a grid across a test chip or within the interstitials in a multi-chip reticle field for monitoring and modeling spatial variations. Sensitivity to long- and short-range pattern density variations should also be considered as they may affect processes such as lithography, RIE, rapid thermal annealing, and CMP. These macros contain resistors, capacitors, MOSFETs, ring oscillators, and a few high-speed experiments and yield monitors. The technology performance macros are migrated from the previous technology generation and care should be taken in standardizing the design so that yield and performance improvements from one technology node to the next can be consistently tracked.

Macros for routine monitoring are measured on silicon wafers in each lot, whether the technology is in the development or manufacturing stage. A macro may be tested on all sites on a wafer or only on a limited number of sites to reduce test time. A wafer site map specifying the (x, y) coordinates of the macros to be tested is created in the test plan, considering the trade-off between test time and sample size. Testing of all sites on a wafer may be needed to obtain wafer-level variations for tuning process tools. It may be prudent to carry out such testing on short-loop lots/test vehicles to keep the test time for actual product wafers to a minimum.

10.3.2 Parameter Naming Convention

A macro naming convention should be established prior to starting design activities. It is convenient to name a macro to indicate its contents. The output parameters of a macro may contain the name of the macro, metal test level, and either DUT location in the macro or DUT description. Using a DUT location in the parameter name is useful for macros sharing a common template. Example formats for macro and parameter names are shown below:

- Macro name: M1_FET_A012_TS2_XXXXX
  This macro is first tested at M1 and is the 12th MOSFET macro designed with template A on test chip TS2. The last set of digits XXX may have additional information such the purpose of the macro. For example, XXX = n_W indicates that the macro is designed to evaluate width effect on n-FETs.
- Parameter name: M1_FET_A012_N10_Vt_M4@1.0 V
  The parameter name indicates V_t measurement at 1.0 V for the 10th n-FET in macro M1_FET_A012 measured at M4 test stop.

These or similar conventions can be exploited in creating test programs and software routines for data analysis by sharing programming code for parameters named with a common set of characters.

10.3.3 Database and Software Tools

The electrical test data collected by ATE are transferred and stored in a relational database. The data obtained from test structures on different test vehicles and technologies are often correlated with data obtained from process tools, metrology tools, and product functional test equipment. It is therefore convenient to have the data collected from these different sources reside in one database or that multiple databases be linked to each other.

If the data volume is small, the data may be imported into a spreadsheet application program such as Microsoft Excel or Lotus symphony for further manipulation, statistical analysis, and preparation of graphical displays and charts. This is very useful when a more detailed, non-standard analysis is needed. Large test data volumes in silicon technology development and manufacturing facilities are loaded into databases, preferably with built-in software packages for data query and display. Software services dedicated to semiconductor technology for yield and performance analysis are offered by several companies. Some of these companies provide customized test structures for specific technology nodes and GRs prescribed by silicon foundries.

There are many software packages available for statistical data analysis and data mining with interactive graphics and user-friendly customization. With the help of computer software, it is easy to create many charts and port these into a presentation. The challenge is not in generating charts and tables but rather in efficiently assimilating the vast amount of data and reaching conclusions that are correctly weighted. Knowledge of the physical nature of a measured or calculated parameter and how it relates to technology process recipes and circuit functions and its inter-relationship to other measured or calculated parameters is extremely valuable. This is complemented by knowledge of statistics and data mining practices and availability and use of software tools. Expertise in each of these areas is developed with learning, training, and experience. The scenario gets more complicated with the incorporation of new features in each technology generation and with the introduction of new product applications.

As an example, consider the case where an unexpected decrease in a critical parameter I_{on} of an n-FET DUT tested at the M1 metal level is observed. Statistical data mining software can quickly (and even automatically) detect the shift and search for any correlation of I_{on} with process conditions gathered from all the tools in the manufacturing sectors and produce an engineering report with suggested actions. The correlation of this shift in I_{on} with corresponding changes in I_{off}, V_t, and circuit delays is obvious. The correlation with a large increase in H0 via resistance and no change in n-FET process recipe may lead to the root cause. A reported correlation of I_{on} dip observed in the test at the M1 level with resistance of the top metal layer is purely accidental.

Two of the commonly used programming languages for statistical analysis of semiconductor data are SAS and R. SAS, originally an acronym for statistical analysis software, is now known for the company that sells software solutions for businesses [10]. The advantage of the SAS programming language is that it can handle large volumes of data very efficiently. SAS-based programs are compact, with a sequence of statements executed in order. One disadvantage of SAS is that the CMOS process and manufacturing team must learn to use SAS for data analysis and report generation or dedicated resources must be added to generate customized SAS programs.

The efficiency of SAS has been extended in the commercially available JMP software tool to provide a user-friendly interface for analysis and graphics [11]. This tool not only can be used by those not familiar with SAS, but also provides the flexibility for SAS users to generate their own programs.

The R programming language, which evolved from the "S" programming language at Bell Laboratories, contains many unique software routines and graphical capabilities [12]. It has a stronger focus on object-oriented programming and new statistical methods can be programmed with relative ease.

SAS and JMP are powerful tools with user-friendly features and software support but have a licensing fee associated with their use. R falls under the general public license (GNU) project and can be downloaded for free. There are a number of other statistical analysis software packages available as well such as SPSS and SYSTAT. Software updates and new offerings are constantly enlarging the scope of statistical data analysis tools.

10.3.4 Number of DUTs to Be Measured

Measurement of all DUTs on all macros on each chip on every wafer is prohibitive from both the wafer throughput and cost perspectives. A test plan is developed to optimize test time while meeting the desired accuracy with which different parameter distributions need to be tracked. The test plan defines the measurements at each test stop, specifying the number of sites (macro locations) per wafer for all the DUTs to be measured and the number of wafers in each lot. The number of sites and number of wafers in a test plan may be different for different DUTs.

Let us consider a parameter x which exhibits a normal distribution. The $\bar{x}$ and s values of a data sample are determined from a limited number of measurements, n (<100). The distribution of $\bar{x}$, obtained from a number of such data samples, is also normal with a mean of μ. The standard deviation of $\bar{x}$, $\sigma(\bar{x})$, in terms of the population's standard deviation σ is given by

$$\sigma(\bar{x}) = \frac{\sigma}{\sqrt{n}}. \tag{10.8}$$

As n increases, $\sigma(\bar{x})$ decreases and $\bar{x}$ gets closer to μ. We would like to know how close $\bar{x}$ is to the μ value for different values of n and what the confidence level is with which $(\mu - \bar{x})$ is known. This information can be obtained from statistical tables [1, 5]. A few representative values of the number of samples required to determine

Table 10.3 Number of observations n to determine $(\mu - \bar{x})$ within a given interval in units of σ, with 90, 95, and 99.7% confidence [5]

$(\mu - \bar{x})$	n for 90% confidence	n for 95% confidence	n for 99.7% confidence
$\pm 1.00\sigma$	3	4	9
$\pm 0.75\sigma$	5	7	16
$\pm 0.50\sigma$	11	16	36
$\pm 0.44\sigma$	14	20	46
$\pm 0.33\sigma$	25	36	83
$\pm 0.25\sigma$	44	62	144
$\pm 0.20\sigma$	68	97	225
$\pm 0.10\sigma$	271	385	900

$(\mu - \bar{x})$ are listed in Table 10.3 for confidence levels of 90, 95, and 99.7%. As an example, for $n = 97$, it can be asserted with a 95% confidence level that $\bar{x}$ lies within $\mu \pm 0.20\sigma$.

The type of information given in Table 10.3 is very useful for designing test structures and for setting up a test plan to determine the optimum number of DUTs to be measured. An estimate of μ and σ for critical parameters is made from models or derived from past experience. The number of DUTs to be measured is determined for the accuracy with which μ must be known for the measured parameter and its σ value. As an example, if the mean value of V_t is 200 mV, and $\sigma(V_t)$ from random variations is 20 mV, then with $n = 62$, V_t can be measured with an accuracy of ±5 mV ($\pm 0.25\sigma$) with 95% confidence. If on the other hand, the $\sigma(V_t)$ for a very narrow MOSFET is ±50 mV, then n has to be increased to 385 to get the same ±5 mV ($\pm 0.10\sigma$) accuracy with 95% confidence.

Once the accuracy and confidence level of μ are estimated, we need to know the sample size for determining σ with a desired accuracy and confidence level. The distribution of σ is a χ^2 (chi-square) distribution, which is an asymmetric distribution for $n < 200$. As n is reduced, the asymmetry in the χ^2 distribution becomes more pronounced with a positive skew, increasing the error on the upper side. At a fixed confidence level, the uncertainty in σ is higher on the upper side than on the lower side. The σ intervals for a 95% confidence level are listed in Table 10.4. Typically, the value of n within a scribe-line macro, designed for characterization of circuit elements, or for sites measured for determining across wafer variation is <100. When data are collected on a large test structure such as a 2D array or memory block or on a test structure on many lots, n may be >1000.

The χ^2 distribution approaches a symmetric shape as $n > 200$ and an approximate expression may be used to determine n for knowing σ within δ% of its value with 95% confidence:

$$n = 2\left(\frac{100}{\delta}\right)^2. \tag{10.9}$$

Table 10.4 Interval limits for $(s-\sigma)$ for 95% confidence, expressed in units of σ, for different values of n [5]

n	$(s-\sigma)$ lower limit	$(s-\sigma)$ upper limit
4	-0.43σ	2.73σ
7	-0.36σ	1.20σ
16	-0.26σ	0.55σ
20	-0.24σ	0.42σ
36	-0.19σ	0.32σ
62	-0.18σ	0.19σ
97	-0.12σ	0.16σ
200	-0.10σ	0.10σ
385	-0.07σ	0.07σ
900	-0.05σ	0.05σ
5000	-0.02σ	0.02σ

From Table 10.3, for a sample size of 36, the value of μ is known within $\pm 0.33\sigma$ with 95% confidence. For the same sample, from Table 10.4, σ is known within 19% of its true value on the lower side of the distribution and within 32% on the upper side. In general, the uncertainty with which σ is known is larger than the uncertainty with which μ is known. Hence, the absolute error in $(\mu - \bar{x})$ and in $(\sigma - s)$ increases with the value of σ.

In the above discussions, it is assumed that measurement accuracy of a parameter is at least an order of magnitude better than the desired accuracy of μ and σ. This ensures that the variability is truly from the physical or electrical properties of the DUTs and not that of the test equipment.

10.3.5 Number of Sites to Be Measured

The criterion for selecting sites or DUT locations on a wafer to be tested is set differently for yield macros than for macros designed for characterization of circuit elements. During technology development, systematic defects may contribute significantly to yield loss. In silicon manufacturing, yield loss is dominated by random distribution of defects on wafers. Generally, all sites are measured to capture the defects which may tend to appear in clusters. The yield over an area A is given by

$$Y = \left(1 + \frac{A \times (\mathrm{DD})}{\alpha_c}\right)^{-\alpha_c}, \tag{10.10}$$

where DD is the defect density and α_c is a clustering parameter [18]. A large value of α_c indicates random defect distribution across the wafers.

The number of sites on a wafer for parametric characterization varies with the accuracy requirements of $\bar{x}$ and s for critical circuit element parameters and whether the variation to be studied is local random, systematic spatial (across a wafer or a reticle field), wafer to wafer within a lot, and temporal lot to lot (over a period of time). Each of these variability components with standard deviations σ_{rdm}, σ_{spt},

σ_{waf}, and σ_{tmp}, respectively, includes random variations arising from test, process, and defects. The grand statistics of a parameter include variations from all sources obtained from a large number of DUTs:

$$(\sigma_{var})^2 = (\sigma_{rdm})^2 + (\sigma_{spt})^2 + (\sigma_{waf})^2 + (\sigma_{tmp})^2 + (\sigma_{res})^2, \tag{10.11}$$

where σ_{res} comprises measurement and defect-induced errors. In the following discussion, we assume that the variations from measurement errors and defects are small or can be filtered from the data (Section 10.4.1).

With estimated values of σ_{rdm}, σ_{spt}, σ_{waf}, and σ_{tmp} from prior knowledge, the number of DUTs and sites on a wafer to be measured for each type of variation for different critical parameters is determined from Tables 10.3 and 10.4. This information is used in designing macros, determining their placement in a reticle field, and generating a test plan. The strategy for accomplishing this is discussed in more detail in the following sections.

Test time can be reduced by isolating the random and systematic across wafer variations and characterizing them independently. If σ_{rdm} and σ_{spt} are either small or nearly constant on all wafers, these can be monitored over a small sub-set of wafers. The impact of random local variations on the determination of across wafer variations is minimized by tracking the local mean parameter value of a number of DUTs placed within a small macro. Across wafer variations are tracked by partitioning the wafer in nearly constant parameter zones and measuring a limited number of sites per zone. The mean parameter value for the wafer is then obtained as a weighted average of the values within zones. Wafer-to-wafer variations are averaged over wafer averages within a lot. This scheme is illustrated in Fig. 10.8.

10.3.5.1 Local Random Variations

Statistics for random variations of a parameter are obtained by measuring a number of nominally identical DUTs in close physical proximity, placed in similar local environments, preferably within one macro.

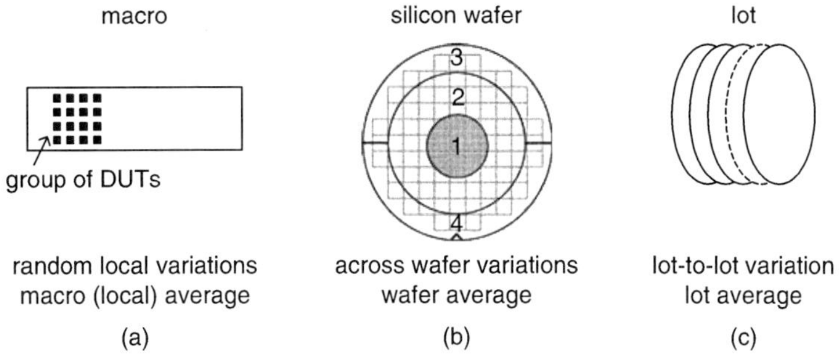

Fig. 10.8 Averaging of parameter values **a** for nominally identical DUTs in a macro, **b** for macros distributed across four (or more) zones in a wafer, and **c** for wafers in a lot

If a macro is designed to measure $\sigma_{\rm rdm}$ with an accuracy of $\pm 10\%$ ($\pm 0.10\sigma$), then the number of DUTs within a site from Table 10.4 is 200. Measuring such a large number of sites consumes silicon area and test time. Hence, measurements of random variability are generally more extensive during the technology development phase than in manufacturing where the process is more or less stable. In the manufacturing phase, only a few sites across a wafer on selected wafers may be periodically sampled for tracking $\sigma_{\rm rdm}$.

Consider $V_{\rm t}$ variation from random dopant fluctuations in a MOSFET. From Eq. (5.8),

$$\sigma_{\rm rdm}\,(V_{\rm t}) = \frac{\rm const}{\sqrt{\rm WL_p}}. \tag{10.12}$$

The value of the constant (~4 mV) may vary with MOSFET engineering. Hence, MOSFET random $V_{\rm t}$ variations are characterized on wafers processed with different recipes. Only a few sites (macros) on a wafer need to be measured. Random variation in H0 resistance, on the other hand, may itself vary across a single wafer and more sites on a wafer need to be measured.

In characterizing spatial and temporal variations discussed in the next sections, random variations are accounted for by using a local mean value of a parameter of a number of DUTs within a macro. The test time is substantially reduced with test structures designed to output parameter values averaged over many DUTs from a single measurement. Nominally identical MOSFETs in a macro may be measured in parallel to obtain a $V_{\rm t}$ value averaged over random dopant fluctuations (RDF) as discussed in Sections 5.1.2 and 5.4.2. Resistances of long wires are averaged over local linewidth variations. Ring oscillators, with stages numbering >25, provide average circuit delay values, eliminating the effect of random local variations in MOSFETs and parasitic circuit elements, as described in Section 6.2.1.1.

10.3.5.2 Spatial Variations

Variations in properties of devices and circuits in different locations on a wafer are present because of the nature of wafer scale processing in silicon technology. Some of the sources of such variations are illustrated in Fig. 10.9a. Spinning of wafers in lithography, CMP, and wafer cleaning steps may introduce radial variations. Gas

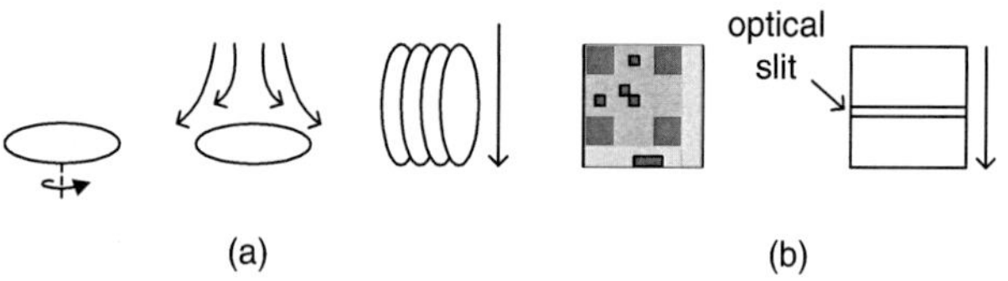

Fig. 10.9 **a** Sources of spatial variations across silicon wafers: spinning, gas flow, and temperature gradient. **b** Sources of variations across a reticle field: circuit density (*gray scale*) and optical exposure

flow and density of reactants in RIE and film deposition processes may be affected by the discontinuity at the wafer edge. Temperature gradients in radial or cross-wafer directions may be present in process chambers. Although every effort is made to mitigate spatial variations at each step, some residual effect is present in all wafers and is generally more pronounced in the chips at the edges of the wafer.

Sources of variations across each reticle field are illustrated in Fig. 10.9b. Circuit density variations, inherent in product design and scribe lines, affect the optical reflectivity and the local temperature during rapid thermal annealing and reactant density in plasma deposition and etching. Linewidth variations in identical circuits placed across a reticle field may be present in optical masks used for patterning or introduced from scanning during optical exposure. In large area product chips, across wafer variations may in turn introduce variations across the chip itself, as is apparent at zone boundaries in Fig. 10.8b.

Ideally, across wafer variation statistics are determined using the mean of the local random distribution per reticle field. A 300 nm silicon wafer typically has $\sim$100 reticle fields. If all sites are measured and across wafer variation is random, from Table 10.3, the mean value is then known within $\pm 0.20\sigma_{\mathrm{spt}}$ with a confidence level of 95%. Note, however, that the spatial effects causing parameter variation are generally systematic and may thus not yield a normal distribution.

A wafer map of the type shown in Fig. 10.1c may be generated to help visualize across wafer variations or a contour map created from discrete measurements. The nature of the spatial variation itself may be changing with time, as in early technology development, in process tuning, or when new process equipment for a critical step is installed. In this case, an algorithm may be used to quantify coefficients of radial and cross-wafer components fitting, for example, a polynomial expansion series [15]. These coefficients can be tracked more easily than viewing many wafer maps.

In order to reduce test time, the wafer is divided into zones of nearly constant parameter values. As an example, a wafer is divided into four regions in Fig. 10.8b. A few sites from each region, measured on many wafers, can provide a wafer average with reasonable accuracy. Information on radial as well as top-to-bottom variation patterns is also obtained from zone averages.

Variations across a product chip arising from across reticle field variations may directly impact product performance and power on all chips. The statistics of this variation are tracked for the chips in different zones as well. The use of ring oscillators to track across product variations is described in Section 6.5. Visualization techniques for across chip variation are covered in Example 3.

10.3.5.3 Temporal Variations

In a silicon manufacturing line, parameter variations of wafer lots processed in any fixed time period (day, week, or month) are monitored to maintain a stable process. As long as the random and spatial variability components remain nearly constant, mean parameter values for each lot are used for tracking temporal variations.

The accuracies with which sample mean, $\bar{x}$, and standard deviation s of a lot are known are dependent on the number of wafers tested in a lot, which is typically 25 or less due to the wafer handling capability of processing equipment. From Tables 10.3 and 10.4, a wafer lot comprising 20 wafers gives $\bar{x}$ within $\sim \pm 0.44\sigma_{\text{waf}}$ of population mean μ and s within $-0.24\sigma_{\text{waf}}$ to $+0.42\sigma_{\text{waf}}$.

A trend chart of the type shown in Fig. 10.1d is a useful way of visualizing temporal variations. Silicon manufacturing lines use Six Sigma or Taguchi or other statistical methods to center the process within specified limits. The Six Sigma process indices C_{p} and C_{pk} are used for monitoring parameter deviation from targets [5]. The process mean μ an upper specification limit USL, and a lower specification limit LSL are defined for each parameter based on model expectations or empirically determined targets. The spread around the process mean over a short period of time over which the data are collected is defined as $\pm 3\sigma_{\text{short}}$. The C_{p} index is defined as

$$C_{\text{p}} = \frac{(\text{USL} - \text{LSL})}{6\sigma_{\text{short}}}. \tag{10.13}$$

The C_{p} index is used when a parameter distribution is perfectly symmetric about the mean. If USL and LSL are the target $+3\sigma$ and -3σ limits, respectively, $C_{\text{p}} = 1$ indicates the parameter and its variation are as specified. A $C_{\text{p}} > 1$ indicates the parameter spread is smaller than target and that the process is well controlled.

Generally, the center of a parameter distribution may drift from the targeted value over time. For some parameters, the distributions may not be symmetric and $(\text{USL} - \mu)$ may be different than $(\mu - \text{LSL})$. A more convenient index is C_{pk}, defined as the smaller of the two values:

$$C_{\text{pk}} = \frac{(\text{USL} - \bar{x})}{3\sigma_{\text{short}}} \tag{10.14}$$

and

$$C_{\text{pk}} = \frac{(\bar{x} - \text{LSL})}{3\sigma_{\text{short}}}. \tag{10.15}$$

As with C_{p}, a C_{pk} value of >1 indicates that the parameter is within the specified limits.

The C_{p} and C_{pk} indices for two different parameters over a period of 3 weeks are summarized in Fig. 10.10. In Fig. 10.10a, the parameter distribution is centered at μ $(= \bar{x})$ and the C_{p} value increases as the spread in the distribution is reduced. In Fig. 10.10b, the C_{pk} values of a different parameter are shown over the same time period. The USL and LSL values for this parameter are not symmetric about μ. The value of $\bar{x}$ drifts with respect to μ, and C_{pk} remains $\leqslant 1$ over this time period.

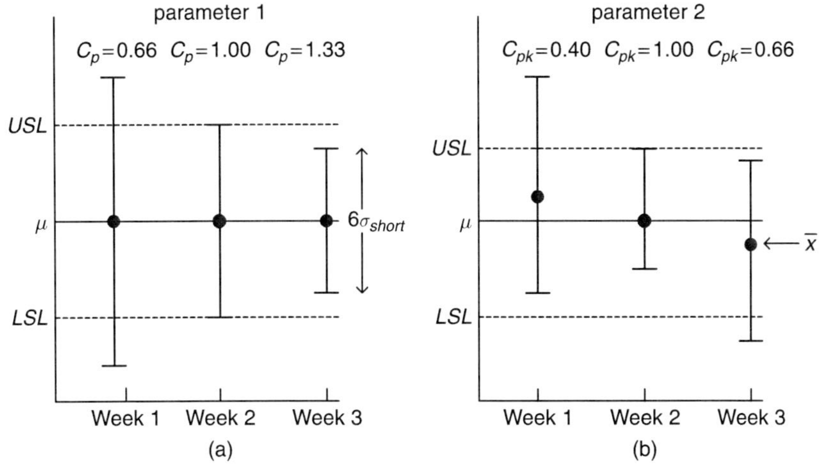

Fig. 10.10 Indices for two parameters showing the movement in $\bar{x}$ (*solid circles*) and $3\sigma_{\text{short}}$ over a period of 3 weeks: **a** C_{p} and **b** C_{pk}

10.3.5.4 Process Recipe Variations (Split Lots)

In CMOS technology development, experiments with process recipes and device engineering are carried out to optimize circuit parameters. Similarly, in silicon manufacturing, process recipe modifications are explored to improve product yield or to tailor selected process steps to a specific product. It is a common practice to split the wafers within a lot into groups and process one group with the "standard" recipe while varying the recipes for the process step to be studied in other groups. All wafers in this lot are processed together with all other process steps being identical. The differences in mean parameter values of interest in each group are compared with the mean values in the "standard" group. In such an exercise, it is essential to identify sufficient number of DUTs to be measured in each group to unambiguously separate the effects of the process recipe change from local random, wafer spatial, and wafer-to-wafer variations within the lot.

A measure of MOSFET performance improvement is an increase in average I_{eff} of n-FETs and p-FETs at constant I_{off}. At least a 15–20% increase in I_{eff} (or I_{on}) and a similar reduction in inverter delay are expected in migrating from one technology node to the next. Let us consider an experiment using a split lot to achieve a 4% increase in average n-FET I_{eff} at a constant I_{off} and with no change in p-FET characteristics. The mean n-FET V_{t} is targeted to be within ±2 mV of the standard process to constrain the I_{off} within ~±5% (Eq. (10.7)). With an n-FET I_{eff} increase of 4%, a reduction of ~2% in an inverter delay τ_{p} is expected. With an ~1:1 correspondence between inverter delay and product cycle time (=1/*f*), this experiment, if successful, would help bring the product performance to its committed frequency target.

To analyze the split lot, we use a MOSFET array macro, testable at the M1 metal level as described in Section 5.4.3. This macro has 4 n-FET and 4 p-FET arrays, each array comprising 30 nominally identical MOSFETs. The MOSFETs in an array can be measured individually or with all 30 connected in parallel. A ring oscillator (RO) macro as described in Section 6.2.2, also testable at the M1 metal level, is situated adjacent to the MOSFET array macro in the reticle field. The change in n-FET I_{eff} is correlated with τ_p values derived from an RO. The split lot is processed up to the M1 metal for early feedback to meet the product ship date.

In this example, there are 20 wafers in a lot with a five-way process split, having 4 wafers in each split. With 100 sites per wafer, there are a total of 400 macros of each type available for testing. The test times for one V_t and one ring oscillator (frequency and power) measurement are each reported to be five times longer than the test time for one I_{eff} measurement. In order to meet the test time budget, we would like to balance the number of V_t, I_{eff}, and RO measurements such that the mean values are known within ±2 mV for V_t, ±0.50% for I_{eff}, and ±0.20% for τ_p.

Example mean and σ components of V_t, I_{eff}, and τ_p, based on prior knowledge, are listed in Table 10.5. We expect σ components to remain fairly constant for all the wafers in the lot. With σ_{var}/μ over four times larger for V_t compared to I_{eff} and τ_p, it is apparent that the number of samples for V_t measurements is an important consideration. For determining μ with a 95% confidence level, the number of samples to be measured n is given by

$$n = \left[\frac{1.96\sigma}{(\mu - \bar{x})} \right]^2 . \tag{10.16}$$

The number of samples n and test time to determine V_t, I_{eff}, and τ_p within the specified accuracy for three different scenarios are listed in Table 10.6. First, we assume that all the DUTs are measured individually. In this case, we need to measure 1104 DUTs for V_t, 236 DUTs for I_{eff}, and 20 DUTs for τ_p. The test time is computed in units of I_{eff} measurements, with 5× test time for V_t and τ_p as mentioned earlier. The test time is reduced by a factor of ~11 if the V_t of 30 n-FETs in the array macro is measured in parallel to get an average local value. Measurement of I_{eff} of 30 n-FETs within a macro in parallel is not feasible because of *IR* drop in the M1 metal wiring. Further test time reduction is possible if a discrete MOSFET macro with five n-FETs of the same design, wired in parallel, is used for I_{eff} measurements. If one or more of the process recipes produce the performance

Table 10.5 The μ and the σ components for n-FET V_t and I_{eff} and for inverter τ_p

Statistics	V_t	I_{eff}	τ_p
μ	200 mV	800 μA/μm	10.000 ps
σ_{rdm}	25 mV	24 μA/μm	0.040 ps
σ_{spt}	20 mV	18 μA/μm	0.020 ps
σ_{waf}	10 mV	9 μA/μm	0.010 ps
σ_{res}	5 mV	2 μA/μm	0.002 ps
σ_{var}	34.0 mV	31.4 μA/μm	0.045 ps
σ_{var}/μ	0.17	0.039	0.004

Table 10.6 Statistics and number of measurements n for n-FET V_t and I_{eff} and inverter τ_p to meet stated accuracy requirements with 95% confidence

	V_t	I_{eff}	τ_p	Test time
μ	200 mV	800 μA/μm	10 ps	
Target $(\mu - \bar{x})$	±2 mV	±4 μA/μm	0.02 ps	
$\sigma_{var}/(\mu - \bar{x})$	17	7.8	2.3	
n	1104	236	20	5860
n	40 (30 parallel)	236	20	538
n	40 (30 parallel)	60 (5 parallel)	20	361

Test time is in units of one I_{eff} measurement

enhancement, local random variations can be measured using the MOSFET array macro for more detailed characterization.

In the above example, a combination of statistics, macro design and placement, and test strategy is used to optimize the split-lot experiment. For this purpose, macro designs are based on known sources of parameter variations from modeling and limited experimental data for the current technology node or data from other technology nodes. Macro designs and their placement, planned with these considerations, provide flexibility in tailoring test resources and in minimizing test time without loss of information.

Sophisticated statistical models for a technology are built once appropriate data become available. These models may be used for further tuning the macro designs, modifying test plans, and to provide design guidelines for future technology generations.

10.4 Data Reduction

The measured data collected from test equipment are manipulated for presentation in a usable format. The data are sanitized and filtered to remove invalid values or values outside a specified range. Parameters of interest are calculated from raw measurements, such as DUT resistance from measured voltage and current values. Parameter values may be normalized to DUT dimensions, model target values, or a reference value to facilitate data analysis. Statistical parameters such as median, σ, C_p, and C_{pk} are convenient ways to summarize large data volumes.

10.4.1 Data Filters

Data filters are applied to remove observations likely to corrupt the dataset or lead to erroneous conclusions. The data removed by filtering can be examined to debug and, if applicable, eliminate the root cause of data corruption. There are four commonly used sets of filters.

The first set of filters removes data reported as tester errors described in Section 9.6. Tester error codes are generated when no contact to the I/O pads is made, an SMU hits compliance, or a tester malfunction occurs. The error code format or values are far removed from any real data and can be automatically filtered. The error count for each type of error points to the problems with the test equipment, probe contacts, or test structure itself.

The second set of filters relates to the integrity of the macro design. As an example, an IDDQ test is carried out for each power supply sector in a macro. An upper limit on the IDDQ value for each sector is specified based on circuit simulations or estimated from published leakage current per unit width of each MOSFET type. An IDDQ value matching the current compliance limit of the test equipment indicates an electrical short in the power grid. An IDDQ value exceeding the specified upper limit of a circuit block such as a decoder or a scan chain, even if below the tester current compliance, is indicative of presence of shorts or opens in the circuit block itself. The data collected from macros failing IDDQ test are rejected. In early technology development, the IDDQ limits may be more relaxed as the experimental hardware may not be centered correctly.

The third set of filters may be applied by setting limits based on the expected range of a parameter. This is not a practical approach for a large number of parameters but may be used in specific cases.

The fourth set of filters is based on examination of the data to remove fliers or outliers. A statistical rule of thumb is to reject any data in the sample outside set upper and lower limits, irrespective of the type of data. The data are divided into four equal parts by quartiles Q1, Q2, and Q3. The first or lower quartile Q1 cuts off the lower 25% of the values; the second quartile Q2 divides the values into two halves; and the third quartile Q3 cuts off the upper 25% of the values. This is graphically illustrated using a box plot in Fig. 10.11. The vertical boundaries of the box are defined by Q1 and Q3, and Q2 is coincident with the median. The cross-bars show the data range containing 90% of the data (from 5 to 95%) and values outside this range are individually shown as dark circles. If the median is not in the center of the box, the data sample is skewed. The mean of the data sample is shown as a cross in the box and may not be coincident with the median. The width of the box can be programmed to be proportional to n or $\log(n)$ to indicate the total number of observations in the data sample.

The range containing the middle 50% of the data is called inter quartile range (IQR = Q3 − Q1) which includes data values within $\pm 0.7\sigma$ of the mean. Filter limits are defined by the lower and upper IQR multipliers, μ_l and μ_u, where

$$\text{Lower filter limit} = \text{Q1} - \mu_\mathrm{l} \times \text{IQR},$$
$$\text{Upper filter limit} = \text{Q3} + \mu_\mathrm{u} \times \text{IQR}.$$

The IQR multiplier values are selected to be in the range of 2.5 to 5, corresponding to $\pm 1.75\sigma$ to $\pm 3.5\sigma$ limits, respectively. Values outside these filter limits, shown as open circles in Fig. 10.11, are removed from the data sample.

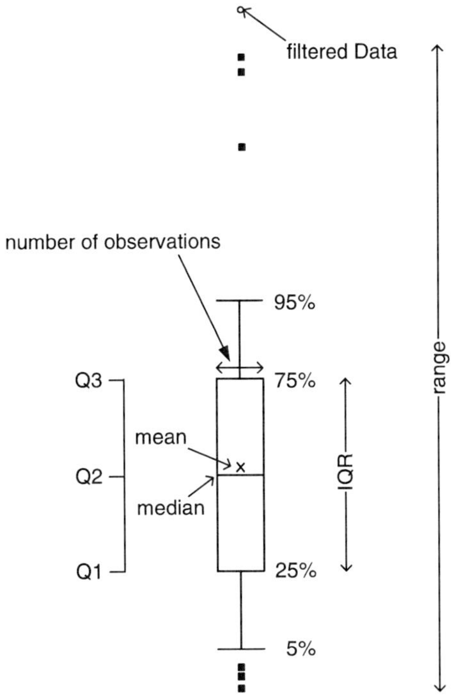

Fig. 10.11 Box plots for graphical illustration of percentile range, sample median, and quartiles

10.4.2 Calculated and Scaled Parameters

Scaling or normalization of DUT parameters to a physical dimension of the DUT is very useful when viewing and tracking data from DUTs of different design dimensions. Metal wire resistances are normalized to the wire length and MOSFET parameters such as I_{ds} and gate capacitances are normalized to device width. The circuit delay/stage, capacitance/stage, and IDDQ/stage are calculated from ring oscillator frequency and power measurements. Test structure design information such as wire lengths, device widths, and number of stages is entered in the database to facilitate such calculations.

Parameters may also be normalized to the nominal model or target value. All parameter values are then dimensionless and targeted to be centered at a value of 1.0. If target μ and σ are both known, the parameter values may be converted to equivalent z value using Eq. (10.4) to render all values to be dimensionless. The mean value should then be 0.0 with ± 3 denoting the "$\pm\, 3\sigma$" values.

Bivariate data of the type shown in Fig. 10.1a can be fitted to equations and the fitting parameters used for describing and comparing data samples or for extrapolation of data outside the measured range. Regression analysis provides a goodness of fit measure, expressed as coefficient of determination R^2 value, where $0 \leqslant R^2 \leqslant 1$.

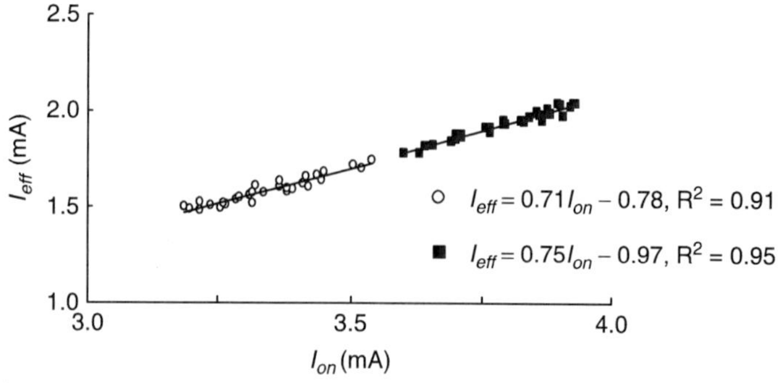

Fig. 10.12 Variation of I_{eff} with I_{on}, obtained with a linear fit to two different data samples

Values of $R^2 > 0.9$ indicated a reasonable correlation between the X and Y parameters, whereas $R^2 < 0.75$ generally indicates very weak or no correlation. An example I_{eff} vs. I_{on} plot on two wafer lots fabricated with different process recipes is shown in Fig. 10.12. A linear fit of the data gives the correlation coefficients and R^2 values. Each data sample is compressed into three parameters (slope, intercept, and R^2) given by the linear fit.

10.4.3 Summary Statistics

Summary statistics are computed for raw, calculated, or normalized parameters. The parameters $\bar{x}$ and s reduce a large number of measured or calculated values of a parameter to two numbers which succinctly describe statistical behavior. In the example shown in Fig. 10.10, C_{p} or C_{pk} values are computed to provide a measure of process centering. The mean C_{pk} and its distribution averaged over all critical parameters summarize the process control status in a manufacturing line.

10.5 Data Analysis Examples

The complexity of CMOS processing requires a high degree of specialization in each of the process modules and in product design, test, and failure analysis [13–18]. Standardization of data analysis and display formats is very helpful, especially when data are shared among many different engineering teams and managers, some of whom may not be familiar with the details of the test structures.

The data analysis examples in this section are specifically targeted toward test structure data from CMOS technology, although a number of visualization techniques described are of more general applicability. The planar nature of CMOS technology with sensitivity to spatial effects across silicon wafers leads to certain preferred methods of analysis and graphical illustrations.

An understanding of the interrelationships of properties of MOSFETs and parasitic elements in CMOS circuits and circuit delay parameters is useful in displaying the data. Target specifications for the parameters are important, but relative values and differential data analysis schemes can also provide insight into the electrical behavior of circuits.

A common requirement in sharing data is that the title, labels, and other text in graphical displays and parameter names as well as values and column headers in tables should provide sufficient detail to the viewer on the source of the data being examined. This information should also allow users to access other data stored elsewhere but related to the experiment under investigation. An example title

IBM4 65 nm M1_FET_A012_TS2 Lot ID: X Wafer ID: Y

indicates that the data are gathered in IBM's silicon fabrication facility #4 for 65 nm technology node on macro M1_FET_A012 placed on test vehicle TS2 on Wafer Y in Lot X. The sub-title

n-FET I_{off} vs. I_{eff} at 0.9 V 25°C $n = 50$, Aug 5, 2010

provides more detail on what is being plotted, test conditions, number of samples, and the date on which measurements were made. Summary statistics, fitting parameter coefficients, and other relevant information may also be included in the graphical display.

Graphics and tables for summarizing data in a hierarchical fashion are shown in Example 1. Graphical displays for showing and validating inter-relationships of parameters are covered in Example 2. Methods of correlating circuit properties in the scribe line with those on-product are described in Example 3. In Example 4, graphical summaries of MOSFET properties extracted from ring oscillator measurements are shown. Finally, in Example 5, graphical correlation of RO data with maximum operating frequency f_{max} of the product is demonstrated.

10.5.1 Example 1: Data Summary

The data collected in a silicon fab are summarized to get a quick hierarchical view of the technology status. A few critical parameters are selected to represent technology or product performance and their C_{pk} values are monitored over time. An example chart for tracking C_{pk} values on a weekly basis is shown in Fig. 10.13a. A box plot of C_{pk} provides a composite summary of how well the process is centered for all selected critical parameters. As discussed in Section 10.3.5.3, a C_{pk} value >1.0 indicates that a parameter is well within specification and parameters for which C_{pk} falls below 1.0 require a corrective action.

Circuit yield is an important consideration in setting parameter specifications, especially in the technology development cycle. Hence, circuit yield must be tracked

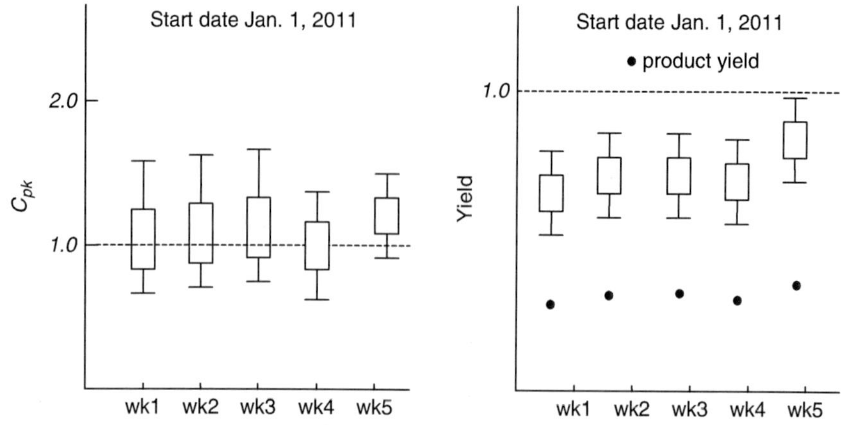

Fig. 10.13 High-level weekly status summary: **a** C_{pk} distribution of critical parameters and **b** yield distribution of selected macros and product yield (*solid circles*)

along with the C_{pk} values. A number of test structures are selected for monitoring yield during manufacturing, from small macros to those covering larger areas and, if available, yield of the product itself. In Fig. 10.13b, a box plot of the composite fractional yield of selected macros is shown for the same time frame and on the same wafers used for tracking C_{pk} values shown in Fig. 10.13a.

Graphical and quantitative views of parameter statistics of a number of parameters of interest are displayed in a tabular format in Fig. 10.14. A box plot of the type illustrated in Fig. 10.11 is rotated sideways for each of the six parameters shown. The table on the right of the box plot gives the number of samples measured along with the variation of sample mean ($\bar{x}$) and standard deviation (s) from their respective target values μ_t and σ_t. From the box plot, it is easy to identify the parameters centered away from the target μ_t or with values outside the $\pm 3\sigma_t$ limits. The target value of $(\bar{x} - \mu_t)/\sigma_t$ is 0.00 and that of s/σ_t is 1.00. As an example, the mean value

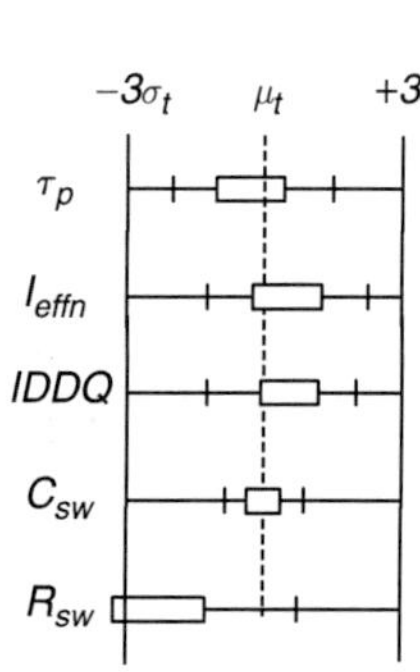

	n	$\frac{(\bar{x}-\mu_t)}{\sigma_t}$	$\frac{s}{\sigma_t}$	$\bar{x}$	s	μ_t	σ_t
τ_p	520	−0.33	0.90	3.90 ps	0.27 ps	4.00 ps	0.30 ps
I_{effn}	600	+0.50	0.90	1.24 mA	0.07 mA	1.20 mA	0.08 mA
IDDQ	560	+0.67	0.83	0.24 μA	0.05 μA	0.20 μA	0.06 μA
C_{sw}	520	0.00	0.50	2.11 fF	0.03 fF	2.11 fF	0.06 fF
R_{sw}	520	−2.30	1.60	1600 Ω	200 Ω	1887 Ω	125 Ω

Fig. 10.14 Box charts and tables providing the statistics on measured parameters. The box charts are normalized to target specifications of μ_t and σ_t

of τ_p is $-0.33\sigma_t$ below target and its standard deviation is 10% below target. The values of $\bar{x}$, s, μ_t, and σ_t are also recorded. The source of data is identified in the table heading (not shown). This format is useful for displaying statistics on many lots, many wafers, or a single wafer.

Spatial variations of parameters across wafers are of special interest in silicon technology. A wafer map of the type shown in Fig. 10.1c or a contour map generated from the measured data is helpful in visualization of this variability. When data are collected on many wafers, a stacked wafer map may be created for summarizing the findings in a single chart. The stacked wafer map shows the statistics at each macro or reticle field location on the wafer for all the wafers in the sample. In Fig. 10.15, a stacked wafer map of 500 wafers is color coded (gray scale) to show the mean f_{max} on reticle field locations on which measurements were made. The mean values and number of yielding chips are also displayed to get a quantitative view. In this example, at a glance, it is apparent that measurements are made on 14 out of 89 locations on the wafer. On an average, the mean f_{max} values are smaller in the top right part of the wafer and the yield in these locations is lower than in the bottom left corner.

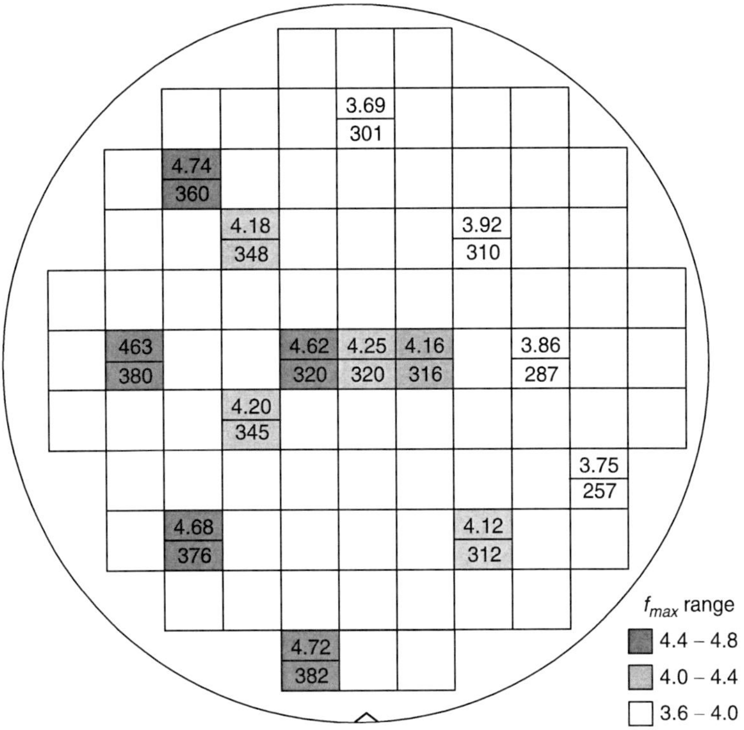

Fig. 10.15 A stacked wafer map of 500 wafers, color coded (*gray scale*) to indicate the range of product chip f_{max}, average f_{max} values and number of yielding chips in measured chip locations

10.5.2 Example 2: Circuit Element Characterization

The relationship between two or more measured or calculated parameters is shown in an *XY* scatter plot also known as a scattergraph. Two example *XY* scatter plots for a technology power-performance metric are shown in Fig. 10.16. In Fig. 10.16a, I_{eff} values for n-FETs and p-FETs are plotted vs. their respective I_{off}, each point representing measurements on one DUT or average values for nominally identical DUTs in a macro. In high-performance CMOS technology, I_{eff} at a fixed I_{off} (= 100 nA/μm) is compared between different silicon foundries or technology nodes [19]. In Fig. 10.16b, inverter delay τ_{p} derived from a ring oscillator test structure is plotted vs. its IDDQ per unit MOSFET width, where the total MOSFET width is $\left(W_{\mathrm{p}} + W_{\mathrm{n}}\right)$. As I_{off} contributions of p-FET and n-FET are from alternate stages in the RO (Section 6.1.1), τ_{p} is compared at IDDQ = 50 nA/μm. The value of I_{off} or IDDQ at which MOSFET current drive or inverter delay is compared is arbitrary and may be different for low-power than for high-performance applications.

In the presence of random variability such as RDF, it is important to consider the total device width of the DUT. Since the variability in MOSFET parameters increases as the width is reduced (Eq. (5.8)), the spread in the data is larger for a single MOSFET compared to an average of 30 MOSFETs as shown in Fig. 10.17. In using an I_{eff} vs. I_{off} plot as a technology metric, an average value of I_{off} over a large number of MOSFETs provides a better representation of product IDDQ.

The data in *XY* scatter plots may be fitted using a known physical relationship or an empirical fit may be made as discussed in Section 10.4.2. Critical MOSFET parameters such as V_{t} and I_{eff} are plotted as a function of L_{p} to assess the impact of channel length variation. These parameters are also plotted against device widths as well as V_{DD} and temperature over the expected application range of products and compared with model predictions.

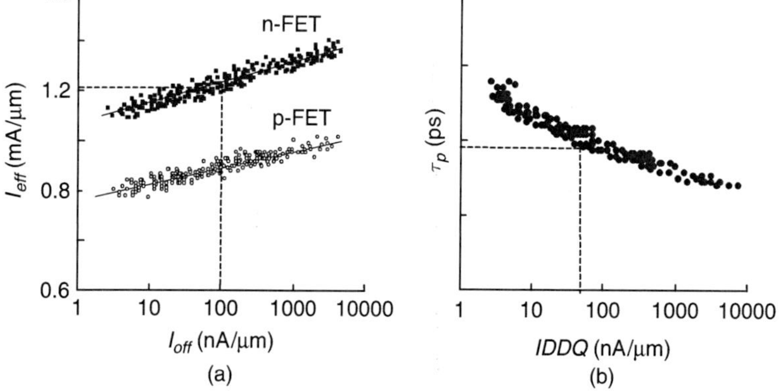

Fig. 10.16 **a** I_{eff} as a function of I_{off} of a p-FET and an n-FET. **b** Inverter delay as a function of IDDQ per unit device width

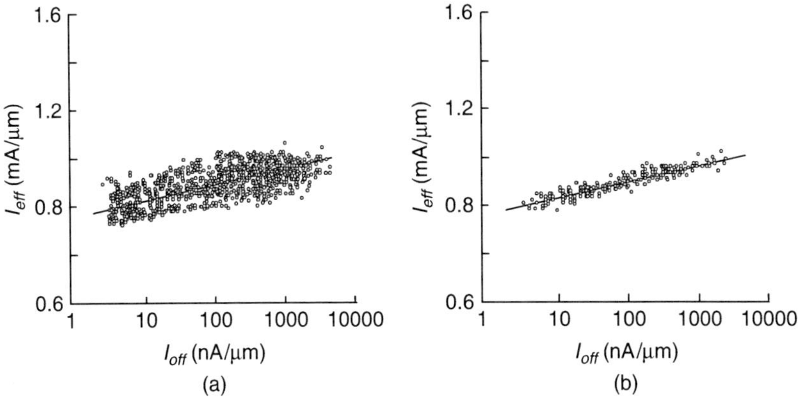

Fig. 10.17 I_{eff} as a function of I_{off} where each point represents **a** a single standard MOSFET and **b** an average of 30 standard MOSFETs

Generally, two to four standard *XY* scatter plots may be placed on a single page without losing clarity of the data being presented. In some cases, a simplified version of an XY plot is used to place 20 or more charts on a single page. The data are normalized to the target specifications or to a reference value and plotted as shown in Fig. 10.18a, b. The identity line (dashed) represents a 1:1 correspondence and the solid circle gives the location of the (1, 1) point when both parameters are at their normalized values. A linear fit of the data is shown as a solid dark line and the correlation coefficient (R^2) indicates the goodness of the fit. If *X* and *Y* parameters have a 1:1 correspondence, data will fall exactly on the identity line. Any deviation

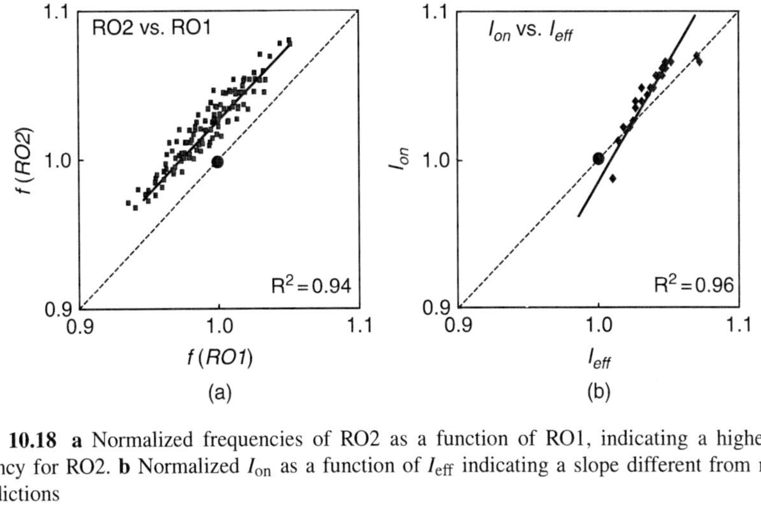

Fig. 10.18 **a** Normalized frequencies of RO2 as a function of RO1, indicating a higher frequency for RO2. **b** Normalized I_{on} as a function of I_{eff} indicating a slope different from model predictions

from the identity line such as an offset or a change in slope is clearly visible. This method of presentation for a large amount of data is demonstrated in Examples 3, 4, and 5.

10.5.3 Example 3: Scribe Line to On-Product Correlation

In a silicon foundry, product designs supplied by customers occupy most of the area in a reticle field. These product designs are carried out using physical ground rules (GRs) and circuit models supplied by the foundry. During silicon wafer fabrication, a common set of electrical test structures placed in the scribe lines of all the wafers in a given technology node are measured for monitoring the process centering and yield. The circuit topologies, local circuit densities, and physical layout of DUTs in the scribe line are generally different than those on the product. In addition, the test equipment and measurement methods used for scribe-line tests in the manufacturing environment are also different than those used in the product test. This may result in differences in measurement accuracy for data from the two sources. Such differences vary in importance from one product to another as products with smaller design margins to meet specifications require a tighter process control.

Although the foundry may have perfectly tailored the process to match the circuit models or tuned it to a specific product request, there is always a question of how well the test structures on the scribe line represent a product design. A comparison of data collected from test structures placed on the scribe line to that obtained from product test must be presented in a way that helps to answer that question in an unambiguous fashion.

It is a good practice to place identical test structure designs on the scribe line and across the product itself whenever possible. Ring oscillators serve this purpose well as frequencies can be easily measured during product functional test. Circuit delays, derived from ring oscillator frequency measurements, provide information on expected maximum product operating frequency. By appropriate design of ROs, information on MOSFET and parasitic circuit element parameters can be extracted as discussed in Section 6.3. Circuit delays, averaged over many logic gates in an RO, minimize the impact of random variations in circuit parameters and are useful for tracking systematic variations across chip, across wafer, and from wafer to wafer and lot to lot. When the RO designs in the scribe line and on-product cannot be truly identical, the same logic gate type such as an inverter ($FO = 3$) may be used.

Here, we show examples of data displays for comparing ring oscillator data collected in the manufacturing line with that from on-product tests [21]. RO frequencies for the on-product and on-scribe-line designs are normalized to target values derived from circuit simulations. Example physical locations of nominally identical ROs on the reticle field are shown in Fig. 10.19a. In this case, there are nine ROs on the product and one each on the scribe lines on the left and right side of the product. The measured RO frequencies are normalized to the target value and plotted vs. that of RO1, which is situated on the top left corner of the product. In Fig. 10.19b

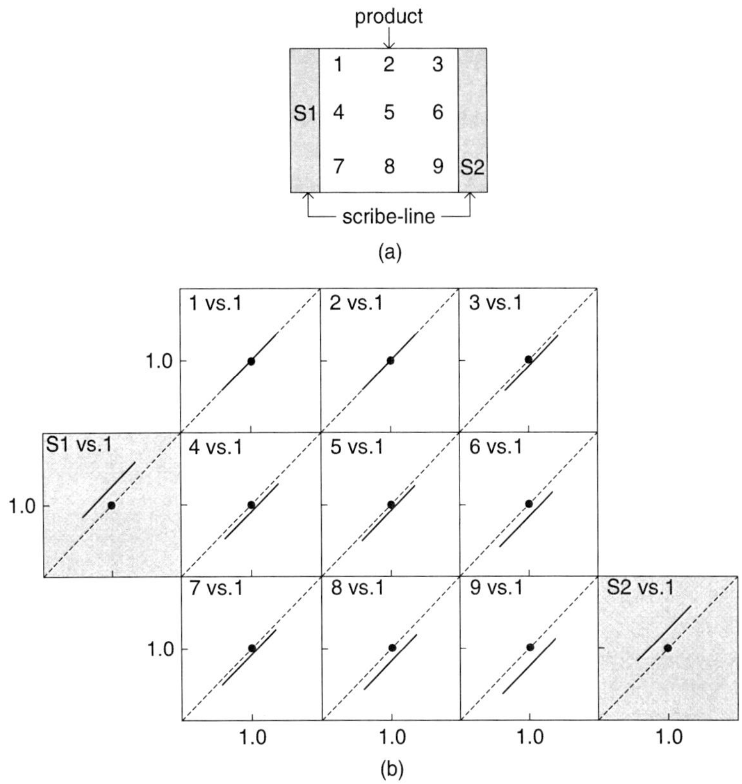

Fig. 10.19 **a** Physical locations of ROs (1–9) embedded within a product and in the scribe line (S1 and S2). **b** Normalized RO frequencies of each RO as a function of RO1 in the same physical arrangement as in the reticle field

the XY plots are placed corresponding to the physical locations of the ROs in the reticle field. As previously described, the dashed lines on each plot indicate 1:1 correspondence of RO frequencies. The solid line is fitted to the measured data or alternatively, individual data points may be shown. The solid circle shows the (1, 1) location, where both RO frequencies match the nominal specifications. Similar charts may be produced by plotting the RO frequencies vs. the mean RO frequency on the product or vs. an RO in the scribe line. If model targets are not available, raw data may be plotted instead.

The data fall exactly on the dashed line on the RO1 vs. RO1 plot in Fig. 10.19b, by definition. The data fall below the dashed line moving to the lower right corner, indicating a systematic across-chip variation, with a reduction in frequency in going from top left to lower right corner of the product. The ROs in the scribe line are shown to have higher frequencies than RO1 on both sides of the product. In this case, a product frequency prediction based on the measurements made in the manufacturing line would be more optimistic.

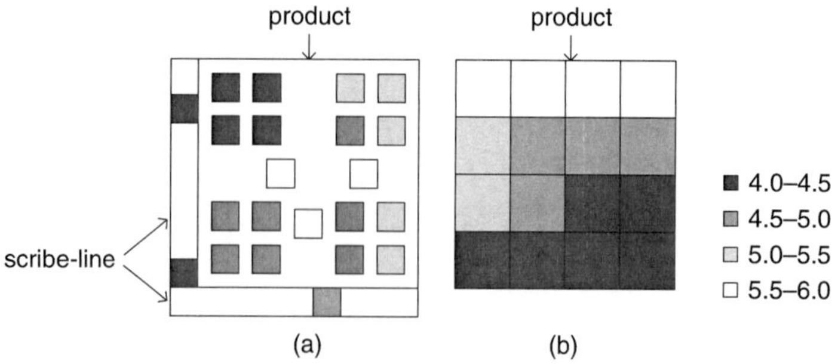

Fig. 10.20 Stacked reticle field maps of frequencies **a** of ROs distributed within product and on scribe lines and **b** of ROs placed on a regular grid on-product

The across reticle field variation may be displayed as a stacked reticle field map for a large number of wafers as shown in Fig. 10.20. In Fig. 10.20a, the color-coded (gray scale) boxes representing RO frequency ranges are centered on the exact locations of the ROs on the product and the scribe lines. If the ROs are on a regular grid on-product, the boxes may be expanded to fill the space as shown in Fig. 10.20b. Different stacked maps may be created for different wafer zones.

A full wafer map showing the frequency ranges of 16 nominally identical ROs across a single product chip located in each reticle field is shown in Fig. 10.21.

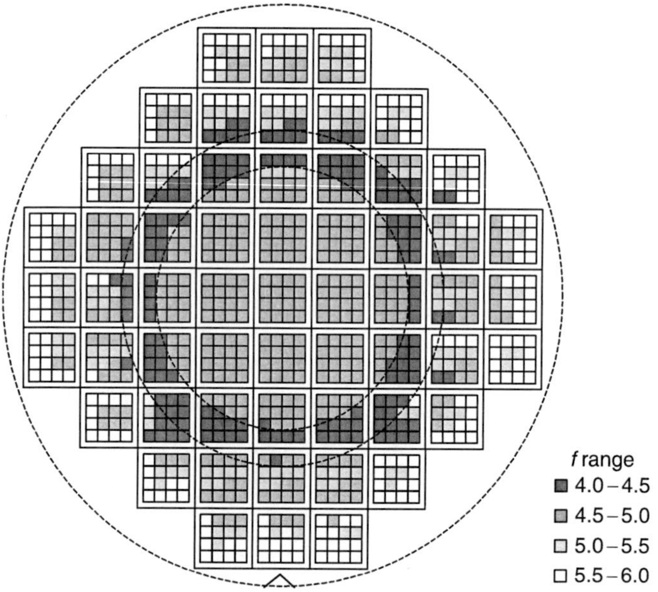

Fig. 10.21 Wafer map showing RO frequency ranges within each reticle field across a wafer, highlighting differences in wafer zones

In the center of the wafer, the RO frequencies across the product are nearly the same. However, there is a systematic radial variation across wafer. Across product variations arising from across wafer variations are clearly visible in the chips away from the center. The scribe-line to on-product correlation may also be different for chips from different zones on the wafers. In this example, product performance may be related to the location of product chips on the wafer as described in more detail in Example 5.

A major advantage of presenting data in this fashion is that measured data are plotted with minimum manipulation. The information is easy to assimilate even for those not familiar with details of test structure design and measurement. Conclusions on differences between scribe-line and on-product data and across chip variation can be drawn and easily agreed upon.

10.5.4 Example 4: Correlation of ROs to Circuit Elements

RO designs for tracking of MOSFET and parasitic parameters are described in Section 6.3. As generally only frequency measurements are possible for RO designs placed on-product, absolute R_{sw} and C_{sw} values of logic gates cannot be extracted. However, the relative values of parameters of MOSFETs and other circuit elements may be tracked by comparing measured RO frequencies for different RO stage designs with a reference (inverter) RO design [20]. Data are displayed for all the ROs on a single page to provide a clear and concise evaluation.

The RO frequencies are normalized to nominal specifications for each RO design. In Fig. 10.22, normalized RO frequencies of a variety of RO designs are plotted against a reference RO comprising a loaded (FO $\sim$3) inverter stage. This loaded inverter stage is not as sensitive to small differences in layouts and parasitic resistances and capacitances as an unloaded (FO = 1) inverter stage. As many as 20 or more such plots can be accommodated in a single page.

A quick view of the plots in Fig. 10.22 indicates that except for Inv(V_{t3}) and Inv(L_{P3}) all other inverter RO frequencies are on target. The NAND and NP (n-passgate) loaded inverters are slower than target, and NOR and PP (p-passgate) loaded inverters are faster than targets. Likely causes of these deviations from targets in NANDs, NORs, and passgates are a stronger p-FET and a weaker n-FET. The inverter with an M2 wire load is slower than target, indicating a problem with M2 layer delineation.

A similar approach is used for displaying data for different process recipes. In Fig. 10.23, measured RO frequencies, normalized to the target values, are plotted against reference RO frequency for two different process recipes, process_1 and process_2. For the majority of ROs, there is no difference in the two process recipes and the fitted lines to the data are coincident. ROs for inverter designs L02, L10, and L15 have a higher than expected frequency for process_2 and only these RO designs need to be investigated. Such graphical displays are useful for RO designs with GR differences and for investigating the effects of layout sensitivities to process changes.

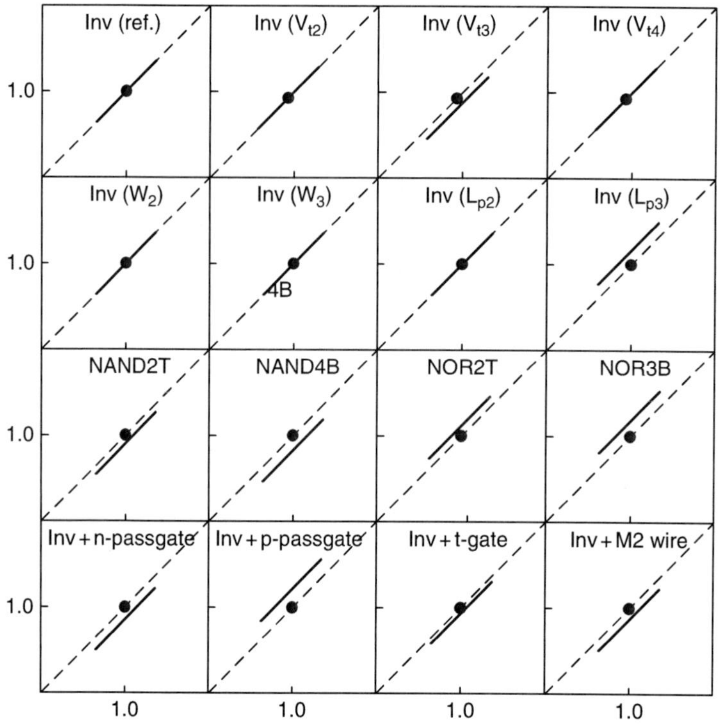

Fig. 10.22 Measured RO frequencies, normalized to target values, of different RO stage designs as a function of a reference RO (inverter stage)

10.5.5 Example 5: Correlation of ROs to Product

In this last example, we show how to link data collected on test structures to product functional data and ultimately how to connect technology process information to product performance [21]. This method of product debug is less complex than detailed analysis used for identification of cycle time limiting paths.

A number of identical ring oscillators (inverter FO $= 3$) are distributed across the product as shown in Fig. 10.24a. In Fig. 10.24b, a stacked wafer map of RO frequencies across the product is shown. The product chips inside the dotted circle have similar across product variation, with ROs at the center top, 02 and 03, exhibiting a higher frequency. The product chips near the wafer edges have higher RO frequencies in locations closest to the outside edge. By correlating the measured frequency f of different ROs to the maximum frequency of operation of the product f_{max} we can gain an insight on the area of the product most critical to its performance. Such information may not be readily available otherwise as timing tools used in product design assume a uniform model across the product.

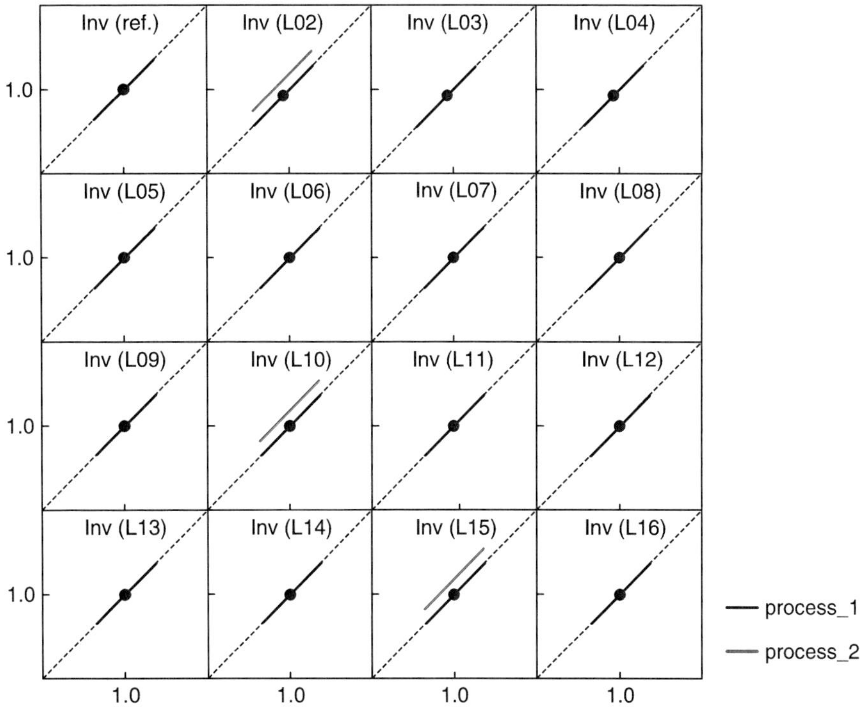

Fig. 10.23 Normalized frequencies of inverter ROs of different physical layouts as a function of normalized frequency of a reference RO for two different process recipes

Correlations of the frequencies of the distributed ROs with f_{max} are shown in Fig. 10.25. The correlation factor for each plot is computed and printed at the bottom right corner of each plot. Here, RO-13, at the bottom left corner of the product, has the highest value of R^2 and hence the best correlation with f_{max}. From the wafer map in Fig. 10.24b, this is also the slow corner in the reticle field in most chips. An improvement in f_{max} may be obtained by tuning the PS level lithography exposure and trimming the channel length in the area in the vicinity of RO-13.

Another technique for linking RO frequencies to product f_{max} makes use of the fact that in CMOS circuits, signal propagation delay sensitivity to V_{DD} varies with circuit topology. The V_{DD} dependence of ROs comprising different logic gate types may be used to determine the circuit composition in frequency limiting paths of the product [21]. Frequencies of a set of ROs of various circuit topologies and product f_{max} are measured at two or more V_{DD} values. The df/dV_{DD} values of the ROs vs. f and df_{max}/dV_{DD} vs. f_{max} are plotted as shown in Fig. 10.26 and compared. The stage design of the RO exhibiting the best match with the product f_{max} data is likely to represent the circuit topology in the frequency limiting paths. In this example, the behavior of the RO-01 circuit is closest to that of the product characteristics. The

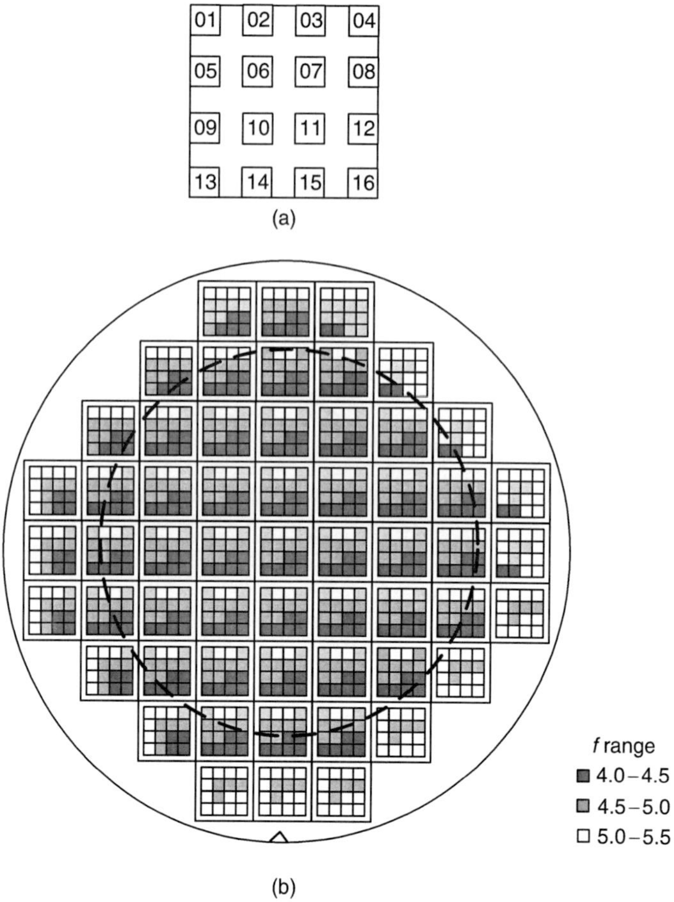

Fig. 10.24 **a** Locations of 16 ROs on-product. **b** Wafer map showing variation in RO frequencies across reticle field

information on physical region and circuit composition of the frequency limiting path obtained from ROs should be further verified by localizing and examining the product design and model predictions.

On-product test structures continue to play a role in monitoring product performance throughout the life of the product and can be used to debug hardware-related product fails reported by customers. Circuit performance degradation with time from BTI and Hot-e effects, generation of defects from electromigration or other aging mechanisms, and product application conditions outside the range of characterization prior to product shipment are some of the possible causes of product fails in the field. Data collected from on-product test structures along with an appropriate graphical display of the data can facilitate the identification of the root cause of such failures.

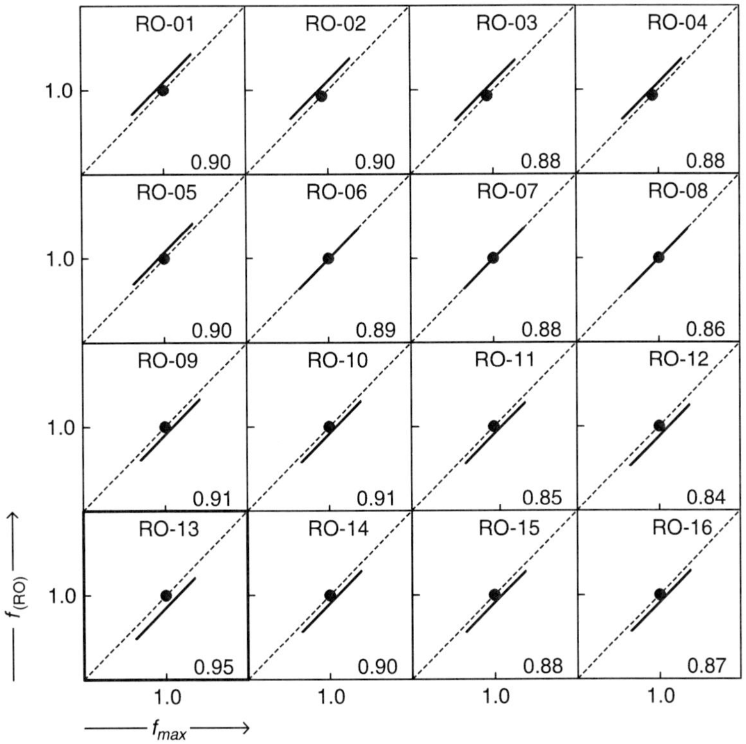

Fig. 10.25 RO frequencies as a function of product f_{max}. The correlation factor is displayed at the bottom right corner of each plot. RO-13 has the highest correlation factor with f_{max}

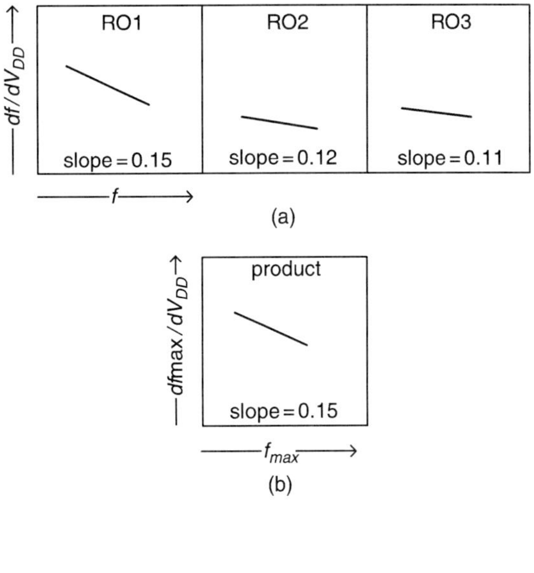

Fig. 10.26 **a** df/dV_{DD} as a function of f of three ROs of different circuit topologies. **b** Product df_{max}/dV_{DD} as a function of f_{max}

References

1. Box GEP, Hunter WG, Hunter JS (1978) Statistics for experimenters: An introduction to design data analysis and model building. Wiley, New York, NY
2. Burr JT (2005) Elementary statistical quality control, 2nd edn. Marcel Dekker, New York, NY.
3. Montgomery DC (2001) Design and analysis of experiments, 5th edn. Wiley, New York, NY
4. Koronacki J, Thomson JR (2001) Statistical process control: the Deming paradigm and beyond, 2nd edn. Chapman and Hall, Boca Raton, FL
5. Joglekar A (2001) Statistical methods for six sigma in R and D and manufacturing. Wiley Interscience, Hoboken, NJ
6. Pande PS, Neuman RP, Cavanagh RR (2000) The six sigma way. McGraw-Hill, New York, NY
7. Tufte E (1983) The visual display of quantitative information. Graphics, Cheshire
8. Tufte E (1997) Visual explanations. Graphics, Cheshire
9. Tufte E (1990) Envisioning information. Graphics, Cheshire
10. Delwiche LD, Slaughter SJ (1998) The little SAS book. SAS Institute, Cary, NC. http://www.sas.com/. Accessed 15 Mar 2011
11. JMP. http://www.jmp.com/. Accessed 15 Mar 2011
12. R. http://www.r-project.org/. Accessed 15 Mar 2011
13. Hannaman DJ, Sayah HR, Allen RA, Buehler MG, Yung M (1990) Fault chip defect characterization for wafer scale integration. Proceedings of the 1990 IEEE international conference on microelectronic test structures, 1990, pp 67–71
14. Montgomery DC (2009) Introduction to statistical quality control. Wiley, New York, NY
15. Ohkawa S, Aoki M, Masuda H (2004) Analysis and characterization of device variations in an LSI chip using an integrated device matrix array. IEEE Trans Semicond Manuf 17:155–165
16. Maxwell P (2006) The design, implementation and analysis of test experiments. Proceedings IEEE international test conference, ITC'06, pp 1–9
17. Pang L-T, Qian K, Spanos CJ, Nikolic B (2009) Measurement and analysis of variability in 45 nm strained-Si CMOS technology. IEEE J Solid-State Circuits 44:2233–2243
18. May GS (2006) Fundamentals of semiconductor manufacturing and process control. Wiley, Hoboken, NJ
19. Tyagi S, Auth C, Bai P, Curello G, Deshpande H, Gannavaram S et al (2005) An advanced low power, high performance, strained channel 65 nm technology. IEDM technical digest, pp 245–247
20. Gattiker A, Bhushan M, Ketchen MB (2006) Data analysis techniques for CMOS technology characterization and product impact assessment. Proceedings of the international test conference, ITC'06, pp 1–10
21. Bhushan M, Gattiker A, Ketchen MB, Das KK (2006) Ring oscillators for CMOS process tuning and variability control. IEEE Trans Semicond Manuf 19:10–18

Appendix A
Standard Physical Layouts and Parameters Used in the Book

Physical ground rules (GRs) and properties of circuit elements used throughout this book are in the range applicable to the 65–45 nm technology nodes. The values are selected for simplifying calculations but do not subscribe to any specific technology node or manufacturing facility.

A.1 Key Physical Layers in CMOS Circuits

The physical cross section of a silicon CMOS circuit with four metal interconnect layers is shown in Fig. A.1. The gates of the n-FET and the p-FET are defined by the PS layer. This PS layer and the n^+ and p^+ regions, capped with a low-resistivity silicided DF layer, are contacted to M1 metal through H0 vias. Metal interconnect layers are denoted by MX ($X = 1, 2, 3,$ and 4) and inter-level vias by HX ($X = 1, 2,$ and 3).

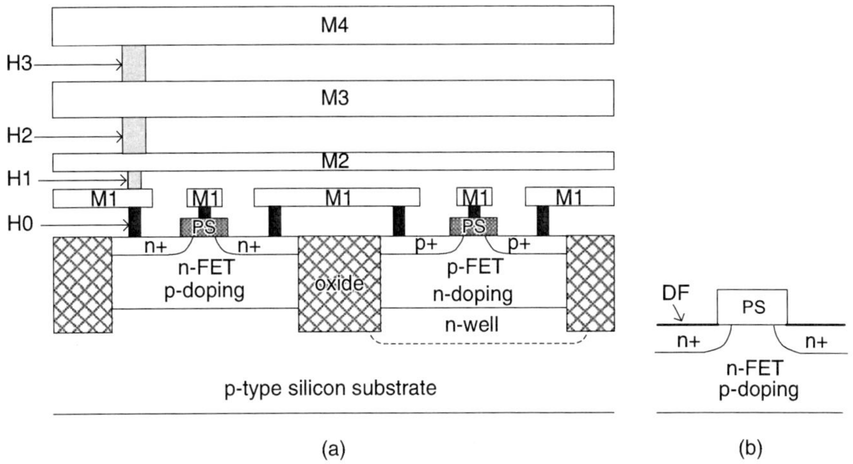

Fig. A.1 **a** Cross section of a CMOS circuit with four metal interconnect layers. **b** n-FET cross section showing silicided PS and DF layers

M. Bhushan, M.B. Ketchen, *Microelectronic Test Structures for CMOS Technology*,
DOI 10.1007/978-1-4419-9377-9, © Springer Science+Business Media, LLC 2011

A.2 MOSFET Parameters

The physical layout dimensions and MOSFET properties, normalized to a width of 1 μm, are listed in Table A.1.

Table A.1 Physical and electrical properties of MOSFETs

Component or parameter	Dimensions (μm)	Properties
MOSFET gate pitch	0.2	Fixed pitch
MOSFET gate length, L_p	0.04	–
MOSFET parameters at 1.0 V	$W = 1.0$	$I_{off} = 100$ nA/μm $V_{tsat} = 0.2$ V $I_{gl} = 1.0$ nA/μm $C_{gT} = 1.0$ fF/ μm
p-FET	$W_p = 1.0$	$I_{on} = 0.8$ mA/μm
n-FET	$W_n = 1.0$	$I_{on} = 1.2$ mA/μm

A.3 Standard Inverter and Circuit Parameters

The layout of a standard inverter is shown in Fig. A.2. The p-FET and the n-FET in the inverter each have two fingers with shared drain terminals. The width of each finger in the p-FET is 0.6 μm ($W_p = 1.2$ μm) and that in the n-FET is 0.4 μm ($W_n = 0.8$ μm). The PS pitch is fixed at 0.2 μm. The inverter height in the vertical direction is 2.0 μm. Physical layout dimensions of an inverter and key parameters used in circuit designs are listed in Table A.2.

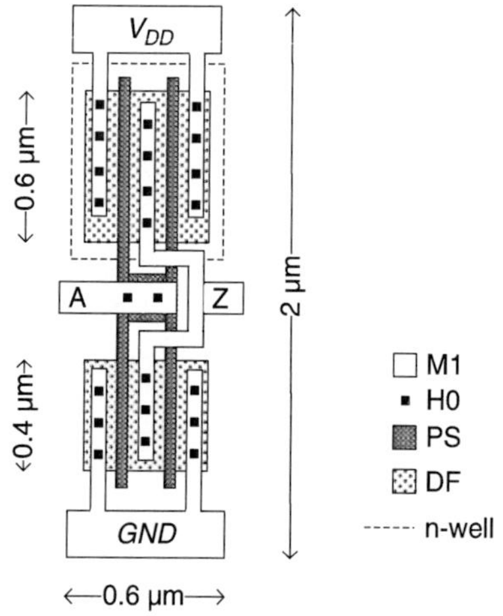

Fig. A.2 Physical layout of a two-finger inverter and layer mapping key

Table A.2 Physical dimensions of a standard inverter and key circuit parameters

Component or parameter	Dimensions (μm)	Properties
Standard inverter (two PS fingers) $W = (W_p + W_n) = 2.0\,\mu m$	$W_p = 1.2$, $W_n = 0.8$	$R_{sw} = 2000\ \Omega\ \mu m$ at 1.0 V $C_{in} = 1.0$ fF/μm $C_{out} = 1.0$ fF/μm $C_{sw} = 4.0$ fF
Logic gate height	2.0	–
RO circuit stage width	3.0	15 PS fingers/stage
Inverter beta for $\tau_{pu} = \tau_{pd}$	–	$W_p/W_n = 1.5$
Inverter delay (FO = 1)	–	$\tau_p = 4.0$ ps at 1.0 V
Inverter delay (FO = 3)	–	$\tau_p = 8.0$ ps at 1.0 V
Pulse rise and fall times, τ_r ,τ_f (FO = 3)	–	20.0 ps
Decoupling capacitor (DECAP)	–	5.0 fF/μm^2

A.4 Properties of Conducting Layers

Physical dimensions, electrical properties and allowed maximum pattern densities of silicided silicon diffusion (DF) and polysilicon (PS) layers, and interconnect metal layers (M1 to M4) are listed in Table A.3.

Table A.3 Physical dimensions and electrical properties of conducing layers

Component or parameter	Dimensions (μm)	Value
PS (polysilicon gate) layer	–	$\rho_{sh} = 10\ \Omega/\square$
DF (silicon diffusion) layer	–	$\rho_{sh} = 10\ \Omega/\square$
M1 and M2 metal layers	–	$\rho_{sh} = 0.20\ \Omega/\square$
M3 and M4 metal layers	–	$\rho_{sh} = 0.10\ \Omega/\square$
M1 and M2 width/space	0.1/0.1	$R_w = 2.00\ \Omega/\mu m$ $C_w = 0.20$ fF/μm $L_w = 0.5$ pH/μm
M3 and M4 width/space	0.2/0.2	$R_w = 0.50\ \Omega/\mu m$ $C_w = 0.20$ fF/μm
Metal area coverage	–	<80%
PS area coverage	–	<50%

A.5 Standard 1×25 Padset Design

Electrical test structures described in the book can be designed with any geometrical arrangement of I/O pads. However, a standard I/O pad configuration is highly desirable for sharing macro templates and probe cards in different stages of technology development as well as among different technologies being processed in the same fabrication facility. The standard I/O pad configuration, used in many of the examples in this book, is a 1×25 linear array as shown in Fig. A.3. A summary of the properties of this padset and of the I/O pads themselves is given in Table A.4. The form factor of the macros designed with this padset is suitable for placement in the scribe line. The number of pads per probe touchdown can be increased to 50 (1×50 padset) by butting two 1×25 padset macros.

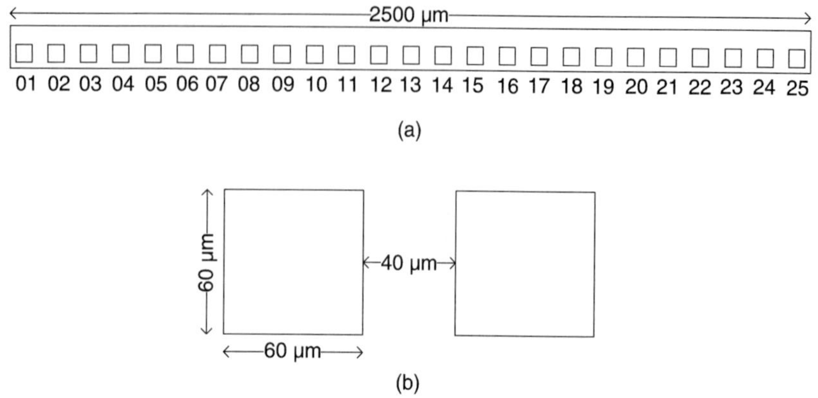

Fig. A.3 **a** I/O pad arrangement in a 1×25 padset macro. **b** Pad placement and dimensions

Table A.4 Properties of standard 1×25 padset used in macro templates

Property	Value
Macro width	2500 μm
Macro height	≥80 μm
Number of I/O pads (linear array)	25
Pad pitch	100 μm
Pad height	60 μm
Pad width	60 μm
Pad space	40 μm
Probe to pad contact resistance	1.0 Ω

Glossary of Symbols

Symbol	Description	Unit
A	Area	cm^2
α	Number of stages in a circuit	None
α_c	Clustering parameter	None
$\boldsymbol{B}$	Magnetic field vector	Gauss
β	MOSFET gain factor	$(\Omega\ V)^{-1}$
c	Speed of light in vacuum (3×10^{10} cm/s)	cm/s
C	Capacitance	F
C_{ch}	MOSFET channel capacitance	F
C_d	Depletion layer capacitance	F
C_{db}	MOSFET drain-to-body capacitance	F
C_{do}	MOSFET gate oxide capacitance in the overlap region	F
C_{down}	Inter-level wire capacitance to layer below	F
C_g	Gate-to-substrate capacitance of a MOS capacitor	F
C_{gb}	MOSFET gate-to-body (substrate) capacitance	F
C_{gs}	Gate-to-source capacitance	F
C_{gT}	MOSFET total gate capacitance	F
C_i	Inversion layer capacitance	F
C_{if}	MOSFET inner fringe capacitance	F
C_{in}	Equivalent input capacitance of a logic gate	F/cm
C_j	Source (drain)-to-body junction capacitance	F
C_{left}	Wire capacitance to adjacent wire to the left	F
C_L	Load capacitance	F
C_{of}	MOSFET outer fringe capacitance	F
C_{out}	Equivalent output capacitance of a logic gate	F/cm
C_{ov}	MOSFET overlap capacitance	F
C_{ox}	Oxide capacitance per unit area	F
C_p	Parasitic interconnect capacitance	F
C_p	Six Sigma index for process control	None
C_{pk}	Six Sigma index for process control	None
C_{pm}	Measured capacitance in parallel mode	F
C_{right}	Wire capacitance to adjacent wire to the right	F
C_{sb}	MOSFET source-to-body capacitance	F
C_{sm}	Measured capacitance in series mode	F
C_{sw}	Switching capacitance of a logic gate	F
C_{up}	Inter-level wire capacitance to layer above	F
C_w	Wire capacitance per unit length	F/cm

Symbol	Description	Unit
d	Film thickness	cm
d_{cl}	Duty cycle	None
D_f	Dissipation factor in impedance element	None
DD	Defect density	cm^{-2}
δ	Accuracy of σ	Variable
δ_m	Measurement error	None
δt	Time difference in signal delays	s
ΔL	Channel length bias	cm
ε	Permittivity	None
ε_{eff}	Effective permittivity	None
ε_o	Vacuum permittivity (=8.85×10^{-14} F/cm)	F/cm
f	Frequency of oscillation	Hz
f_{max}	Maximum frequency of oscillation	Hz
FO	Fan out	None
g	Inductance correction factor	None
GND	Ground potential	V
γ	Area efficiency of a defect monitor	None
γ	MOSFET body-effect coefficient	None
h	Dielectric thickness	cm
H	History effect	%
H_t	History effect for 1SW–2SW transitions	%
H_{tpd}	PD history effect for 1SW–2SW transitions	%
H_{tpu}	PU history effect for 1SW–2SW transitions	%
I	Current	A
I_{DD}	Power supply current	A
IDDQ	Quiescent current of a circuit	A
IDDA	Active current of a circuit	A
I_{dlin}	MOSFET drain-to-source linear current	A
I_{ds}	MOSFET drain-to-source current	A
I_{dsat}	MOSFET drain-to-source saturation current	A
I_{eff}	MOSFET effective current = $(I_{hi} + I_{lo})/2$	A
I_f	Forced current	A
I_g	MOSFET gate current	A
I_{gl}	MOSFET gate oxide leakage current	A
I_{hi}	MOSFET I_{ds} at $V_{ds} = V_{DD}/2$, $V_{gs} = V_{DD}$	A
I_{lo}	MOSFET I_{ds} at $V_{ds} = V_{DD}$, $V_{gs} = V_{DD}/2$	A
I_m	Measured current	A
I_{off}	MOSFET drain-to-source leakage current	A
I_{offn}	n-FET drain-to-source leakage current	A
I_{offp}	p-FET drain-to-source leakage current	A
I_{on}	MOSFET on current	A
k	Dielectric constant relative to SiO_2	None
K	One element in a set of N	None
K	Thermal conductivity	W/cm/°C
κ	Cell (bin) size in a histogram and in history macro	Variable
l	Wire length	cm
L	Inductance	H
L_{eff}	MOSFET effective channel length	cm
L_{ov}	MOSFET gate overlap length	cm
L_p	MOSFET gate length	cm
L_w	Inductance per unit length	H/cm

Symbol	Description	Unit
m	Number of elements	None
μ	Sample mean	Variable
μ_{eff}	Carrier mobility	cm^2/V s
μ_l	Lower filter multiplier	None
μ_o	Permeability of free space	H/cm
μ_t	Target sample mean	Variable
μ_u	Upper filter multiplier	None
μ_w	Magnetic permeability of a wire	H/cm
n	Number of data points or samples	None
n_{sq}	Number of squares in a film	None
N	Number of elements	None
N_c	Number of elements in a matrix column	None
N_r	Number of elements in a matrix row	None
η	Divide-by-factor	None
$p(x)$	Probability density of value x	None
pc	Pre-switch state of PD-SOI gate	None
P	Power dissipation	W
P_{ac}	Active power dissipation	W
P_d	Power density	W/cm^2
P_{max}	Maximum power dissipation	W
P_{off}	Standby power dissipation	W
P_{sc}	Short-circuit power	W
Φ	Magnetic flux	Wb
Q_c	Charge transfer during switching of a CMOS circuit	C
Q1	First quartile in descriptive statistics	Variable
Q2	Second quartile in descriptive statistics	Variable
Q3	Third quartile in descriptive statistics	Variable
χ	I_{ds} ratio in a set of parallel MOSFETs with V_C and V_g	None
r_{sw}	Switching resistance of a logic gate	Ω
R	Resistance	Ω
R_m	Measured resistance	Ω
R_p	Parasitic probe resistance	Ω
R_p	Parasitic parallel resistance of a capacitor	Ω
R_{pm}	Measured parallel resistance of a capacitor	Ω
R_s	Parasitic series resistance	Ω
R_{sd}	Source–drain series resistance	Ω
R_{sm}	Measured series resistance of a capacitor	Ω
R_{sw}	Specific switching resistance of a logic gate	Ω cm
R_{th}	Thermal resistance	°C/W
R_w	Wire resistance per unit length	Ω/cm
ρ	Resistivity	Ω cm
ρ_{sh}	Sheet resistance	Ω/□
s	Sample variance	Variable
s	Spacing between wires in the same metal layer	cm
S	Primary CMOS technology scaling factor	None
S_κ	Secondary CMOS technology scaling factor	None
SS	MOSFET subthreshold slope	V/decade
σ	Standard deviation	Variable
σ_t	Target standard deviation	Variable
σV_t	Standard deviation of V_t	V
t	Time	s

Symbol	Description	Unit
t_{eq}	Dielectric thickness equivalent to SiO_2	cm
t_{ox}	Oxide thickness	cm
T	Temperature	°C
T_c	Clock period	s
TCR	Temperature coefficient of resistance	Ω/°C
T_d	Time delay	s
T_p	Period of oscillation	s
T_s	Setup time of a latch	s
T_w	Pulse width	s
τ	Delay of a logic gate	s
τ_{1pd}	1SW PD delay of a logic gate	s
τ_{1pu}	1SW PU delay of a logic gate	s
τ_{2pd}	2SW PD delay of a logic gate	s
τ_{2pu}	2SW PU delay of a logic gate	s
τ_f	Signal fall time	s
τ_l	Delay of a latch	s
τ_p	Average of pull-down (PD) and pull-up (PU) delays	s
τ_{pd}	Pull-down (PD) delay	s
τ_{pm}	Measured value of τ_p	s
τ_{pu}	Pull-up (PU) delay	s
τ_r	Signal rise time	s
v	Speed of light	cm/s
V	Voltage	V
V_C	Clamp voltage	V
V_{ds}	MOSFET drain-to-source voltage	V
V_{DD}	Power supply voltage	V
V_{DDC}	Common power supply voltage	V
V_{DDE}	Experiment power supply voltage	V
V_f	Forced voltage	V
V_g	MOS capacitor gate voltage	V
V_{in}	Voltage of an input circuit node	V
V_m	Measured voltage	V
V_{out}	Voltage of an output circuit node	V
V_t	MOSFET threshold voltage	V
V_{tlin}	MOSFET threshold voltage in linear mode	V
V_{tsat}	MOSFET threshold voltage in saturation mode	V
w	Wire width	cm
W	MOSFET width	cm
W	Logic gate width $(= W_p + W_n)$	cm
W_n	n-FET width	cm
W_p	p-FET width	cm
ω	Angular frequency	Radians/s
x	Observed values in a data sample	Variable
y	Transformed variable	Variable
Y	Yield	None
z	Transformed variable	None
Z	Impedance	Ω

Acronym

ATE	Automated test equipment
BTI	Bias temperature instability
C4	Controlled collapse chip connections
CAD	Computer-aided design
CBCM	Charge-based capacitance measurement
CMOS	Complementary metal oxide semiconductor
CMP	Chemical mechanical polishing
DECAP	DEcoupling CAPacitor
DIBL	Drain-induced barrier lowering
DFM	Design for manufacturing
DOE	Design of experiments
DRC	Design rule checker
DUT	Device under test
ESD	Electrostatic discharge
FD-SOI	Fully depleted silicon-on-insulator
GIDL	Gate-induced drain leakage
GR	Ground rule
HK	High-k (gate dielectric material)
IFVM	Current force voltage measure
ILD	Inter-level dielectric
I/O	Input/output
IQR	Inter quartile range
LSL	Lower specification limit
LSSD	Level-sensitive scan design
LVS	Layout vs. schematic
MEMS	Micro electro-mechanical system
MOS	Metal oxide -semiconductor
MOSFET	Metal oxide semiconductor field-effect transistor
MPS	Multiple serpentine
n-FET	n-type MOSFET (NMOS)
NIST	National Institute of Standards and Technology
p-FET	p-type MOSFET (PMOS)
PD	Pull down
PD-SOI	Partially depleted silicon-on-insulator
PICA	Pico-second imaging circuit analysis
PLC	Power line cycle
PU	Pull up
RDF	Random dopant fluctuation

RIE	Reactive ion etching
RO	Ring oscillator
ROI	Return on investment
SCE	Short-channel effect
SMU	Source measure unit
SoC	System-on-chip
SOI	Silicon-on-insulator
SPC	Statistical process control
SPICE	Simulation program with integrated circuit emphasis
SRAM	Static random access memory
STI	Shallow trench isolation
TFI	Thin film interposer
USL	Upper specification limit
VFIM	Voltage force current measure
VLSI	Very large-scale integration

Index

K

L

M

N